SUSTAINABLE WATER MANAGEMENT AND TECHNOLOGIES

SUSTAINABLE WATER TECHNOLOGIES

VOLUME II

GREEN CHEMISTRY AND CHEMICAL ENGINEERING

Series Editor: Sunggyu Lee

Ohio University, Athens, Ohio, USA

Proton Exchange Membrane Fuel Cells: Contamination and Mitigation Strategies
Hui Li, Shanna Knights, Zheng Shi, John W. Van Zee, and Jiujun Zhang

Proton Exchange Membrane Fuel Cells: Materials Properties and Performance
David P. Wilkinson, Jiujun Zhang, Rob Hui, Jeffrey Fergus, and Xianguo Li

Solid Oxide Fuel Cells: Materials Properties and Performance
Jeffrey Fergus, Rob Hui, Xianguo Li, David P. Wilkinson, and Jiujun Zhang

Efficiency and Sustainability in the Energy and Chemical Industries: Scientific Principles and Case Studies, Second Edition
Krishnan Sankaranarayanan, Jakob de Swaan Arons, and Hedzer van der Kooi

Nuclear Hydrogen Production Handbook
Xing L. Yan and Ryutaro Hino

Magneto Luminous Chemical Vapor Deposition
Hirotsugu Yasuda

Carbon-Neutral Fuels and Energy Carriers
Nazim Z. Muradov and T. Nejat Veziroğlu

Oxide Semiconductors for Solar Energy Conversion: Titanium Dioxide
Janusz Nowotny

Lithium-Ion Batteries: Advanced Materials and Technologies
Xianxia Yuan, Hansan Liu, and Jiujun Zhang

Process Integration for Resource Conservation
Dominic C. Y. Foo

Chemicals from Biomass: Integrating Bioprocesses into Chemical Production Complexes for Sustainable Development
Debalina Sengupta and Ralph W. Pike

Hydrogen Safety
Fotis Rigas and Paul Amyotte

Biofuels and Bioenergy: Processes and Technologies
Sunggyu Lee and Y. T. Shah

Hydrogen Energy and Vehicle Systems
Scott E. Grasman

Integrated Biorefineries: Design, Analysis, and Optimization
Paul R. Stuart and Mahmoud M. El-Halwagi

Water for Energy and Fuel Production
Yatish T. Shah

Handbook of Alternative Fuel Technologies, Second Edition
Sunggyu Lee, James G. Speight, and Sudarshan K. Loyalka

Environmental Transport Phenomena
A. Eduardo Sáez and James C. Baygents

Resource Recovery to Approach Zero Municipal Waste
Mohammad J. Taherzadeh and Tobias Richards

Energy and Fuel Systems Integration
Yatish T. Shah

Sustainable Water Management and Technologies, Two-Volume Set
Daniel H. Chen

Sustainable Water Management
Daniel H. Chen

Sustainable Water Technologies
Daniel H. Chen

GREEN CHEMISTRY AND CHEMICAL ENGINEERING

SUSTAINABLE WATER MANAGEMENT AND TECHNOLOGIES

SUSTAINABLE WATER TECHNOLOGIES

VOLUME II

edited by

Daniel H. Chen

CRC Press is an imprint of the
Taylor & Francis Group, an **informa** business

CRC Press
Taylor & Francis Group
6000 Broken Sound Parkway NW, Suite 300
Boca Raton, FL 33487-2742

Printed on acid-free paper
Version Date: 20160531

International Standard Book Number-13: 978-1-4822-1510-6 (Hardback)

Library of Congress Cataloging-in-Publication Data

Names: Chen, Daniel H., 1949- author.
Title: Sustainable water technologies / Daniel H. Chen.
Description: Boca Raton : Taylor & Francis, CRC Press, 2017. | Includes bibliographical references and index.
Identifiers: LCCN 2016013727 | ISBN 9781482215106 (alk. paper)
Subjects: LCSH: Water conservation. | Water--Purification.
Classification: LCC TD388 .C436 2017 | DDC 628.1028/6--dc23
LC record available at https://lccn.loc.gov/2016013727

Visit the Taylor & Francis Web site at
http://www.taylorandfrancis.com

and the CRC Press Web site at
http://www.crcpress.com

Printed and bound in the United States of America by Sheridan

Contents

Preface

Water is the fundamental building block for human civilization, economic development, and the well-being of all living species. While the world population and the economy continue to grow, the availability of water and other natural resources remain nearly constant. Water shortages inevitably lead to conflicts between competing interests (irrigation, municipal, industrial, energy, environmental), regions (arid vs. wet), and nations (water scarce vs. water rich, developed vs. developing). Facing a looming water crisis, society not only needs to make significant scientific/engineering efforts to advance cost-effective water monitoring/treatment/reuse/integration technologies but also needs to tackle strategy/management issues such as water resources planning/governance, water infrastructure planning/adaption, proper regulations, and water scarcity/inequality as an integrated part of the solution or approach toward water sustainability. For this reason, the *CRC Sustainable Water Management and Technologies* addresses both cornerstone areas: management and technology. This book set presents the best practices as a foundation and proceeds to stress emerging technologies and strategies that facilitate water sustainability for future generations. Timely water topics like unconventional oil and gas development, global warming with changing precipitation patterns, integration of water and energy sustainability, and green manufacturing are discussed. The book is intended for a global audience that has a concern and interest about water quality, supply, resources conservation, and sustainable use.

Water, energy, and climate interactions are the most pressing issues for the 21st century. Water is currently treated as if in infinite supply, yet this is far from the case and use is drastically up worldwide owing to population growth and the pursuit of higher living standards. Water consumption in the production of everyday products such as coffee, beef, and plastics will eventually be priced in. The shale gas revolution is a welcome change in the energy front because of its low greenhouse emissions and relative cleanness for power generation. But the impact of hydraulic fracturing (fracking) on surface water and groundwater quality is of concern. This handbook provides expert assessments on this subject.

The handbook covers the basic principles, best practices, and latest advances in sustainable water management/technology with emphases on the following:

1. Emerging nanotechnology, biotechnology, geographical information system/global position system (GIS/GPS), and membrane technology applications
2. Sustainable processes/products to protect the environment/human health, to save water, energy, and materials
3. Best management practices for water resource allocation, groundwater protection, and water quality assurance, especially for rural, arid, and underdeveloped regions of the world
4. Timely issues such as the impact of shale oil/gas development, adapting water infrastructure to climate change, energy–water nexus, and interaction among water, energy, and ecosystems

This handbook is composed of two books: one is *Sustainable Water Management* and the other is *Sustainable Water Technologies*, the latter devoted to technologies for water resources monitoring, water efficiency (conservation, treatment, reclamation, recycle, reuse, and integration), and water quality (safe for drinking, landscaping, groundwater recharging, and industrial purposes).

This handbook is intended as a technical reference for environmental/civil/chemical engineers, water scientists/risk managers/regulators, academics, and advocacy groups that have responsibilities or interests in water resources, quality, and sustainability. It is also my hope that this handbook will facilitate young science, engineering, and social science students to learn the basics of water technology and management and then to develop the aspiration and skill set to contribute to the solution of this water sustainability issue facing mankind in the 21st century.

The information contained herewithin is the result of professional experience, literature review, and skillful analysis by the leading experts of the field (in alphabetical order): Frank Anscombe, Daniel Attoh, John A. Byrne, Ramesh C. Chawla, Daniel H. Chen (Editor), Liwen Chen, Hyeok Choi, Tapas K. Das, Dionysios D. Dionysiou, Rachel Fagan, Polycarpos Falaras, Pilar Fernández-Ibáñez, Lucas Gregory, Changseok Han, Leslie D. Hartman, Jude O. Ighere, Natalie Johnson, Carey W. King, Teik Thye Lim, Hebin Lin, Cindy Loeffler, Helen H. Lou, Willy Giron Matute, Declan E. McCormack, Mark McFarland, Mohamed K. Mostafa, Dorina Murgulet, John Nielsen-Gammon, Che-Jen Lin, Kevin O'Shea, Robert W. Peters, Jennifer L. Peterson, Suresh C. Pillai, Qin Qian, Walter Rast, Larry A. Redmon, Kelly T. Sanders, Preetam Kumar Sharma, Virender K. Sharma, Saqib Shirazi, Richard Stumpf, Jeffrey A. Thornton, Ross Tomson, Yen Wah Tong, Mike Twardowski, Kevin Urbanczyk, Kevin Wagner, Judy Westrick, Ralph Wurbs, Y. Jeffrey Yang, and Hesam Zamankhan. I sincerely appreciate their dedication and contributions.

I wish to express my gratitude to Kevin Wagner, Dion Dionysiou, and Carey King for identifying many of the chapter authors for the book. I also thank Robert Peters, Tapas Das, Dorina Murgulet, Ross Tomson, Kevin Urbanczyk, Liwen Chen, and Yen Wah Tong for contributing multiple chapters. Finally, a heartfelt thank you is extended to Allison Shatkin of Taylor & Francis/CRC Press for the initiation and production of this handbook.

Daniel H. Chen, PhD

Editor

Daniel H. Chen, PhD, PE, is a professor at the Dan F. Smith Department of Chemical Engineering and holds the title of University Scholar at Lamar University, Beaumont, Texas. He is the director of the Photocatalysis Lab and deputy director of the Process Engineering Center. Dr. Chen's areas of expertise are air/water pollution, nanostructured catalytic materials, and process modeling. He has published 4 book chapters and 37 articles and has given numerous presentations throughout the United States and worldwide. Since 2006, he has worked on a series of field sampling and water treatment projects aimed at improving the water quality of the Rio Grande Basin in south Texas. Dr. Chen holds a PhD degree in Chemical Engineering from Oklahoma State University. He has previously served as a shift supervisor in the Hualon-Teijin Polyester Plant, Taiwan; as a project engineer at Chemical Engineering Consultants, Inc., Stillwater, Oklahoma; as a US Department of Energy research fellow at the National Energy Technology Laboratory in Morgantown, West Virginia; and as a US Army/Battelle faculty associate at Aberdeen Proving Ground in Maryland. He also has served as session chair/co-chair of Sustainable Water Use and Management, Reaction Engineering for Combustion and Pyrolysis, and Liquid Phase Reaction Engineering at the American Institute of Chemical Engineers (AIChE). Dr. Chen is a registered professional engineer in the state of Texas.

Contributors

John Anthony Byrne
Nanotechnology and Integrated BioEngineering Centre
Ulster University
Newtownabbey, United Kingdom

Ramesh C. Chawla
Department of Chemical Engineering
Howard University
Washington, D.C.

Liwen Chen
Dan F. Smith Department of Chemical Engineering
Lamar University
Beaumont, Texas

Hyeok Choi
Department of Civil Engineering
The University of Texas at Arlington
Arlington, Texas

Tapas K. Das
University of Wisconsin at Stevens Point
College of Natural Resources
Stevens Point, Wisconsin

Dionysios D. Dionysiou
Environmental Engineering and Science Program
Department of Biomedical, Chemical and Environmental Engineering
University of Cincinnati
Cincinnati, Ohio

Rachel Fagan
Centre for Research in Engineering Surface Technology (CREST)
Dublin Institute of Technology
Dublin, Ireland

Polycarpos Falaras
Institute of Nanoscience and Nanotechnology
National Center for Scientific Research "Demokritos"
Athens, Greece

Pilar Fernández-Ibáñez
Plataforma Solar de Almería CIEMAT
Almería, Spain

Changseok Han
Environmental Engineering and Science Program
Department of Biomedical, Chemical and Environmental Engineering
University of Cincinnati
Cincinnati, Ohio

Jude O. Ighere
Department of Chemical Engineering
Howard University
Washington, D.C.

Siddharth Jain
Department of Chemical and Biomolecular Engineering
National University of Singapore
Singapore, Singapore

Natalie Johnson
Department of Environmental and Occupational Health
School of Public Health
Texas A&M University
College Station, Texas

Teik Thye Lim
Division of Environmental and Water Resources Engineering
School of Civil and Environmental Engineering
Nanyang Technological University
Singapore, Singapore

Che-Jen Lin
Department of Civil and Environmental Engineering
Lamar University
Beaumont, Texas

Helen H. Lou
Dan F. Smith Department of Chemical Engineering
Lamar University
Beaumont, Texas

Declan E. McCormack
Centre for Research in Engineering Surface Technology (CREST)
Dublin Institute of Technology
Dublin, Ireland

Mohamed K. Mostafa
Department of Civil, Construction, and Environmental Engineering
University of Alabama at Birmingham
Birmingham, Alabama

Dorina Murgulet
Department of Physical and Environmental Sciences
Center for Water Supplies Studies
Texas A&M University–Corpus Christi
Corpus Christi, Texas

Mallikarjuna N. Nadagouda
Water Quality Management Branch
National Risk Management Research Laboratory
United States Environmental Protection Agency
Cincinnati, Ohio

Kevin O'Shea
Department of Chemistry and Biochemistry
Florida International University
Miami, Florida

Robert W. Peters
Department of Civil, Construction, and Environmental Engineering
University of Alabama at Birmingham
Birmingham, Alabama

Suresh C. Pillai
Nanotechnology Research Group
Department of Environmental Science
Institute of Technology Sligo
Sligo, Ireland

Qin Qian
Department of Civil Engineering
Lamar University
Beaumont, Texas

Bangxing Ren
Environmental Engineering and Science Program
Department of Biomedical, Chemical and Environmental Engineering
University of Cincinnati
Cincinnati, Ohio

Peyton C. Richmond
Dan F. Smith Department of Chemical Engineering
Lamar University
Beaumont, Texas

George Em. Romanos
Institute of Nanoscience and Nanotechnology
National Center for Scientific Research "Demokritos"
Athens, Greece

Virender K. Sharma
Department of Environmental and Occupational Health
School of Public Health
Texas A&M University
College Station, Texas

Saqib Shirazi
San Antonio Water System
San Antonio, Texas

Zexin Tian
Dan F. Smith Department of Chemical Engineering
Lamar University
Beaumont, Texas

Ross Tomson
Tomson Technologies, LLC
Houston, Texas

Yen Wah Tong
Department of Chemical and Biomolecular Engineering
National University of Singapore
Singapore, Singapore

Kevin Urbanczyk
Department of Biology, Geology and Physical Sciences
Rio Grande Research Center
Sul Ross State University
Alpine, Texas

Ingo Wolf
Department of Chemical and Biomolecular Engineering
National University of Singapore
Singapore, Singapore

Cen Zhao
Department of Chemistry and Biochemistry
Florida International University
Miami, Florida

1 Water Transport

Qin Qian

CONTENTS

1.1 INTRODUCTION

Freshwater, which makes up 2.15% of the Earth's water, is one of the most basic and important human needs (United Nations Environment Programme [UNEP, Vandeweerd et al. 1997]). It has been under unprecedented pressure and requires urgent attention and sustainable water resources management. Problems regarding water resources are expected to worsen because of population growth and climate change. Population growth is aggravated by continued rural–urban migration and may lead to more pollution on surface water sources. Ample evidence demonstrates that climate change will increase hydrologic variability and, in turn, will lead to extreme weather events such as the recent droughts in the western parts of the United States and floods caused by hurricanes and severe storms. Issues on sustainable management and protection of freshwater are often the limiting factor for development since water supply and wastewater systems are the main themes on the development agenda.

With urbanization, freshwater resource degradation caused by wastewater from stormwater runoff and sewage is a major environmental concern in many cities around the world (Vandeweerd et al. 1997). In some cities, sewerage facilities are also utilized to transport stormwater runoff, which makes the city even more vulnerable to flooding. There has been flooding caused by stormwater runoff in most big cities in China in the past 5 years. As we know, freshwater resources include groundwater and surface water from rivers, lakes, reservoirs, and water bodies along oceans. In recent years, groundwater is being depleted faster than it is being replenished and worsening water quality degrades the environment and adds to costs. Nutrients such as organic matter, phosphorus, and nitrogen that originated mainly from agricultural runoff as well as from industrial and urban discharges come into contact with surface water and degrade water quality. The dynamics in streams and rivers are different from those in lakes and reservoirs. Eutrophication of lakes and reservoirs is becoming more serious around the world.

To better manage freshwater in different water bodies, this chapter will focus on water transport principles and related water transport characteristics in pipes, as well as natural water bodies, such as rivers, streams, estuaries, bays, harbors, lakes, and reservoirs. To probe into the environmental concerns regarding water resources, the transport process of various substances, such as contaminants, pollutants, and artificial tracers, is also included.

1.2 WATER TRANSPORT PRINCIPLES

Natural water, as a universal solvent, from various sources contains dissolved gases, minerals, and organic and inorganic substances. Water can be considered as an incompressible fluid because the physical properties of water vary slightly under different pressure and temperature (Eckart 1958). The density of water reaches a maximum of 1000 kg/m^3 at 4°C, while it is 999 kg/m^3 at 20°C. The viscosity of water is 1.002 E^{-3} N s/m^2 at 20°C and decreases with an increase in temperature. In general, water exhibits Newtonian fluid behavior because the shearing stress is linearly related to the rate of angular deformation (Batchelor 1967). The Navier–Stokes equation, named after L.M. Navier (1785–1836) and Sir G.G. Stokes (1819–1903), provides a complete mathematical description of the water flow velocity at a given point in space and time.

1.2.1 The Navier–Stokes Equation

The Navier–Stokes equation applies Newton's second law to Newtonian fluids in a control volume and can be expressed as in Cartesian coordinates (x, y, and z) as

$$\begin{aligned}
\rho\left(\frac{\partial u}{\partial t}+u\frac{\partial u}{\partial x}+v\frac{\partial u}{\partial y}+w\frac{\partial u}{\partial z}\right)&=\frac{-\partial p}{\partial x}+\rho g_x+\mu\left(\frac{\partial^2 u}{\partial x^2}+\frac{\partial^2 u}{\partial y^2}+\frac{\partial^2 u}{\partial z^2}\right)\\
\rho\left(\frac{\partial v}{\partial t}+u\frac{\partial v}{\partial x}+v\frac{\partial v}{\partial y}+w\frac{\partial v}{\partial z}\right)&=\frac{-\partial p}{\partial y}+\rho g_y+\mu\left(\frac{\partial^2 v}{\partial x^2}+\frac{\partial^2 v}{\partial y^2}+\frac{\partial^2 v}{\partial z^2}\right)\\
\rho\left(\frac{\partial w}{\partial t}+u\frac{\partial w}{\partial x}+v\frac{\partial w}{\partial y}+w\frac{\partial w}{\partial z}\right)&=\frac{-\partial p}{\partial z}+\rho g_z+\mu\left(\frac{\partial^2 w}{\partial x^2}+\frac{\partial^2 w}{\partial y^2}+\frac{\partial^2 w}{\partial z^2}\right),
\end{aligned} \tag{1.1}$$

where ρ = density (M/L^3); x, y, z = coordinate system (L); u, v, w = velocity at the x, y, and z directions, respectively (L/T); p = pressure (F/L^2); μ = viscosity (FT/L^2); and

g_x, g_y, g_z = local acceleration of gravity (L/T^2). Note that the values depend on the orientation of gravity with respect to the chosen set of coordinates.

The water flow velocity can be described by combining Equation 1.1 with the conservation of mass equation for incompressible fluid:

$$\frac{\partial u}{\partial x}+\frac{\partial v}{\partial y}+\frac{\partial w}{\partial z}=0. \tag{1.2}$$

Note that the Navier–Stokes equation can be written in cylindrical coordinates (r, θ, z) as

$$\begin{aligned}
&\rho\left(\frac{\partial v_r}{\partial t}+v_r\frac{\partial v_r}{\partial r}+\frac{v_\theta \partial v_r}{r\partial\theta}-\frac{v_\theta^2}{r}+v_z\frac{\partial v_r}{\partial z}\right)\\
&\quad=\frac{-\partial p}{\partial r}+\rho g_r+\mu\left(\frac{1}{r}\frac{\partial}{\partial r}\left(r\frac{\partial v_r}{\partial r}\right)-\frac{v_r}{r^2}+\frac{1}{r^2}\frac{\partial^2 v_r}{\partial\theta^2}-\frac{2}{r^2}\frac{\partial v_\theta}{\partial\theta}+\frac{\partial^2 v_r}{\partial z^2}\right)\\
&\rho\left(\frac{\partial v_\theta}{\partial t}+v_r\frac{\partial v_\theta}{\partial r}+\frac{v_\theta \partial v_\theta}{r\partial\theta}+\frac{v_r v_\theta}{r}+v_z\frac{\partial v_\theta}{\partial z}\right)\\
&\quad=\frac{-1\partial p}{r\partial\theta}+\rho g_\theta+\mu\left(\frac{1}{r}\frac{\partial}{\partial r}\left(r\frac{\partial v_\theta}{\partial r}\right)-\frac{v_\theta}{r^2}+\frac{1}{r^2}\frac{\partial^2 v_\theta}{\partial\theta^2}+\frac{2}{r^2}\frac{\partial v_r}{\partial\theta}+\frac{\partial^2 v_\theta}{\partial z^2}\right)\\
&\rho\left(\frac{\partial v_z}{\partial t}+v_r\frac{\partial v_z}{\partial r}+\frac{v_\theta \partial v_z}{r\partial\theta}+v_z\frac{\partial v_z}{\partial z}\right)\\
&\quad=-\frac{\partial p}{\partial z}+\rho g_z+\mu\left(\frac{1}{r}\frac{\partial}{\partial r}\left(r\frac{\partial v_z}{\partial r}\right)+\frac{1}{r^2}\frac{\partial^2 v_z}{\partial\theta^2}+\frac{\partial^2 v_z}{\partial z^2}\right).
\end{aligned} \tag{1.3}$$

Although four unknowns (u, v, w, p) are well posed in four equations, they are not amenable to exact mathematical solutions because of nonlinear partial differential terms, i.e., $u\,\partial u/\partial x$, $w\,\partial v/\partial z$, in the Navier–Stokes equation. It is generally difficult to analytically solve nonlinear partial differential equations. In some instances, the exact solution can be obtained with some flow assumptions and initial/boundary condition formulation, for example, the steady, laminar flow between fixed parallel plates, the steady laminar flow in circular tubes (e.g., pipe flow), and the Couette flow. Refer to *An Introduction to Fluid Dynamics* by Batchelor (1967) for a detailed derivation and application of the Navier–Stokes equation.

1.2.2 Transport Processes in Water

To quantify chemical or biological effects on water quality, water transport is implicitly involved with the transport processes of heat, mass, and momentum. The *gradient flux law* (Potter et al. 2002) describes that "the transport of momentum, energy and mass occurs due to the gradient of momentum concentration, energy concentration and mass concentration, respectively." The transport of momentum states is as in the Navier–Stokes equation. By applying the Reynolds transport theorem to mass and heat, the mass and heat transport equation can be derived. The law of conservation expresses that accumulation in the control volume = input − output + generation − degradation. The three-dimension mass transport equation can be obtained by considering mass flux components in the three coordinate directions:

$$\frac{\partial C}{\partial t}+u\frac{\partial C}{\partial x}+v\frac{\partial C}{\partial y}+w\frac{\partial C}{\partial z}=D\left(\frac{\partial^2 C}{\partial x^2}+\frac{\partial^2 C}{\partial y^2}+\frac{\partial^2 C}{\partial z^2}\right)+r_{\mathrm{g}}-r_{\mathrm{d}}, \tag{1.4}$$

where C = the concentration of the substance (M/L^3), D = mass diffuse coefficient or mass diffusivity (L^2/T), r_{g} = rate of generation of the substance (M/L^{3}T), and r_{d} = rate of degradation of the substance (M/L^{3}T).

The energy equation has a form similar to the mass transport equation. The major difference is in the source term. The form of the energy equation for water is

$$\frac{\partial T}{\partial t}+u\frac{\partial T}{\partial x}+v\frac{\partial T}{\partial y}+w\frac{\partial T}{\partial z}=\frac{K}{\rho c_v}\left(\frac{\partial^2 T}{\partial x^2}+\frac{\partial^2 T}{\partial y^2}+\frac{\partial^2 T}{\partial z^2}\right)+\frac{S}{\rho c_v}, \tag{1.5}$$

where T = the temperature (Θ), K = thermal conductivity (ML/(T^{3}Θ)), c_v = specific heat capacity (L^2/(T^{2}Θ)), and S = thermal energy production or dissipation (ML^2/T^2).

Most water flows encountered in the environment are turbulent flow, which has identifiable structures called eddies (Potter et al. 2002). The existence of a large range of eddy sizes is attributed to the flow domain scale (width or depth of the river or lake) of a few millimeters, that is, *Kolmogorov microscale*. The larger eddies are usually unstable and transfer some of the kinetic energy to a smaller scale to generate smaller eddies. Such a process is known as energy cascade, which can be repeated until the smallest length scale generates the highest-frequency eddies at the end of transfer. In terms of transport processes in water flow, the concentration of the substance, water temperature, and velocity components are composed of the mean (the overbar notation) and fluctuating (the prime notation) components:

$$\begin{aligned} &C(t)=\bar{C}+C'(t), \quad T(t)=\bar{T}+T'(t), \\ &u(t)=\bar{u}+u'(t), \quad v(t)=\bar{v}+v'(t), \quad w(t)=\bar{w}+w'(t). \end{aligned} \tag{1.6}$$

By following Reynolds decomposition, Equation 1.6 can be incorporated into Equations 1.4 and 1.5 to yield the turbulent mass and heat transport equation.

To simplify the decomposition process, only the x component is considered, and then Equation 1.6 can be rewritten as $C(t) = \bar{C} + C'(t)$, $u(t) = \bar{u}\ u'(t)$, $v(t) = v'(t)$, $w(t) = w'(t)$. By substituting into Equation 1.4, the following equation can be obtained:

$$\frac{\partial(\bar{C}+C')}{\partial t}+(\bar{u}+u')\frac{\partial(\bar{C}+C')}{\partial x}+(v')\frac{\partial(\bar{C}+C')}{\partial y}(w')\frac{\partial(\bar{C}+C')}{\partial z} = D\left(\frac{\partial^2(\bar{C}+C')}{\partial x^2}\right)+\bar{r}_g+r'_g-(\bar{r}_d+r'_d), \tag{1.7}$$

where $\bar{r}$ is the time-averaged source/sink term and r' is the fluctuation in the source/sink term. By averaging over time, Equation 1.7 can be written as

$$\begin{aligned}
&\overline{\frac{\partial(\bar{C}+C')}{\partial t}}+(\bar{u}+u')\overline{\frac{\partial(\bar{C}+C')}{\partial x}}+(v')\overline{\frac{\partial(\bar{C}+C')}{\partial y}}+(w')\overline{\frac{\partial(\bar{C}+C')}{\partial z}} \\
&= D\overline{\left(\frac{\partial^2(\bar{C}+C')}{\partial x^2}\right)}+\bar{r}_g+\overline{r'_g}-\left(\bar{r}_d+\overline{r'_d}\right) \\
&\frac{\partial\bar{\bar{C}}+\partial\bar{C}'}{\partial t}+\bar{\bar{u}}\frac{\partial\bar{\bar{C}}}{\partial x}+\bar{\bar{u}}\frac{\partial\bar{C}'}{\partial x}+\bar{u}'\frac{\partial\bar{\bar{C}}}{\partial x}+u'\overline{\frac{\partial C'}{\partial x}}+\bar{v}'\frac{\partial\bar{\bar{C}}}{\partial y}+v'\overline{\frac{\partial C'}{\partial y}}+\bar{w}'\frac{\partial\bar{\bar{C}}}{\partial z}+w'\overline{\frac{\partial C'}{\partial z}} \\
&= D\left(\frac{\partial^2\bar{\bar{C}}}{\partial x^2}+\frac{\partial^2\bar{C}'}{\partial x^2}\right)+\bar{\bar{r}}_g+\bar{r}'_g-\left(\bar{\bar{r}}_d+\bar{r}'_d\right).
\end{aligned} \tag{1.8}$$

Applying the definition of averaging rules $\bar{a}' = 0$ and $\bar{\bar{a}} = \bar{a}$ to Equation 1.8,

$$\frac{\partial\bar{C}}{\partial t}+\bar{u}\frac{\partial\bar{C}}{\partial x}+u'\overline{\frac{\partial C'}{\partial x}}+v'\overline{\frac{\partial C'}{\partial y}}+w'\overline{\frac{\partial C'}{\partial z}}=D\frac{\partial^2\bar{C}}{\partial x^2}+\bar{r}_g-\bar{r}_d. \tag{1.9}$$

Employing Equation 1.9 at the x direction results in

$$\frac{\partial\bar{C}}{\partial t}+\bar{u}\frac{\partial\bar{C}}{\partial x}=D\frac{\partial^2\bar{C}}{\partial x^2}-\frac{\partial\overline{u'C'}}{\partial x}+\bar{r}_g-\bar{r}_d. \tag{1.10}$$

The terms in Equation 1.10 from left to right can be described as the mean storage of substance concentration, the advection of the substance by the mean flow, the mean molecular diffusion of the substance, the turbulence diffusion or eddy diffusion, and the net mean body source for additional substance processes. The term $\overline{u'C'}$ represents the net mass flux attributed to turbulent advection. It can be computed if the turbulence field is fully calculated. However, this is quite complex and computationally intensive and even prohibitive for many flows (Csanady 1973).

Alternatively, the turbulent flux term has been related to the mean flow and represented as a gradient diffusion process and expressed in the following equation:

$$\overline{u'C'} = -D_{tx}\frac{\partial \bar{C}}{\partial x}, \tag{1.11}$$

where D_{tx} is the turbulent diffusion or eddy diffusion coefficient at the x direction (L^2/T). The typical values of the turbulent diffusion in oceans, lakes, and rivers are five orders of magnitude larger than the molecular diffusion coefficient (10^{-5} cm^2/s) (Potter et al. 2002). The molecular diffusion term in Equation 1.10 can be ignored. Assuming the identical turbulent diffusion coefficient D_t (L^2/T) at three directions, the turbulent mass transport equation in three dimensions can be obtained:

$$\frac{\partial \bar{C}}{\partial t} + \bar{u}\frac{\partial \bar{C}}{\partial x} + \bar{v}\frac{\partial \bar{C}}{\partial y} + \bar{w}\frac{\partial \bar{C}}{\partial z} = D_t\left(\frac{\partial^2 \bar{C}}{\partial x^2} + \frac{\partial^2 \bar{C}}{\partial y^2} + \frac{\partial^2 \bar{C}}{\partial z^2}\right) + \bar{r}_g - \bar{r}_d. \tag{1.12}$$

Following similar procedures to derive a heat transport equation in turbulent flow,

$$\frac{\partial \bar{T}}{\partial t} + \bar{u}\frac{\partial \bar{T}}{\partial x} + \bar{v}\frac{\partial \bar{T}}{\partial y} + \bar{w}\frac{\partial \bar{T}}{\partial z} = \alpha_t\left(\frac{\partial^2 T}{\partial x^2} + \frac{\partial^2 T}{\partial y^2} + \frac{\partial^2 T}{\partial z^2}\right) + \frac{\bar{S}}{\rho c_v}, \tag{1.13}$$

where α_t is the turbulent diffusion coefficient for heat (L^2/T). Note that the overbars in Equations 1.12 and 1.13 are often omitted.

Equations 1.12 and 1.13 provide the basis for studying transport processes of mass and heat in natural water bodies. These equations require a significant amount of information and effort for application toward rivers, lakes, and coastal waters. With advances in computer techniques, hydrodynamics and water quality models have been developed with various algorithms and have made big progress: from being a steady-state model to becoming a dynamic model; from being zero-dimensional to becoming one-dimensional, two-dimensional, and three-dimensional models; and from being a point source model to becoming coupling point and nonpoint source models. These models become effective tools to simulate and predict the solute transport in a water environment. However, the different predications among models sometimes are big because of the different theories and numerical algorithms. It is essential to select suitable models that can fulfill the objectives and meet the available budget for data collection and analysis. Therefore, the selected main hydrodynamic and water quality models are listed in Table 1.1 based on their characteristics. Numerous case studies have been included in the model manual, which provides useful guidance for the model application. In fact, nature is too complex to be fully described with a set of differential equations. Therefore, simple, probabilistic models are a better approach. For example, Stow et al. (2003) concludes predictive accuracy was not improved by using the more process-oriented spatially detailed models than the aggregate probabilistic model by comparing three estuarine water quality models for total maximum daily load development in Neuse River estuary.

TABLE 1.1
Selected Main Hydrodynamic and Water Quality Models

Models and Current Version	Description	Reference
CE-QUAL-W2, version 3.7.1, 2014	Longitudinal-vertical two-dimensional hydrodynamics and water quality in stratified and nonstratified systems, nutrient-dissolved oxygen–organic matter interactions, fish habitat, selective withdrawal from stratified reservoir outlets, hypolimnetic aeration, multiple algae, epiphyton/periphyton, zooplankton, macrophyte, CBOD, and generic water quality groups.	Portland State University, Water Quality Research Group, http://www.ce.pdx.edu/w2/
HEC-RAS, version 4.1, 2010	One-dimensional steady flow, unsteady flow, sediment transport/mobile bed computation, and water temperature and water quality (dissolved nitrogen, dissolved phosphorus, algae, dissolved oxygen and carbonaceous biological oxygen demand) modeling.	US Army Corps of Engineers, Hydrologic Engineering Center, http://www.hec.usace.army.mil/software/hec-ras/
QUAL 2K version 2.11b8, 2009	One-dimensional, nonuniform steady flow water quality model for a river and stream to predict the conventional pollutants (nitrogen, phosphorus, dissolved oxygen, BOD, sediment oxygen demand, algae). pH, periphyton, pathogens from point and nonpoint loads.	United Sates Environmental Protection Agency (USEPA) http://epa.gov/athens/wwqtsc/html/qual2k.html
WASP version 7.52, 2013	One-, two-, and three-dimensional for variety of pollutant types. WASP also can be linked with hydrodynamic and sediment transport models that can provide flows, depths velocities, temperature, salinity and sediment fluxes.	United States Environmental Protection Agency (USEPA) http://epa.gov/athens/wwqtsc/html/wasp.html
EFDC model, version 1.0, 2002	Dynamically coupled transport equations for turbulent kinetic energy, turbulent length scale, salinity, and temperature, to provide the hydrodynamic model linking to WASP for water quality model.	United States Environmental Protection Agency (USEPA) http://epa.gov/athens/wwqtsc/html/efdc.html
GEMSS, 2002	Three-dimensional hydrodynamic and transport modules embedded in a geographic information and environmental data system, including thermal analysis, water quality, sediment transport, particle tracking, oil and chemical spills, entrainment, and toxics.	*Waterbody Hydrodynamic and Water Quality Modeling: An Introductory Workbook and CD-ROM on Three-Dimensional Waterbody Modeling* by John Eric Edinger (2002), ASCE

(*Continued*)

TABLE 1.1 (CONTINUED)
Selected Main Hydrodynamic and Water Quality Models

Models and Current Version	Description	Reference
CCHE1D, CCHE2D, up to date	One- and two-dimensional hydrodynamic, sediment transport, water quality model for channel networks.	The University of Mississippi, National Center for Computational Hydroscience and Engineering, http://www.ncche.olemiss.edu/sw_download
PC_QUASAR, 1997	One-dimensional dynamic water quality and flow model for river networks to predict river flow, pH, nitrate, temperature, *Escherichia coli*, biochemical oxygen demand, dissolved oxygen and conservative pollutant or tracer.	Centre for Ecology and Hydrology, http://www.ceh.ac.uk/index.html
MIKE 11, 1993; MIKE 21, 1996; MIKE 31, 1996	Water quality simulation in rivers, estuaries, and tidal wetlands, including one-, two-, and three-dimensional models.	Denmark Hydrology Institute, http://www.mikebydhi.com/products
CH3D-IMS; CH3D-SSMS, up to date	Curvilinear-grid hydrodynamics three-dimensional model suitable for application to coastal and near shore waters with complex shoreline and bathymetry	University of Florida, http://aces.coastal.ufl.edu/CH3D/

1.3 WATER TRANSPORT IN PIPES

Water transport in pipes includes full water flows and partially full water flows in closed pipes. Since partially full water flows are one of the open channel flows, this chapter will focus on the full water flow, that is, pressure pipe flow. Such flow systems include any pipe network carrying an incompressible, single-phase, Newtonian fluid in full pipes such as industrial cooling systems and municipal water utilities (i.e., water supply system, sewage system, and stormwater system). The principles of hydraulics and water quality analysis in the pipe systems will be reviewed.

1.3.1 Pipe Flow

Pipe flows can be thought of as an energy gradient applied on the water and are described by energy equations. For a given flow rate (Q), the total energy of pipe flow at any location can be calculated by the potential head or elevation (z), the pressure head (p/γ), and the velocity head ($V^2/2g$). The flow velocity is the section mean velocity, which can be defined as the flow rate (Q) divided by the cross-sectional area (A):

$$V = \frac{Q}{A}. \tag{1.14}$$

Flow velocity from Equation 1.14 can be used to determine the Reynolds number (Re):

$$\mathrm{Re} = \frac{\rho VD}{\mu} = \frac{VD}{\nu}, \tag{1.15}$$

where Re = Reynolds number, D = pipeline diameter (L), ρ = fluid density (M/L^3), μ = absolute viscosity (M/L/T), and ν = kinematic viscosity (L^2/T).

The *Reynolds number* defines the ratio between inertial and viscous forces in a fluid. The ranges of the Reynolds number that define the three pipe flow regimes are as follows: *laminar* < 2000, *transitional* 2000–4000, and *turbulent* > 4000.

Because of the shear stresses along the walls of a pipe, the velocity profile over the pipe diameter is not uniform. It is equal to zero at the pipe wall and increases with distance from the pipe wall, reaching the maximum along the centerline of the pipe. The shape of the velocity profile will vary depending on whether the flow regime is laminar or turbulent.

Mathematically, the velocity profile in a laminar flow is shaped like a parabola and the maximum velocity is twice as large than the average flow velocity, which can be expressed as in the Navier–Stokes equation (Equation 1.1). Although turbulent flow is characterized by more eddies producing random variation in the velocity profile, the velocity distribution is more uniform than laminar flow. It follows the general form of a logarithmic curve and becomes flatter as the Reynolds number increases. Therefore, the average velocity is bigger than half of the maximum velocity.

Water flowing in pipes contains kinetic energy, potential energy, and pressure energy. The pump supplies the energy to the pipe flow, while the turbine accepts the energy from the pipe flow. The one-dimensional steady flow form of the energy equation is expressed as

$$\frac{p_1}{\gamma} + \alpha_1 \frac{V_1^2}{2g} + z_1 + h_\mathrm{p} = \frac{p_2}{\gamma} + \alpha_2 \frac{V_2^2}{2g} + z_2 + h_\mathrm{t} + h_\mathrm{L}, \tag{1.16}$$

where Z = elevation above datum (L), p = pressure (M/L/T^2), γ = fluid specific weight (M/L^2/T^2), V = velocity (L/T), g = gravitational acceleration constant (L/T^2), h_p = head supplied by a pump (L), h_t = head supplied to a turbine (L), and h_L = head loss between sections 1 and 2 (L).

The line that plots total head (H), $H = \frac{p_1}{\gamma} + \alpha_1 \frac{V_1^2}{2g} + z_1$, versus distance through a system is called the *energy grade line*. The sum of the elevation head and pressure head yields the *hydraulic grade line*. The power (P) is directly supplied to the flow by pump (P_p) or the power is supplied directly to the turbine (P_t) as

$$P_\mathrm{p} = Q\gamma h_\mathrm{p};\ P_\mathrm{t} = Q\gamma h_\mathrm{t}. \tag{1.17}$$

TABLE 1.2
Friction Equations as a Function of Pipe Length (*L*) and Diameter (*D*), Flow Rate (*Q*), and Friction Factor (*f* for Darcy–Weisbach, C_{HW} for Hazen–Williams, and *n* for Manning's Equation)

Equation	BG System	SI System
Darcy–Weisbach	$h_f = \frac{0.0252 fL}{D^5} Q^2$	$h_f = \frac{0.0826 fL}{D^5} Q^2$
Hazen–Williams	$h_f = \frac{4.73L}{D^{4.87} C_{HW}^{1.85}} Q^{1.85}$	$h_f = \frac{10.7L}{D^{4.87} C_{HW}^{1.85}} Q^{1.85}$
Manning	$h_f = \frac{4.64 n^2 L}{D^{5.33}} Q^2$	$h_f = \frac{10.3 n^2 L}{D^{5.33}} Q^2$

If power is related to the electrical or mechanical energy of the pump or turbine, an efficiency factor must be included in the calculation. Head losses (h_L) are generally the result of friction along the pipe walls (i.e., *major losses*) and turbulence attributed to changes in streamlines through fitting and appurtenances (i.e., *minor losses*).

Major losses can be calculated using the Darcy–Weisbach formula, the Hazen–Williams equation, and Manning's equation. The Darcy–Weisbach formula is a more physically based equation developed using dimensionless analysis. With appropriate fluid viscosities and densities, Darcy–Weisbach can be used to find the head loss in a pipe for any Newtonian fluid in any flow regime.

However, the Hazen–Williams and Manning formulas are empirically based expressions and generally only apply to water under turbulent flow conditions. The Hazen–Williams formula uses a pipe carrying capacity factor (*C*). Higher *C* factors represent smoother pipes, while lower *C* factors describe rougher pipes. Manning's equation calculates loss expression more typically associated with open channel flow. Table 1.2 presents these three equations in two unit configurations.

Minor losses occur at valves, tees, bends, reducers, and other *appurtenances* within the piping system. Head loss caused by minor losses can be computed by multiplying a *minor loss coefficient* by the velocity head ($V^2/2g$). Minor loss coefficients are found experimentally, and data are available for many different types of fittings and appurtenances as from sources such as a hydraulic engineering design manual.

1.3.2 Pipelines and Pipe Networks

In practice, pipelines and pipe networks are designed for a given project when a number of pipes are connected together to transport water. Series pipes, parallel pipes, branching pipes, elbows, valves, meters, and other appurtenances are presented together. Every element in pipelines and networks is influenced by its neighbors. The entire system is interrelated such that the condition of one element must be consistent with the condition of all other elements. These interconnections are defined by the conservation of mass and the conservation of energy. The principle of the *conservation of mass* dictates that the fluid mass entering any pipe will be equal to the mass leaving the pipe. The conservation of mass equation is applied to all junction nodes and tanks in a network, and one equation is written for each of them. The principle of the *conservation of energy* dictates that the energy difference between two points must be the same regardless of the path that is taken. For real water distribution systems, one continuity equation must be developed for each node in the system, and one energy equation must be developed for each pipe (or loop). To solve the real water distribution system problem, powerful numerical techniques have been developed, for example, WaterCAD developed by Bentley Systems and EPANET developed by USEPA (http://www.epa.gov/nrmrl/wswrd/dw/epanet.html).

1.3.3 Water Quality Analysis in Pipe Flow

The governing equation for water quality analysis in pipe flow is based on the principle of conservation of mass coupled with reaction kinetics. Advective transport, mixing at pipe junctions and storage facilities, and decay/reactions are the fundamental physical and chemical processes in water quality analysis in pipe flows. Water quality analysis is an extension by using the network hydraulic solution, such as flow rates in pipes and the flow paths. The equations describing transport through pipes, mixing at nodes, chemical formation and decay reactions, and storage and mixing in storage facilities are adapted from Rossman et al. (1993) and Rossman and Boulos (1996). Assuming that there is no intermixing of mass between adjacent parcels of water traveling down in a pipe (i.e., ignoring longitudinal dispersion), a dissolved substance will travel down the length of a pipe with the same average velocity as water while at the same time reacting (either growing or decaying) at some given rate.

Advective transport in pipes is represented with the following equation:

$$\frac{\partial C_i}{\partial t} = -u_i \frac{\partial C_i}{\partial x} + r(C_i), \tag{1.18}$$

where C_i = concentration (M/L^3) in pipe i as a function of distance x (L) and time t (T), u_i = flow velocity (L/T) in pipe i, which can be calculated as (Q_i/A_i), and r = rate of reaction (M/L^3/T) as a function of concentration.

The mixing of fluid at junctions is assumed to be complete and instantaneous. Thus, the combined concentration at pipe i described by Equation 1.18 can be written by performing a mass balance on concentrations entering and leaving a junction node. For a specific node j, the concentration for flow leaving node j can be written as

$$C_{\text{out},j} = \frac{\sum_{i \in \text{in},j} Q_i C_i + Q_{i,\text{source}} C_{i,\text{source}}}{\sum_{i \in \text{out},j} Q_i + Q_{i,\text{source}}}, \tag{1.19}$$

where $C_{\text{out},j}$ = concentration leaving the junction node j (M/L^3), $\sum_{i \in \text{out},j} / \sum_{i \in \text{in},j}$ = set of pipes leaving/entering node j, Q_i = flow rate entering the junction node from pipe i (L^3/T), C_i = concentration entering the junction node from pipe i (M/L^3), $Q_{i,\text{source}}$ = flow rate source at junction node j (L^3/T), and $C_{i,\text{source}}$ = concentration source at junction node j (M/L^3).

When the pipe system connects to the storage facilities such as tanks and reservoirs, complete mixing is also assumed at the junction (Rossman and Grayman 1999). In addition, the internal concentration could be affected by reactions, such that a mass balance of concentrations entering or leaving the tank or reservoir (k) can be expressed as

$$\frac{\mathrm{d}C_k}{\mathrm{d}t} = -\frac{Q_i}{V_k}(C_i(t) - C_k) + r(C_k), \tag{1.20}$$

where C_k = concentration within tank or reservoir k (M/L^3), V_k = volume in tank or reservoir k (L^3), and $r(C_k)$ = reaction term (M/L^3/T).

The reaction term $r(C_k)$ can be considered as *bulk flow reactions* while a substance moves down in a pipe or resides in storage and as *pipe wall reactions* while dissolved substance reacts with material such as corrosion products or biofilm that are on or close to the wall. The rate of bulk flow reactions can generally be described as a power function of concentration:

$$r(C_k) = kC^n, \tag{1.21}$$

where k = a reaction constant (1/T) and n = the reaction order.

A reaction constant (k) is a function of temperature. It can be positive (to indicate a formation reaction) or negative (to state a decay reaction). Zero-, first-, and second-order reactions are commonly used. Some examples of different reaction rate expression have been implemented in EPLANT and WaterCAD. For first-order kinetics, the rate of a pipe wall reaction can be expressed as (Rossman et al. 1994)

$$r(C_k) = \frac{2k_\text{w}k_\text{f}C}{R(k_\text{w} + k_\text{f})}, \tag{1.22}$$

where k_w = wall reaction rate constant (L/T), k_f = mass transfer coefficient (L/T), $k_\text{f} = \text{Sh}\frac{D}{d}$, Sh is Sherwood number, D = the molecular diffusivity, and d = pipe diameter; and R = pipe radius.

Allowing incomplete mixing at pipe junctions in water distribution networks, bulk advection mixing (BAM) has been developed by Ho and Khalsa (2009) by combining momentum transfer and separation of impinging fluid streams within a cross junction. The model has been implemented in the new version, EPANET-BAM, with a mixing parameter, s, which allows the user to select the bulk advective mixing model (s = 0) and the existing complete mixing model (s = 1), or a result that is linearly scaled between the results of the two models. More information can be found at http://ascelibrary.org/doi/abs/10.1061/41024%28340%2987. Recently, a new computer model, ADRENT (Li 2006), has considered stochastic water demands and unsteady mass dispersion to predict the spatial and temporal distribution of disinfectant in a pipe network. The model demonstrates better agreement with field observations in locations with prevalent laminar flow than the prediction from the EPANET model.

1.4 WATER TRANSPORT IN NATURAL WATER BODIES

1.4.1 Water Transport in Rivers and Streams

It is important to understand and learn about water transport in rivers and streams because they are a dynamic combination of water, sediment, aquatic organisms, and riparian vegetation, where human beings like to settle down (Wampler 2012). More than 3.5 million miles of rivers/streams in the United States covering an extremely rich and diverse ecosystem have been considered from physical, chemical, and biological perspectives. The main physical characteristics of rivers and streams are geometry, slope, bed roughness, tortuosity, flow velocity, dispersion mixing characteristics, water temperature, suspended solid, and sediment transport (Thomann and Mueller 1987). The important chemical contents related to water quality management are dissolved oxygen, pH, total dissolved solids and chlorides, and chemicals that are potentially toxic (Thomann and Mueller 1987). Bacteria and viruses, fish populations, rooted aquatic plants, and biological slimes are the biological characteristics, which are of specific significance for water quality (Thomann and Mueller 1987).

1.4.1.1 Water Flow in Rivers and Stream

It is quite clear for anyone who has been fishing, canoeing, or paddling in rivers and streams that the distinguishing feature of water transport is more or less rapid in a downstream direction because the main driving force is gravity. Rivers and streams are examples of open channel flow, in which the upper free surface of the liquid is open to the atmosphere. Factors contributing to water flow in rivers and streams include, among others, precipitation in the watershed, stream flow, droughts and floods, groundwater, and snow melting. Advances in computer development and data monitoring allow the development and application of a number of models in hydrology to predict the flow rate. Hydrology models consider the parameter variations in space and time by incorporating various equations to describe hydrologic transport processes through numerical methods. The most public models are HEC-HEM (HEC 2006), USEPA SWMM (http://www2.epa.gov/water-research/storm-water-management-model-swmm), and the SCS TR20 (USDA, http://www.hydrocad.net/tr-20.htm) to simulate rainfall–runoff for single storm events. The more in-depth information can be referred to the textbook entitled *Hydrology and Floodplain Analysis* (Bedient et al. 2008).

The most important hydrological aspects for water quality are flow rate, flow velocity, and geometry. Flow velocity can be measured directly by current meters as introduced by the US Geology Survey (USGS) (http://water.usgs.gov/edu/streamflow2.html)

or indirectly by tracking the time for objects in the water to travel a given distance as demonstrated by USEPA (http://water.epa.gov/type/rsl/monitoring/vms51.cfm). The flow velocity varies with width and depth because of frictional effects in the river and stream system. Therefore, the mean velocity must be estimated at a specific cross section by following the procedures and instructions from USGS and USEPA. The average flow velocity can also be obtained from the open channel equation developed by Chezy in 1775 and Manning in 1890 (Chow 1959):

$$V = \frac{\kappa}{n} R_h^{2/3} S_0^{1/2}, \tag{1.23}$$

where V = flow velocity (L/t), κ = constant, 1 for SI unit and 1.49 for BG unit; n = roughness coefficient, $R_h = A/P$, hydraulic radius (L), A = area (L^2) and P = wet perimeter in a cross section (L), and S_0 = slope of the channel bed (L/L).

Equation 1.23 can be applied only for steady flows with a small constant slope (<10°) and reasonably long, straight reaches. In addition, empirical estimation of the mean velocity in a reach has been presented in the literature based on the observed data accounting for pools, riffles, dead zones, and so on.

The flow rate at a specific location can be calculated with the mean flow velocity and the cross section area as

$$Q = VA, \tag{1.24}$$

where Q = flow rate (L^3/T), V = flow velocity (L/T), and A = cross-section area at a specific location (L^2).

The USGS has established a large number of flow gaging stations throughout the United States. A correction between flow rate and river stage (height) is developed for each control section to provide the flow rate at 15- to 60-min intervals. The flow rate varies seasonally depending on watershed location and degree of upstream regulation controlled by dams, reservoirs, or locks. Empirical estimate flows of various durations directly from meteorological data, climate, and watershed characteristics have been carried out by extensive statistical analysis with USGS gage flow data.

For example, Thomas and Benson (1970) studied the relationship between river flows and watershed characteristics in eastern, central, southern, and western regions; Rifai et al. (2000) developed the low flow characteristics for Texas streams by considering basin contributing drainage area, channel length and slope, basin shape factor, mean annual precipitation, predominant hydrologic soil group, and the 2-year 24-h precipitation. Please note that low flows are often of interest in water quality studies. The relationship between flow rate and river morphometry–hydraulic geometry has also been developed. River morphometry indicates the form or shape of a river, including its width, depth, slope, meander, and floodplain, and the grain size distribution of the bed. The empirical relationship, which estimates velocity, depth, and width, and takes the form of a power function with flow as the independent variable, has been developed and examined in various rivers by Leopold and Maddock (1953).

Hydraulic design applications, such as bridge, culvert, dam, and channelization, can change the water transport pattern in rivers and streams. Such applications can be completed by HEC-RAS, CCHE1D, CCHE2D, and CE-QUAL-W2 as listed in Table 1.1 to predict the flow velocity field under different design criteria.

1.4.1.2 Turbulent Transport in Rivers and Streams

The transport of mass and heat in natural rivers and streams is dependent on the vertical and horizontal turbulence of very different length scales. The horizontal turbulence with the length scale comparable to the width of the flow is a dominative and effective transport process at the longitudinal direction, while the vertical turbulence is limited by the depth of the stream and is smaller than the horizontal turbulence. Turbulent motions of many different length scales coexist in natural streams and rivers. The horizontal turbulence at the longitudinal direction can be significantly larger than the vertical dispersion and is generally required by simulating separately from vertical transport by applying large eddy simulations (Chu and Babarutsi 1988). Sometimes, the primary interest is to obtain cross-sectional average solutions for the contaminant concentration or water temperature profiles by integrating the transport equation over a cross section. Alternatively, it is often possible to estimate the turbulence from dye experiment measurements under certain conditions. The rapid rate of mixing, called dispersion of dye, is noted at the longitudinal direction, that is, a larger scale direction and much larger than the turbulent diffusion. In addition, the nonuniform velocity distribution can enhance dispersion mixing. Measurements indict that the dispersion coefficient (K) is much larger than the turbulent diffusion. To account for the effects on the cross-sectional averaged concentration attributed to the variation in velocity while crossing the channel cross section, the longitudinal dispersion coefficient (K) has been formulated by Potter et al. (2002) as

$$K = \alpha_k \frac{U^2 b^2}{K_z}, \tag{1.25}$$

where K = longitudinal dispersion coefficient (L^2/T), α_k = constant with the range from 0.001 to 0.016, U = mean flow velocity at longitudinal direction, b = channel width, and K_z = transverse dispersion coefficient (L^2/T).

The transverse dispersion coefficient accounts for the effects on the depth-averaged tracer concentration of depth variation in the transverse velocity. It is proportional to the local shear velocity (U_*) and the water depth (h) and can be calculated as (Potter et al. 2002)

$$K_z = \alpha_z U_* h, \tag{1.26}$$

where $U_* = \sqrt{gHS}$, average shear velocity, g = acceleration due to gravity, H = mean depth, and S = channel slope, and α_z = a constant ranging from 0.3 to 0.9.

The dispersion coefficients in rivers and streams, along with cross-sectional averaged velocity, are required to solve the transport equations for the water quality model. The cross-sectional average (laterally) mass and heat transport can be rewritten from Equations 1.12 and 1.13 as

$$\frac{\partial b\bar{C}}{\partial t}+\frac{\partial b\bar{U}\bar{C}}{\partial x}+\frac{\partial b\bar{W}\bar{C}}{\partial z}=\left(\frac{\partial\left(bK\frac{\partial\bar{C}}{\partial x}\right)}{\partial x}+\frac{\partial\left(bK_z\frac{\partial\bar{C}}{\partial z}\right)}{\partial z}\right)+\overline{br_{\mathrm{g}}}-\overline{br_{\mathrm{d}}}. \tag{1.27}$$

By defining E as the longitudinal dispersion coefficient for heat, the heat transport equation is

$$\frac{\partial b\bar{T}}{\partial t}+\frac{\partial b\bar{U}\bar{T}}{\partial x}+\frac{\partial b\bar{W}\bar{T}}{\partial z}=\left(\frac{\partial\left(bE\frac{\partial T}{\partial x}\right)}{\partial x}+\frac{\partial\left(bE_z\frac{\partial T}{\partial z}\right)}{\partial z}\right)+\frac{bS}{\rho c_v}. \tag{1.28}$$

By combining the mass or heat transport equation with the hydrodynamic model to solve the water quality problem has been studied extensively. Such examples can be found under the Reference column in Table 1.1.

1.4.2 Water Transport in Estuaries, Bays, and Harbors

The coastal regime of estuaries, bays, and harbors is a fascinating, diverse, and complex water system between the free-flowing river and the ocean. The estuaries region provides a home (nursery and adult habitat) for fish; controls flood, salt incursion, and pollution from the surrounding coastal environment; and serves as a recreation area and a transportation route. A better understanding of the basic flow circulation and associated transport mechanism is crucial to managing the ecosystem and preventing deterioration. The tides moving in and out of estuaries and the associated water density effects attributed to salinity are of particular importance in describing the water quality of estuaries, bays, and harbors.

Tides refer to the movement of water rising above and falling below the mean sea level, and the associated horizontal water movement into and out of an estuary is called tidal currents (Defant 1958). They are caused by the combined effects of the gravitational force exerted by the moon and the sun on the waters of the earth. These motions occur on a regular cyclical basis in terms of the regularity of the lunar and solar cycles. Some estuaries experience two almost even high tides and two low tides per day, called a *semidiurnal tide*. Some locations experience only one high and one low tide per day, called a *diurnal tide*. Some locations experience a *mixed tide*, which has two uneven tides a day, or sometimes only one high tide and one low tide per day. A typical oscillation of the tidal velocity is approximately sinusoidal, although many other oscillations may result in a variety of cyclical patterns. The variation in tides and currents is recorded by tide gauges, which can be downloaded from the National Oceanic and Atmospheric Administration (http://tidesandcurrents.noaa.gov/gmap3/). An important characteristic of estuarine flow is the net tidal flow over a tidal cycle or a given number of cycles, which is a significant parameter to estimate the distribution of estuarine water quality. The net flow at any location is the sum of the upstream flow from the up-basin drainage area, the flow from the well-defined tributary, the flow from the waste input, and the incremental flow between these latter point flow inputs and local runoff.

The complicated circulation pattern in the fully developed estuarine region resulted from the vertical stratification of freshwater inflow. The density differences attributed to salinity and temperature in different regions of an estuary or bay combine with a complex circulation pattern, making the transport phenomenon a complicated process. It is an important task to examine the relationship between temperature, salinity, and the density of water. Various forms of equations have been developed. For example, Eckart (1958) developed the equation as

$$\rho = \frac{5890 + 38T - 0.375T^2 + 3S}{5890.72 + 37.774T - 0.33625T^2 - 1.706S - 0.1TS}, \tag{1.29}$$

where S = salinity (M/L^3) and T = temperature (°C).

Crowley (1968) gave the equation of state as

$$\rho = 1 + \{10^{-3}[(28.14 - 0.0735T - 0.00469T^2) + (0.802 - 0.002T)(S - 35)]\}. \tag{1.30}$$

Note that S is expressed in parts per thousand.

The three-dimensional flow fields in estuaries can be expressed as a Reynolds-averaged Navier–Stokes equation by employing the eddy-viscosity concept and including the Coriolis effect (Sheng 1983):

$$\begin{gathered}
\frac{\partial u}{\partial x} + \frac{\partial v}{\partial y} + \frac{\partial w}{\partial z} = 0 \\
\rho\left(\frac{\partial u}{\partial t} + \frac{\partial uu}{\partial x} + \frac{\partial vu}{\partial y} + \frac{\partial wu}{\partial z}\right) = \rho f v + \frac{-\partial p}{\partial x} + \frac{\partial}{\partial x}\left(\mu_H \frac{\partial u}{\partial x}\right) + \frac{\partial}{\partial y}\left(\mu_H \frac{\partial u}{\partial y}\right) + \frac{\partial}{\partial z}\left(\mu_v \frac{\partial u}{\partial z}\right) \\
\rho\left(\frac{\partial v}{\partial t} + \frac{\partial uv}{\partial x} + \frac{\partial vv}{\partial y} + \frac{\partial wv}{\partial z}\right) = -\rho f u + \frac{-\partial p}{\partial y} + \frac{\partial}{\partial x}\left(\mu_H \frac{\partial v}{\partial x}\right) + \frac{\partial}{\partial y}\left(\mu_H \frac{\partial v}{\partial y}\right) + \frac{\partial}{\partial z}\left(\mu_v \frac{\partial v}{\partial z}\right) \\
\rho\left(\frac{\partial w}{\partial t} + \frac{\partial uw}{\partial x} + \frac{\partial vw}{\partial y} + \frac{\partial ww}{\partial z}\right) = \frac{-\partial p}{\partial z} - g + \frac{\partial}{\partial x}\left(\mu_H \frac{\partial w}{\partial x}\right) + \frac{\partial}{\partial y}\left(\mu_H \frac{\partial w}{\partial y}\right) + \frac{\partial}{\partial z}\left(\mu_v \frac{\partial w}{\partial z}\right).
\end{gathered} \tag{1.31}$$

The Coriolis terms (*fv and –fu*) are included in the basic equation to reflect the effect of earth's rotation on fluid motion.

Following mass and heat transport equations, salinity and temperature transport equations are expressed as

$$\frac{\partial S}{\partial t} + \frac{\partial uS}{\partial x} + \frac{\partial vS}{\partial y} + \frac{\partial wS}{\partial z} = \frac{\partial}{\partial x}\left(D_H \frac{\partial S}{\partial x}\right) + \frac{\partial}{\partial y}\left(D_H \frac{\partial S}{\partial y}\right) + \frac{\partial}{\partial z}\left(D_v \frac{\partial S}{\partial z}\right) \tag{1.32}$$

$$\frac{\partial T}{\partial t}+\frac{\partial uT}{\partial x}+\frac{\partial vT}{\partial y}+\frac{\partial wT}{\partial z}=\frac{\partial}{\partial x}\left(D_{\mathrm{H}}\frac{\partial T}{\partial x}\right)+\frac{\partial}{\partial y}\left(D_{\mathrm{H}}\frac{\partial T}{\partial y}\right)+\frac{\partial}{\partial z}\left(D_{\mathrm{v}}\frac{\partial T}{\partial z}\right), \quad (1.33)$$

where D_{H} = the horizontal diffusion coefficient and D_{v} = the vertical turbulent mixing coefficient.

With the results from temperature and salinity, density can be estimated from Equations 1.29 and 1.30. A similar transport equation can be formulated for modeling transport of pollutants.

Hydrodynamic and salinity transport modeling is important in studies on estuaries to provide the evolution of the ecosystem. It is a tool for predicting fluid flow, salinity/temperature distribution, and concentration level in the study area by applying a numerical scheme. The model requires appropriate initial and boundary conditions as described by the hydrodynamic quantities of tide, wind, and river inflows, as well as transport quantities of salinity, temperature, and concentration. The development of a comprehensive three-dimensional hydrodynamic and salinity/temperature transport model started in 1973. CH3D-IMS, CH3D-SSMS, and EFDC models (as listed in Table 1.1) have been developed, calibrated, and validated to predict the future ecosystem in the different estuaries and bays in the United States.

1.4.3 Water Transport in Lakes and Reservoirs

Lakes and reservoirs vary from small backyard ponds to the magnificent and monumental large lakes, such as Lake Superior, one of the Great Lakes. The study of the physical, chemical, and biological behavior of lakes is *limnology*. This section focuses on the water transport process in lakes and reservoirs. A lake or reservoir is an open system with an upper free surface boundary exposed to the atmosphere, relatively low flow velocity, and development of significant vertical gradients in temperature and other water quality variables. The physical relationships for a lake are typical in area–depth and volume–depth curves. The time for emptying the lake or reservoir is defined as *detention time*. It can be calculated as the volume divided by the flow rate out of the lake or reservoir. To determine the hydrologic balance of a lake, the change in volume, surface inflow and outflow, and precipitation can be measured easily with the designed flow measurement tools. However, evaporation loss and exchange with groundwater sometimes are challenging. The evaporation loss depends on the incoming solar radiation, wind, air temperature, and water temperature. It can be estimated by measurement of water loss in evaporation pans or using a water balance equation. The gain and loss from groundwater are a function of the regional hydraulic head, specific discharge, and stream lines, which can be complex systems depending on the location of the lake or reservoir.

The hydrodynamics of lake and reservoir is derived from wind shear, solar radiation, heat losses, inflows, and outflows. Lakes and reservoirs in general are not well mixed and usually imply vertical stratification by density gradients. Stratification can be caused by thermal effects and the inflow of water with varying densities. Solar radiation causes the heat exchange between water and air at the water surface to increase the water temperature at the upper layer of the lake. The degree of the stratification plays a significant role in water quality by isolating pollutants in regions. For example, the water in the bottom of the lake is cut off from the atmosphere to reduce oxygen supply and this can kill the fishes in the lake and have an adverse effect on water quality. Quantifying turbulent transport phenomena is one of the major challenges in lake and reservoir water quality analysis. Hydrodynamic and transport models have been studied for several decades. The two-dimensional laterally averaged transport equation can be expressed as

$$\begin{aligned}
&\frac{\partial ub}{\partial x}+\frac{\partial wb}{\partial z}=q_{\mathrm{t}}\\
&\frac{\partial u}{\partial t}+u\frac{\partial u}{\partial x}+w\frac{\partial u}{\partial z}=\frac{-\partial(bp)}{\rho b\partial x}+\frac{\partial}{b\partial x}\left(bK_x\frac{\partial u}{\partial x}\right)+\frac{\partial}{b\partial z}\left(bK_z\frac{\partial u}{\partial z}\right)+\frac{q_{\mathrm{t}}}{b}(u_{\mathrm{t}}-u)\\
&\frac{\partial w}{\partial t}+\frac{\partial uw}{\partial x}+\frac{\partial ww}{\partial z}=\frac{-\partial(bp)}{\rho b\partial z}-g+\frac{\partial}{b\partial x}\left(bK_x\frac{\partial w}{\partial x}\right)+\frac{\partial}{b\partial z}\left(bK_z\frac{\partial w}{\partial z}\right).
\end{aligned}\tag{1.34}$$

Similarly, the mass and heat transport equations in lake can be expressed as

$$\begin{aligned}
&\frac{\partial C}{\partial t}+u\frac{\partial C}{\partial x}+w\frac{\partial C}{\partial z}=\frac{\partial}{b\partial x}\left(bD_x\frac{\partial C}{\partial x}\right)+\frac{\partial}{b\partial z}\left(bD_z\frac{\partial C}{\partial z}\right)+\frac{q_{\mathrm{t}}}{b}(C_{\mathrm{t}}-C)\\
&\frac{\partial T}{\partial t}+u\frac{\partial T}{\partial x}+w\frac{\partial T}{\partial z}=\frac{\partial}{b\partial x}\left(bD_x\frac{\partial T}{\partial x}\right)+\frac{\partial}{b\partial z}\left(bD_z\frac{\partial T}{\partial z}\right)+\frac{q_{\mathrm{t}}}{b}(T_{\mathrm{t}}-T),
\end{aligned}\tag{1.35}$$

where q_{t} = rate of mass addition by lateral flows. The additional heat attributed to solar radiation or heat loss to the sediment has also been accounted for in the model study. Case studies can be developed using QUAL2E, CE-QUAL-W2, and WASP as listed in Table 1.1.

1.5 SUMMARY AND CONCLUSION

This chapter introduced water transport principles, various substances, and heat transport characteristics in pipes and natural water bodies, such as rivers and streams, estuaries, bays and harbors, lakes, and reservoirs. The gradient flux laws combine a fluid property with the state variable gradient. The principles of conservation of heat, mass, and momentum are combined with the flux laws and provided the basis for the advection–diffusion (dispersion) transport in the environment. Three different scales of transport processes are introduced. A substance that is spread by dispersion is significantly larger than that spread by turbulence diffusion. Turbulence diffusion is also much larger than molecular diffusion. However, it is not always necessary to ignore molecular diffusion. Hydrodynamic and water quality models have been developed to solve the transport equation of heat, mass, and momentum in natural water bodies. This chapter has presented the typical transport procedures and useful water quality models in rivers and streams, estuaries, bays and harbors, lakes, and reservoirs.

REFERENCES

Batchelor G. K. 1967. *An Introduction to Fluid Dynamics*. Cambridge Mathematical Library series. Cambridge University Press.

Bedient, P. B., W. C. Huber, and B. E. Vieux. 2008. *Hydrology and Floodplain Analysis*. 4th edition. Prentice Hall, Inc., NJ.

Centre for Ecology and Hydrology. http://www.ceh.ac.uk/index.html (accessed September 30, 2014).

Chow, V. T. 1959. *Open-Channel Hydraulics*. McGraw-Hill, New York.

Chu, V. H. and S. Babarutsi. 1988. Confinement and bed-friction effects in shallow turbulent mixing layers. *J. Hydraulic Engineering*, 114: 1257–1274.

Crowley, W. P. 1968. A global numerical ocean model: Part I. *J. Comp. Phys.*, 3: 111–147.

Csanady, G. T. 1973. *Turbulent Diffusion in the Environment*. Geophysics and astrophysics monographs 3. D. Reidel Publishing Company, Boston.

Defant, A. 1958. *Ebb and Flow, the Tides of Earth, Air and Water*. University of Michigan Press, Ann Arbor, MI.

Denmark Hydrology Institute. http://www.mikebydhi.com/products (accessed September 30, 2014).

Eckart, C. 1958. Properties of water. Part II—The equation of state of water and seawater at low temperature and pressure. *Am. J. Sci.*, 256: 225–240.

Edinger, J. E. 2002. *Waterbody Hydrodynamic and Water Quality Modeling: An Introductory Workbook and CD-ROM on Three-Dimensional Waterbody Modeling*, American Society of Civil Engineering.

HEC 2006. http://www.hec.usace.army.mil/software/hec-hms/ (accessed September 30, 2014).

Ho, C. and S. Khalsa. 2009. EPANET-BAM: Water quality modeling with incomplete mixing in pipe junctions. *Water Distribution Systems Analysis 2008*: 1–11. doi: 10.1061/41024(340)87.

Leopold, L. B. and T. Maddock. 1953. The hydraulic geometry of stream channels and some physiographic implications. Geological Survey Professional Paper 252. Washington, D.C.

Li, Z. 2006. Network water quality modeling with stochastic water demands and mass dispersion, PhD diss., University of Cincinnati.

Portland State University, Water Quality Research Group, http://www.ce.pdx.edu/w2/ (accessed September 30, 2014).

Potter, M., D. Wiggert, and T. Shih. 2002. Mechanics of Fluids, 3rd edition, *Cengage Learning*, 862 pp.

Rifai, H. S., S. M. Brock, K. B. Ensor, and P. B. Bedient. 2000. Determination on low-flow characteristics for Texas streams, *J. Water Resour. Plan. Manage.*, 126: 310–319.

Rossman, L. A., P. F. Boulos, and T. Altman. 1993. Discrete volume-element method for network water-quality models. *J. Water Resour. Plan. Manage.*, 119(5): 505–517.

Rossman, L. A., R. M. Clark, and W. M. Grayman. 1994. Modeling chlorine residuals in drinking-water distribution systems. *J. Environ. Eng.*, 120(4): 803–820.

Rossman, L. A. and P. F. Boulos. 1996. Numerical methods for modeling water quality in distribution systems: A comparison. *J. Water Resour. Plan. Manage.*, 122(2): 137–146.

Rossman, L. A. and W. M. Grayman. 1999. Scale-model studies of mixing in drinking water storage tanks. *J. Environ. Eng.*, 125(8): 755–761.

Sheng, Y. P. 1983. *Mathematical modeling of three-dimensional coastal currents and sediment dispersion: Model development and application.* Technical Report CERC-83-2, U.S. Army Engineer Waterways Experiment Station. Vicksburg, MS.

Stow, C. A., C. Roessler, M. E. Borsuk, J. D. Bowen, and K. H. Reckhow. 2003. Comparison of estuarine water quality models for total maximum daily load development in Neuse River Estuary, *J. Water Resour. Plan. Manage.*, 129(4): 307–314. doi: 10.1061/(ASCE)0733-9496.

The University of Mississippi, National Center for Computational Hydroscience and Engineering, http://www.ncche.olemiss.edu/sw_download (accessed September 30, 2014).
Thomann, R. V. and J. A. Mueller. 1987. *Principles of Surface Water Quality Modeling and Control.* Harper Collins Publishers Inc., New York.
Thomas, D. M. and M. A. Benson. 1970. *Generalization of Streamflow Characteristics from Drainage-Basin Characteristics*. U.S. Geological Survey WS Paper 197t, U.S. Governmental Printing Office, Washington, D.C.
US Army Corps of Engineers, Hydrologic Engineering Center, http://www.hec.usace.army.mil/software/hec-ras/ (accessed September 30, 2014).
USDA. http://www.hydrocad.net/tr-20.htm (accessed September 30, 2014).
USEPA. http://www2.epa.gov/water-research/storm-water-management-model-swmm (accessed September 30, 2014).
USEPA. http://epa.gov/athens/wwqtsc/html/efdc.html (accessed September 30, 2014).
USEPA. http://epa.gov/athens/wwqtsc/html/qual2k.html (accessed September 30, 2014).
USEPA. http://epa.gov/athens/wwqtsc/html/wasp.html (accessed September 30, 2014).
USEPA. http://water.epa.gov/type/rsl/monitoring/vms51.cfm (accessed September 30, 2014).
USEPA. http://www.epa.gov/nrmrl/wswrd/dw/epanet.html (accessed September 30, 2014).
USGS. http://water.usgs.gov/edu/streamflow2.html (accessed September 30, 2014).
Vandeweerd, V., M. Cheatle, B. Henricksen, M. Schomaker, M. Seki, and K. Zahedi. 1997. Global Environment Outlook (GEO)—UNEP Global State of Environment Report.
Wampler, P. J. 2012. Rivers and streams—Water and sediment in motion. *Nature Education Knowledge*, 3(10): 18.

2 Groundwater Contaminant Transport Mechanisms and Pollution Prevention

Dorina Murgulet

CONTENTS

2.1 GROUNDWATER FLOW

2.1.1 Basic Concepts of Groundwater Flow

Groundwater is stored in and moves slow through stratigraphic formations located at varying depths with different hydrogeologic characteristics (i.e., permeabilities). These water-bearing formations are called aquifers and can be open to direct recharge from the surface (i.e., unconfined aquifers) and completely or partially isolated from surface recharge (i.e., confined or semiconfined aquifers). An aquifer may be composed of one or multiple layers of unconsolidated silt, sand and gravel, sandstone or cavernous limestone, fractured granite, or basalt with sizable openings or rubbly lava flows. In terms of water storage, groundwater is the largest single resource of freshwater available for human consumption worldwide. Groundwater, which occupies empty spaces (i.e., pores) between sediment grains or within fractures in rocks or unconsolidated sediment in the subsurface, is inevitably replenished (i.e., recharge) rainfall seeps and from surface infiltration of lake and river water; it is important to precisely delineate the location of major areas of recharge to limit contamination from surface sources such as waste disposal sites, application of fertilizers, and so on [1]. The amount of water that accumulates as groundwater and is available for extraction is dependent on the porosity and permeability of the subsurface materials. If all the water is removed from a saturated aquifer material through drying (in a laboratory), the volume removed represents the *total porosity* of the material, which is the fraction of voids within the total volume of solids (Figure 2.1) [2].

Permeability is a measure of how easily water can flow through rock or sediment. Generally, high permeability formations are characterized by high porosity; however, this is not the case for clays, which often have very high porosity but low permeability (could be as much as 50%); individual clay particles are very small (usually on the order of 10^{-6} meters [1 μm]) in diameter, and consequently, the pores are abundant but very small, hindering flow through the clay by two processes (Figure 2.1). First, the surface area of the clay particles is much greater than that of the pore space volume producing high frictional resistance to flow between clay crystals. Second, clay particles have weak electrostatic forces that bind water to their particles, holding it in place. Layers of sediment with very low permeabilities such as clays are impermeable and act as barriers to groundwater flow, otherwise known as confining beds. Groundwater is lost through discharge where water leaves the aquifer; this can occur naturally as springs, phreatophyte wetlands, playas (i.e., dry lakes that

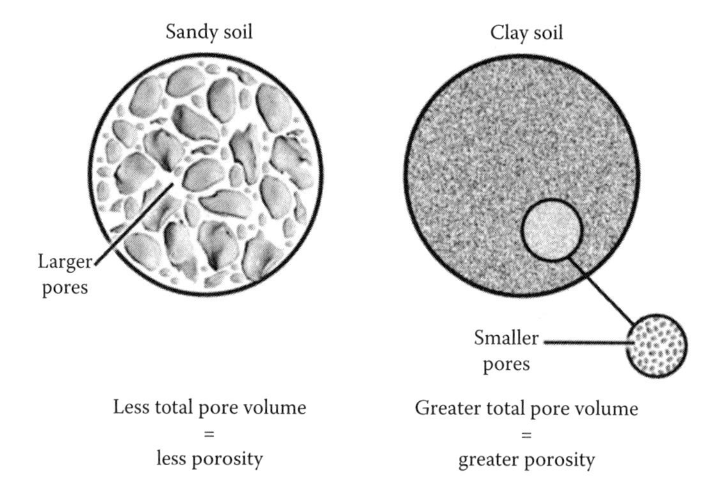

FIGURE 2.1 Pore space in sandy versus clay formations. (From The COMET Program; *Basic Hydrologic Science Course Understanding the Hydrologic Cycle Section Five: Groundwater* [cited August 24, 2014]; Available from: http://www.goesr.gov/education/comet/hydro/basic/HydrologicCycle/print_version/05-groundwater.htm.)

release water to the atmosphere through evaporation directly from the ground), or as baseflow to rivers and other surface waters or anthropogenically through production/pumping using wells. Increasing groundwater pumping to meet the growing water needs can lead to depletion and mining of this resource, specifically when extraction rates exceed recharge.

Groundwater flow is governed by differences in hydraulic pressures and the resulting gradients, and in general, it closely mimics the surface topography. Groundwater flows from areas of higher hydraulic pressure or potential energy (i.e., recharge areas) to areas of lower hydraulic pressure (i.e., discharge), or in the direction of decreasing pressure/energy or total head; the rate and direction of flow are highly dependent on permeability. Preferentially, water will flow through higher-permeability zones. Groundwater flows through small pore spaces in the subsurface and, thus, flow rates are relatively slow as opposed to surface waters in the form of streams. The fundamental law that dictates water flow through a porous medium (i.e., groundwater flow) was developed by Henry Darcy [3]. Darcy's law states that discharge Q of water through a column of sand is proportional to the cross-sectional area A of the sand column, the media hydraulic conductivity K, and the difference in piezometric head between the ends of the column, $h_1 - h_2$, and inversely proportional to the length of the column L. That is,

$$Q = KA\frac{h_2 - h_1}{L} r Q = KAi. \tag{2.1}$$

The fraction $\frac{h_2 - h_1}{L}$ defines the hydraulic gradient i over the length of the column. Most often, it is more convenient to refer to the specific discharge q [LT^{-1}], given simply by Q/A rather than to the total discharge Q. Darcy's law if then written in terms of specific discharge and the hydraulic gradient:

$$q = \frac{Q}{A} = -K\frac{h_2 - h_1}{L}. \tag{2.2}$$

There is a fundamental relationship between the specific discharge q and the groundwater velocity v. It is important to note that, in the case of a pipe filled with sediment, specific discharge is different from the water velocity. That is because the cross-sectional area open to flow is dependent on the size of pore spaces and their interconnectedness. The cross-sectional area open to flow when the pipe is filled with sediment is significantly smaller, but for the same specific discharge, water will travel faster than if the pipe was empty. Therefore, the velocity through the porous medium is higher than the specific discharge. The effective area to flow is An_e, where n_e is the effective porosity of rock. Hence, groundwater velocity is calculated by normalizing the specific discharge to the pore space from which water can be extracted:

$$v = \frac{q}{n_e} \tag{2.3}$$

Groundwater storage is directly related to the total *effective porosity* of an aquifer, which is the amount of void spaces in a unit volume of rock where water can be stored and extracted. Good aquifers may have up to 30% effective porosity but higher than this have been observed. The volume of water that can be extracted from aquifers is dependent on the amount of effective porosity. A characteristic closely related to this is the *specific yield* (Sy) of the aquifer, which represents the volume of water per unit volume of aquifer that can be extracted by pumping [4]. Effective porosity and specific yield are most often considered to be equivalent. The specific yield is an extremely important aquifer characteristic that is used to estimate the total volume of available groundwater from the saturated thickness (ST) (Figure 2.2) [2]:

$$\text{Volume} = \text{Area} \times \text{ST} \times \text{Sy} \tag{2.4}$$

Hydraulic conductivity, which is a constant of proportionality, accounts for the porosity and permeability of medium and certain characteristics of the fluid such as viscosity and density: it offers a quantitative comparison of permeability for different rocks [1]. Hydrologists apply Darcy's law to determine the direction of groundwater flow, to evaluate flow rates, and to monitor groundwater availability [3]. Darcy's law is only applicable to laminar flow. Laminar flow typically occurs when the viscosity of fluid is higher and the velocity is lower; fluid basically moves slowly in layers, along a defined direction without significant mixing among layers. On the other hand, turbulent flow occurs in environments with high velocities and considerable mixing occurs [3]. Most groundwater flows through sediment in a laminar fashion although turbulent flow may be encountered nearby strong discharge points and in karstified aquifers.

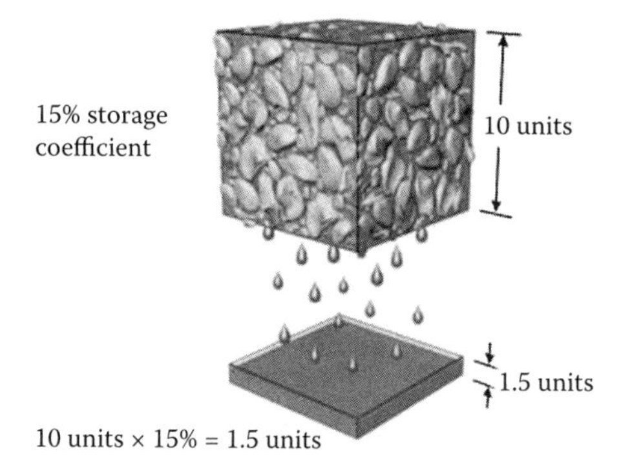

FIGURE 2.2 Illustration of water yield from a given volume of saturated aquifer material. In this example, the storage coefficient is 0.15, meaning that 15% of the total volume of water in the aquifer material will drain freely by gravity. The remaining aquifer volume (i.e., 85% in this example) is composed of water that does not drain by gravity and aquifer material such as rock, sand, gravel, or clay. If the given volume of sample is completely drained (only achievable through drying of sample in laboratory), resulting in a 10-unit drop in the water level, only 1.5 depth units (or 15% of the 10 units) of water are produced. (From The COMET Program; *Basic Hydrologic Science Course Understanding the Hydrologic Cycle Section Five: Groundwater* [cited August 24, 2014]; Available from: http://www.goesr.gov/education/comet/hydro/basic/HydrologicCycle/print_version/05-groundwater.htm.)

2.1.2 Aquifer Vulnerability

An aquifer is a water-bearing hydrostratigraphic unit composed of unconsolidated materials or fractured rocks, with hydraulic conductivity and transmissivity high enough to allow water to flow to a well or any natural form of discharge (i.e., groundwater can be extracted easily). Layers of low permeability or transmissivity, acting as confining units, hamper the movement of water in and out from an aquifer (i.e., cross-flow between aquifers; direct recharge to aquifers from local sources). Depending on the degree of confinement, these units are also known as an aquitard (i.e., acts as a semipervious membrane) or aquicludes (i.e., acts as an impervious membrane) [3]. There are two major types of aquifers: confined and unconfined (Figure 2.3) [5]. The upper boundary of an unconfined aquifer is called the water table and water is under atmospheric conditions (i.e., pressure head is zero); there is a direct connection between the groundwater and the atmosphere, and a low permeable confining layer or aquitard does not exist at the upper boundary. A confined aquifer is overlain by a confining unit and the water level rises to above the top of the aquifer; the level to which it rises is called the potentiometric surface. Perched aquifers are found within the unsaturated zone where there is a clay or impermeable lens to serve as a barrier to water percolating from the surface to the water table [3].

Highly permeable and unconfined aquifers are generally at risk of contamination from surface and near-surface sources. Overdeveloped aquifers where production

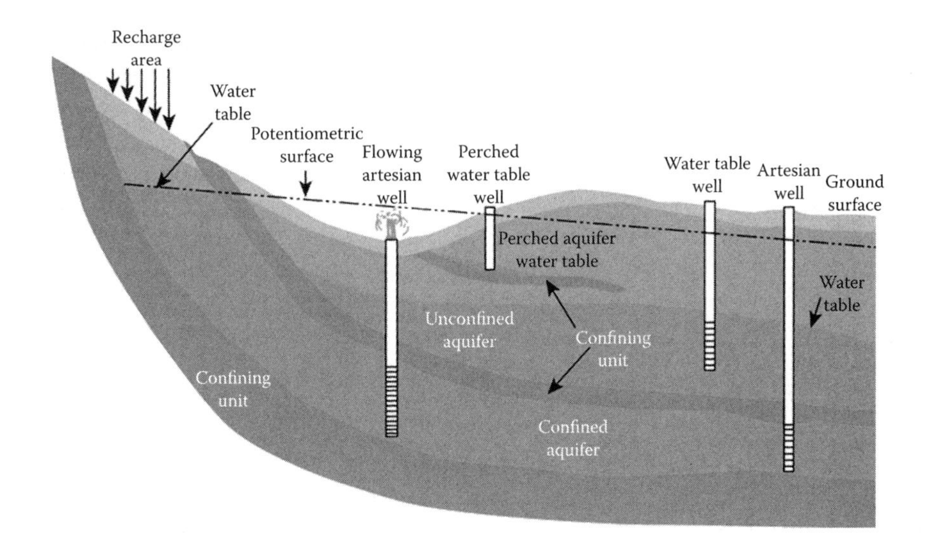

FIGURE 2.3 Cartoon showing different types of aquifers: confined, unconfined, and perched. Recharge zones are typically at higher altitudes but can occur wherever water enters an aquifer, such as from rain, snowmelt, river and reservoir leakage, or irrigation. Discharge zones can occur anywhere and not only through wells such as depicted in the diagram, but as springs near the stream and in wetlands at low altitude, wells, and high-altitude springs. (From NGWA Press Publication, *Ground Water Hydrology for Water Well Contractors*; Chapter 14, 1999 [cited July 14, 2014]; Available from: http://www.ngwa.org/fundamentals/hydrology/pages/unconfined-or-water-table-aquifers.aspx.)

rates exceed recharge are also at risk of contamination. This risk of contamination is referred to as aquifer vulnerability. For this reason, aquifer protection is essential for sustainable use of groundwater resources [6]. Vulnerability, in this sense, is defined as the sensitivity of an aquifer to an imposed contaminant load, which is determined by the intrinsic characteristics of the aquifer. For instance, hydraulic conductivity of the overlying strata is a key parameter to assessing aquifer vulnerability [6]. It is important to note that "vulnerability" does not directly lead to a "pollution risk." The characteristics that make an aquifer vulnerable are intrinsic and are fully dependent on the pollutants that may or may not infiltrate. This notion of intrinsic vulnerability needs to be taken into consideration. If the aquifer's vulnerability is deemed low, but the contaminant load is large or highly concentrated, the pollution risk can still be high. On the other hand, if an aquifer is classified as highly vulnerable, but there is no significant contaminant load, the risk of pollution is low [7]. Aquifer vulnerability is either a qualitative or quantitative measure of how much the physical and biochemical characteristics of the subsurface contribute to or inhibit the transport of pollutants into or within the aquifer itself. There are different methods available to evaluate aquifer vulnerability, each utilizing different indicators including physical and chemical measurements, integrated hydrological modeling, and statistical methods. Some of the models developed to evaluate aquifer vulnerability are DRASTIC, GOD, EPIK, and AVI [8]. Details on these methods are not presented in this chapter. In addition, there has been intensive effort to meld and adapt certain methods to focus on specific contaminants at particular locations.

Infiltration and percolation of meteoric water and contaminants from the land surface to the water table are governed by several factors including drainage and conductivity capacity of the soils, water table elevations, regional topography, and the land covers [9,10]. It has been demonstrated that aquifer vulnerability is strongly influenced by the hydrogeological characteristics and flow dynamics of the aquifers, as well as the specific type of contaminants (i.e., toxicity, physicochemical properties, and fate/transport characteristics) identified for a region [11–13]. Aquifers are often most vulnerable to contamination where direct connection with the land surface (i.e., unconfined or water-table aquifer) via outcrops exists. In most developed and urbanized areas (i.e., with impervious surfaces covering most of the area), recharge rates will be drastically altered and typically significantly reduced [14]. On the other hand, for some regions, surface runoff or artificial recharge (i.e., irrigation or channelized runoff) can enhance recharge to aquifers [15–18]. However, this phenomenon is typically isolated to very small and specific areas of the aquifers and the contribution to recharge is limited when compared to the regional extent of most aquifer systems. In any case, given the complexity and variability of the geology and dynamics of the aquifers involved, a vulnerability index does not always accurately predict contaminant distribution [9].

2.2 GROUNDWATER CONTAMINATION

2.2.1 Sources of Contamination (Natural and Anthropogenic)

Groundwater contamination is defined as dissolved solutes in water such that it is rendered unfit for human use or harmful to ecosystem health [3]. Since the 1970s, there has been increasing attention paid to hazardous waste sites in the United States.

Before that, little was known about the vulnerability of groundwater that supplies freshwater to millions of people in the United States and around the world. It was common practice for industrial waste to be dumped randomly at the locations where they were being produced. Disposal practices such as injection wells and underground storage tanks were considered to be safe, but significant and sometimes severe leakage became a significant problem [19]. Certain state-level restrictions were being imposed and enforced before the Federal Government finally enacted hazardous waste restriction laws when, in 1976, it passed the Resource Conservation and Recovery Act, known as the RCRA [20]. The RCRA regulates the disposal of various forms of waste. In 1980, the Federal Government passed the CERCLA, also known as Superfund. Through this act, the criteria to make a National Priorities List (NPL) for probable hazardous waste sites were implemented [21].

Sources of groundwater contamination are abundant and widespread in the United States. They can include underground and aboveground storage tanks, injection wells, landfills, pesticides, pipelines, radioactive waste disposal sites, acid mine drainage, saltwater intrusions, wastewater lagoons, and possibly thousands of accidental spills [19]. Most of the listed sources are relatively small and can be considered point sources, depending on how widespread they are. For example, landfills and farmlands sprayed with pesticides can sometimes be quite extensive and would be classified as nonpoint sources. Another example of a nonpoint source is polluted precipitation. Beyond just anthropogenic sources as listed above, there are numerous naturally occurring contaminants that may pollute groundwater. In fact, most natural waters have dissolved constituents such as lead and arsenic, which, even though are in trace amounts (i.e., a few micrograms per liter), may pose serious contamination problems in groundwater [22]. A large range of dissolved constituents occur naturally in groundwater as a result of interaction with the atmosphere, the surficial environment, host rock, and residence times. The main dissolved chemical components commonly found in groundwater are bicarbonate, sulfate, chloride, potassium, sodium, calcium, and magnesium [23]. Most of these elements are found in rocks and minerals and are either weathered into the soil and surface waters and gradually seep into groundwater or are leached from rock as a result of chemical exchanges between groundwater and minerals (i.e., water-rock interaction) [24]. An in-depth knowledge of chemical and physical weathering, fluid mechanics, and flow characteristics according to Darcy's law is required to sufficiently understand and prevent groundwater contamination.

2.2.1.1 Nonaqueous Phase Liquids in Groundwater

Nonaqueous phase liquids (NAPLs) are hydrocarbons that exist as a separate, immiscible phase when in contact with water or air. In the recent years, NAPL contamination has become an issue of increasing environmental concern. NAPLs are contaminants that, like oil, are mostly immiscible in water. Their very distinct physiochemical characteristics set them apart from other types of harmful contaminants; NAPL spills can result in the formation of large toxic plumes of particular concern that pose a high contamination problem [25]. Petroleum products and chlorinated solvents enter the subsurface as nonaqueous-phase solutions. Sources of NAPLs include petroleum storage tanks, accidental spills, and leakage from natural deposits. There are two classes of NAPLs: light NAPLs (LNAPLs), which have densities

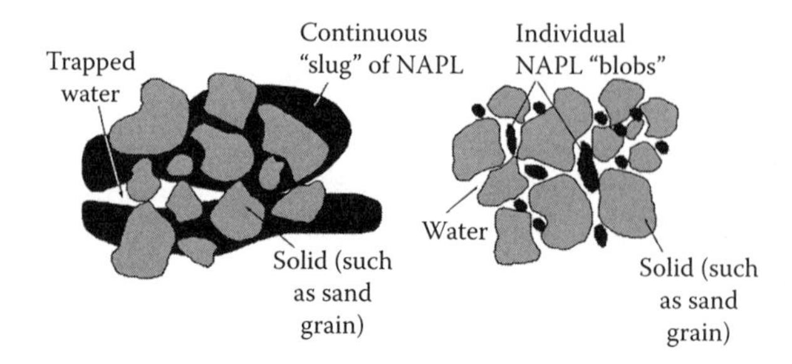

FIGURE 2.4 Mobile versus residual NAPL.

less than that of water, and dense NAPLs (DNAPLs), which have densities greater than that of water. Differences in the chemical and physical properties of NAPLs and water prevent mixing of the two fluids and lead to formation of a physical interface between the two [26]. These contaminants, however, can partially dissolve into water at very slow rates. Migration of NAPLs is a complex process governed by gravity, buoyancy, capillary forces, soil and aquifer texture, and NAPL chemical and physical properties (i.e., light vs. dense phase). Furthermore, distribution in the subsurface is dictated by four major processes: volatilization (Henry's law), dissolution (solubility), sorption (revealed by tailing effects in pump and treat [P&T] remediation systems), and biodegradation (mostly for the aqueous phase). Nonaqueous-phase liquids in the subsurface can be distributed as pools or "continuous slug" of significant size and thickness and relatively high solute saturations (Figure 2.4) [27]. The solute is found within the sediment pores as either mobile-phase (or free-phase) NAPL or as residual (trapped) NAPL (Figure 2.4). Mobile-phase NAPL is a continuous mass of solute that can flow, provided there is a hydraulic gradient. Depending on the pore sizes and other physical and chemical characteristics of the solute (density, viscosity) and subsurface (i.e., interfacial tension, wettability), a portion of the flowing NAPL will remain trapped between the soil particles and may not easily move in response to hydraulic differences (i.e., hydraulic gradients) (Figure 2.4).

2.2.1.1.1 DNAPLs and LNAPLs

The transport of NAPLs, both dense and light, is largely determined by their physical and chemical characteristics. NAPLs are mostly immiscible in groundwater, and therefore flow separately within the groundwater system. When released in the environment, NAPLs (i.e., LNAPL or DNAPL) migrate downward under the force of gravity and move through the unsaturated zone where a fraction of the hydrocarbon will be retained by capillary forces as residual globules in the soil pores. This generally depletes the plume mass, and if only a small volume has been released, NAPL transport will cease. On the other hand, if sufficient NAPL is released, a small part will be retained in the unsaturated zone while the largest part will migrate downward to the water table. Because of the large differences in densities, DNAPL (heavier than water) will migrate until it encounters a physical barrier (e.g., low permeability strata) while LNAPL (lighter than water) will float on top of the water table owing to buoyancy forces (Figure 2.5) [28].

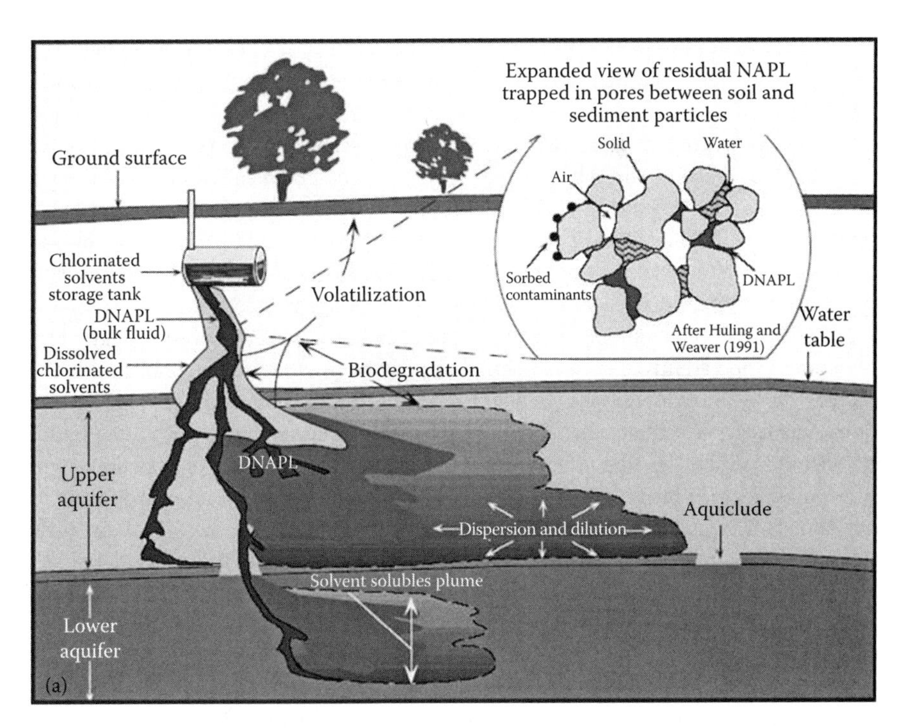

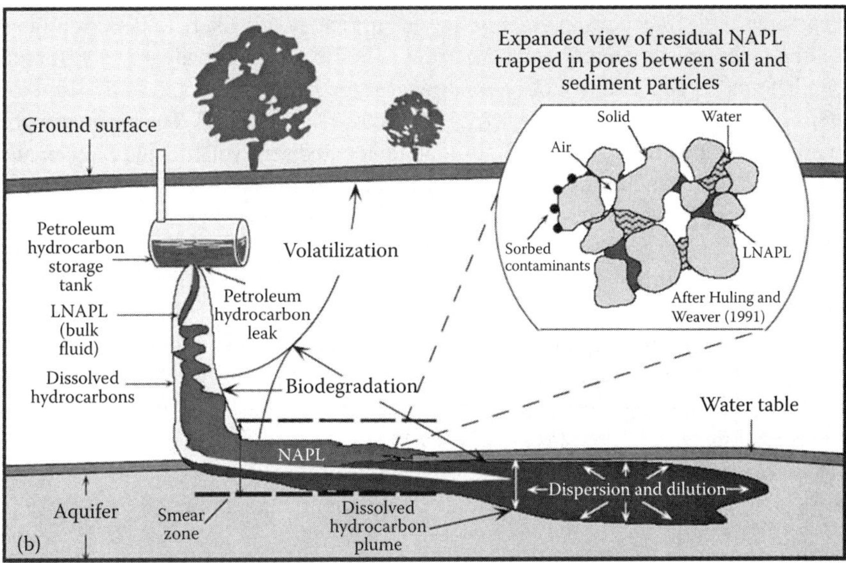

FIGURE 2.5 DNAPLs (a) such as dry-cleaning solvents sink in water and accumulate at the top of a confining layer while LNAPLs (b) such as gasoline float on water. (From Stewart S., *Groundwater Remediation*. Environmental Geoscience: Environmental Science in the 21st Century—An Online Textbook [cited August 31, 2014]; Available from: http://oceanworld.tamu.edu/resources/environment-book/groundwaterremediation.html.)

LNAPLs may move laterally as a continuous layer along the upper boundary of the water-saturated zone owing to gravity and capillary forces [25]. Common forms of LNAPLs are petroleum hydrocarbon liquids such as oil, gas, petrochemicals such as benzene and benzene derivatives, toluene, and xylene [29]. Polychlorinated biphenyl, known as PCB, is a highly toxic synthetic organic chemical compound that was banned in the United States in 1979. Before this, they were heavily used as dielectric and coolant fluids [30]. LNAPLs commonly affect groundwater quality at many sites across the world; the release of petroleum products is the most common LNAPL-related groundwater contamination problem. These products are characteristically composed of a mixture of organic chemicals with varying degrees of water solubility. Under ideal conditions, some additives such as methyl tertiary-butyl ether and alcohols are highly soluble, while benzene, toluene, ethylbenzene, and xylenes are slightly soluble; other components such as *n*-dodecane and *n*-heptane components are list soluble. The physical and chemical properties that affect the transport and fate of LNAPL compounds make them potential long-term sources for continued groundwater contamination at many sites [31]. However, because LNAPLs are generally confined to the top of the water table, remediation efforts are largely more successful compared to those of DNAPLs.

DNAPLs are heavier than water and therefore sink through the pores and cracks of the geologic formations to the bottom of aquifers; this makes characterization of the plume's shape and extent more difficult. DNAPLs have been identified at numerous hazardous waste sites and their presence is suspected at many more. DNAPLs are largely undetected as a result of the complexity of physical and chemical variables influencing their transport and fate in the subsurface; this is likely to be a significant limiting factor in site remediation. A partial list of the most prevalent DNAPL compounds found at superfund sites by a national screening is presented in the 1991 US Environmental Protection Agency (EPA) Groundwater issue [31]. These compounds, generally included in the halogenated/nonhalogenated semivolatile and halogenated volatile chemical categories, are typically found in the following wastes and waste-producing processes: solvents, wood-preserving wastes (creosote, pentachlorophenol), coal tars, and pesticides. The most frequently cited group of these contaminants to date are the chlorinated solvents. Some examples of DNAPLs are polycyclic aromatic hydrocarbons or PAH, transformer oil, PCBs, coal tars, chlorinated solvents, mercury, and creosote [32]. PAHs are forms of NAPL that are ubiquitous in the environment and are mostly anthropogenic but also occur naturally; these compounds are highly carcinogenic even at relatively low levels and thus demand much attention and regulation. Natural sources include forest fires, volcanic eruptions, and degraded biological materials. Human sources include coke production, burning of coal refuse banks, automobiles, wood gasifiers, and commercial incinerators [29].

2.2.1.1.2 Dissolved Phase Constituents

The extent to which NAPLs spread out in the subsurface is governed by the interfacial tension between the NAPL and the water. NAPLs are usually a mixture of several chemical compounds that react differently given their different physical and chemical characteristics. Accidental spills and leaks of fuel oils and other hydrocarbons from surface or near-surface sources often result in accumulations of NAPLs in the subsurface, which, when hydrologically connected with an aquifer (i.e., the water table), can constitute a succeeding source of contamination to groundwater. For more precise evaluation of hydrocarbon in aquifers for improved remediation efforts, an understanding of their partitioning from the NAPL source to groundwater is necessary [33]. Partitioning of dissolved phase in groundwater and the chemical composition of hydrocarbons is controlled by major factors such as the NAPL composition and the partitioning characteristics of individual NAPL constituents [34]. The concentration of the dissolved NAPL phase is dependent on the partitioning equilibrium of the constituents (ideally descried by Raoult's law), which are the nonmiscible organic phase and aqueous phase [35–37].

$$C_{eq,i} = X_i \, S_{1,i}, \tag{2.5}$$

where $C_{eq,i}$ (in moles per liter) is the concentration of the solute i in the aqueous phase at equilibrium with the organic phase, $S_{1,i}$ (in moles per liter) is the aqueous solubility of the pure liquid compound i at the considered temperature, and X_i is the molar fraction of compound i in the organic phase at equilibrium [38]. Furthermore, concentrations of NAPL in the dissolved phase are also kinetically driven (i.e., movement of NAPL) and are dependent on the rate of molecular diffusion of the NAPL solutes and the interface (i.e., between the dissolved and NAPL phases) [29,38,39]. Temperature and pH govern the rate of diffusion [36]. The relative concentrations of hydrocarbons in the dissolved phase are expected to be different from those of the source NAPL because of the different partitioning properties of most hydrocarbons in an NAPL. Groundwater will have higher concentrations of soluble hydrocarbons than the NAPL composition. Hydrocarbons such as volatile aromatic compounds, benzene, toluene, ethylbenzene, and xylenes (BTEX), typically found in petroleum products such as gasoline and diesel fuel products, are present in the dissolved phase in much higher concentration relative to those of the less-soluble, higher-molecular-weight hydrocarbons, such as C3- or C4-alkyl benzenes (e.g., 1,2,4-trimethylbenzene) or PAHs [31]. Comparable distributions of BTEX are common in many sources. Aqueous solubility of hydrocarbon is the controlling factor in their partitioning between NAPL and groundwater [34,38,39]. Gasoline-range petroleum NAPLs contain many volatile hydrocarbons besides BTEX [38,39] that are commonly found in groundwater; depending on the source, these other hydrocarbons are paraffins, isoparaffins, alkylated monoaromatics (aromatics), naphthenes, and olefins (PIANO). Some of the PIANO products are also found in coal tar and coal-tar by-product NAPLs [33].

2.3 CONTAMINANT TRANSPORT AND MASS TRANSFER

2.3.1 Aquifer Characteristics and Solute Transport

As presented at the beginning of this chapter, there are multiple aquifer characteristics that dictate flow in the subsurface. This section will build upon the information presented earlier in this chapter to better portray the impact of aquifer hydraulic properties on groundwater flow and transport of solutes. For best predictions of solute transport, a thorough characterization of the hydrologic system and its hydraulic

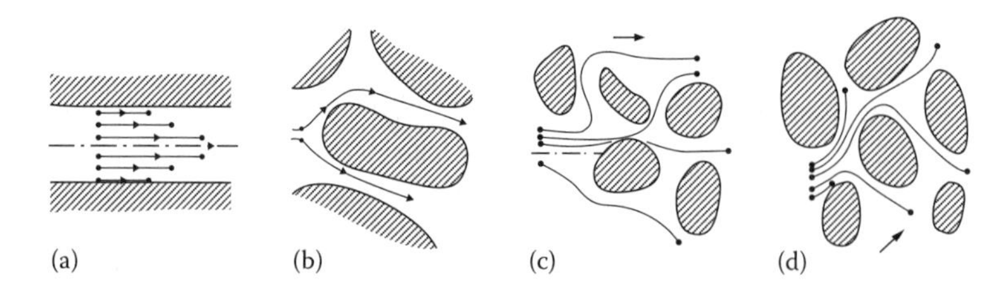

FIGURE 2.6 General aspects of solute dispersion in an aquifer: (a) macroscopic scale; increasing tortuosity of groundwater flow paths (b to d) attributed to microscopic variances in hydraulic conductivities. Solute spreading can be influenced by preferential flowpaths that arise from heterogeneities at a decimeter scale. (From Zheng C. and Gorelick S.M., *Ground Water*, 2003, 41(2):142–155; Troisi S. and Vurro M., *Water Resources Management*, 1987, 1(4):305–312; doi: 10.1007/BF00421882.)

properties is necessary. Since the composition of any given aquifer varies, hydraulic characteristics will differ as well. Important characteristics to consider when evaluating groundwater flow and contaminant transport are the heterogeneous and anisotropic nature of an aquifer or groundwater reservoir. Most field measurements reveal the existence of these properties and their prevalence lies on a thorough understanding of the different depositional geological environments. An aquifer is considered homogeneous if its hydraulic properties are the same at any given point in space. If properties vary in space, the aquifers are said to be heterogeneous. Furthermore, in an anisotropic medium, physical and mechanical properties such as hydraulic conductivity change along different axes (Figure 2.6) [40,41] while in an isotropic matrix, these properties are identical regardless of direction. The homogeneity and isotropy are closely related. For instance, geologic formations with hydraulic properties that are independent of position and direction are homogeneous and isotropic; where hydraulic properties are dependent on position and vary directionally, the formation is heterogeneous and anisotropic. Geologic formations can also be homogeneous and anisotropic and homogeneous and isotropic. Most aquifers are heterogeneous and anisotropic, having hydraulic conductivities that are different from one location to another and higher values in the horizontal direction compared to the vertical direction [1]. For large-scale evaluations, these differences are most often attributed to rock units having distinct stratigraphic layers (i.e., heterogeneity attributed to large-scale variation of hydraulic conductivities); to simplify calculations, isotropy and homogeneity are assumed for each individual layer. However, for small-scale aquifer calculations, heterogeneous and anisotropic conditions must be considered for more precise estimations of flow and solute transport rates, fate of contaminant, and discharge locations. Precise estimations of in situ hydrologic properties are important for more reliable predictions of contaminant fate and transport in heterogeneous materials.

2.3.1.1 Advective Transport

Solute transport is the movement of dissolved substances in flowing groundwater [3]. Different physical and chemical processes affect the movement of solutes through an aquifer. Solute movement through an idealized aquifer (i.e., homogeneous and isotropic) is dictated by advective flow. Advective flow is a relatively simple transport mechanism in which a substance or a solute is moved downgradient by the bulk flow of water; contaminants will move with the general direction of groundwater flow. This is the primary means of transport for any solute, including contaminants [42].

2.3.1.2 Dispersive Transport

Spreading of contaminants as they move through the groundwater system is controlled by the diffusion and dispersion processes. Through spreading, contaminants are diluted in some parts of the system and concentrated in other parts beyond what would be expected through advection alone. Through diffusion, both dissolved ionic and molecular species move via concentration gradients, from areas of higher concentration to areas of lower concentration. On the other hand, dispersion is the process through which solutes mix during advective transport as a result of random, localized variations in flow speed caused by flow variations within the pores and by nonuniform hydraulic conductivity distribution (i.e., heterogeneities) (Figures 2.6 and 2.7); it usually causes more mixing than simple molecular diffusion [43]. While dispersion happens strictly because of variations in flow velocities, diffusion will happen regardless of flow [40].

Advective and dispersive processes cause contaminants to spread into multiple areas of the groundwater system, making assessment and remediation more difficult. Characterization and understanding of contaminant spreading become challenging especially when the source is unknown (i.e., nonpoint source). Generally, contaminants will flow through the path of least resistance, which usually is highly irregular [3]. For instance, groundwater will tend to flow faster through course-grained channels (i.e., gravel and sands) and very slow though fine-grained layers or lenses (i.e., clays) [44]. Furthermore, an effect of small-scale heterogeneity on the direction of flow has been observed by many studies. Preferential flowpaths and flow barriers caused by small-scale aquifer heterogeneities (Figure 2.6) appear to govern solute transport [40,41,45]. As the contaminant plume spreads because of dispersive and

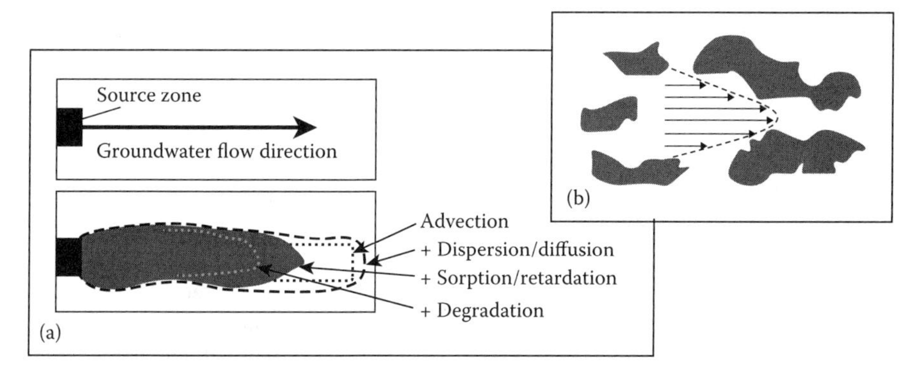

FIGURE 2.7 Diagram illustrating transport processes of contaminants in groundwater (a) and the velocity variation within an individual pore (b). Contaminant plume is diluted because of dispersive spreading which is attributed to aquifer heterogeneities (i.e., variations in hydraulic conductivity and porosity). (From EUGRIS, portal for soil and water management in Europe; Contaminant Hydrology [cited August 21, 2014]; Available from: http://www.eugris.info/FurtherDescription.asp?e=70&Ca=2&Cy=0&T=Contaminant%20hydrology.)

advective forces, it can take different shapes depending on the physical and chemical characteristics of the aquifer and solute [46].

If we consider a portion of the groundwater flow system and a certain mass of solute as a tracer, a concentration gradient will exist. Molecular diffusion will occur in the system if the velocity is zero, while mechanical dispersion is the process that will spread the tracer if the velocity is not zero. The latter phenomenon will likely to cause a complex behavior to occur in the aquifer [47]. Furthermore, in some cases, the spreading may occur as a result of the "tortuosity" of the system, but at the macroscopic level, the main flow has a more simplified direction (Figure 2.6). Although the impact of these processes is not visible at the macroscopic level, they must be considered for better prediction of contaminant spreading. In order to accurately predict the spreading of a solute in an aquifer, it is utterly important to employ the hydrodispersion coefficient expressed by the following equations:

$$D_{ij} = D_{\mathrm{L}} \frac{u_i u_j}{u^2} T \frac{u_i u_j}{u^2} {}_0\tau, \; i \neq j; \; i = 1,2; \; j = 1,2 \tag{2.6}$$

$$D_{\mathrm{L}} = \alpha_{\mathrm{L}} u, \tag{2.7}$$

$$D_{\mathrm{T}} = \alpha_{\mathrm{L}} u, \tag{2.8}$$

where D_{ij} = dispersion coefficient (in meters squared per second), D_{L} = longitudinal dispersion coefficient (in meters squared per second), D_{T} = transverse dispersion coefficient (in meters squared per second), τ = tortuosity (dimensionless), u_i, = x component of velocity vector (in meters per second), u_j = y component of velocity vector (in meters per second), D_0 = molecular diffusion (in meters squared per second), x, y = horizontal coordinates (in meters), α_{L} = longitudinal dispersivity (in meters), and α_{L} = transverse dispersivity (in meters).

2.3.1.3 Contaminant Fate/Chemical Reactions

Aside from advection, dispersion, and physical characteristics of the subsurface, many chemical processes also dictate contaminant transport. Numerous chemical reactions such as various chemical equilibrium (i.e., adsorption), ion exchange, redox, precipitation/dissolution, and biodegradation, among others, may occur along flowpaths. Water–rock interactions can be quite ubiquitous; minerals in the rocks will react with solutes in the groundwater and change its composition. These reactions happen according to the initial equilibrium or disequilibrium conditions of the groundwater solution with the rock. The flow of solutes is then affected by changes in ion concentration, mobility of solutes, or changes in pH caused by the chemical exchange between minerals and groundwater [48]. Adsorption, defined as the adhesion of atoms, ions, or molecules from a fluid to a surface, creates a thin layer of the adsorbate (i.e., gas or liquid) on the adsorbent (i.e., insoluble, rigid particles such as sediment grains). Adsorption is relevant to solute transport especially in the context of retardation or alteration of solute transport. When solutes are adsorbed onto an aquifer material (e.g., clay), concentrations in the aqueous phase decrease and the velocity of contaminant transport/migration is experiencing differing degrees of retardation. For organic contaminants, the extent of retardation will depend on the fraction of organic carbon (f_{oc}) of the aquifer materials; higher f_{oc} values are related to more abundant sites available for adsorption [49]. The extent of adsorption and retardation depends on the chemical equilibrium achieved in the aquifer at any given location. If a contaminant has a relatively higher solubility, it will stay dissolved and continue to flow with the groundwater. A contaminant that has a relatively low solubility will be prone to adsorption and, consequently, may move orders of magnitude slower [3]. Additional parameters that govern solute transport in the subsurface are persistence, volatility, and molecular size. Persistence refers to the residence time or the presence of a substance in an aquifer. A good example of contaminants with long residence times is the PCBs. Contaminants that do not break down easily will persist in the environment and pose long-term problems. Volatility refers to the tendency of a substance to change from a solid or liquid to a gas; the higher the volatility, the greater the likelihood of loss to the atmosphere. Lastly, apart from the aquifer physical characteristics (i.e., permeability), the molecular size also dictates how easily a solute can move from pore to pore; the smaller the molecule size, the easier the solute will move through the aquifer [50].

Colloid transport is another important aspect of groundwater contaminant studies. A colloid is a homogenous, noncrystalline substance in the form of a particle no greater than 10 μm in diameter. These particles can exist in a suspended dispersion in the groundwater. Colloids may consist of small, precipitated mineral particles, NAPLs, or bacteria among others. These small particles have high relative surface area to which solutes will partition to. Depending on their size and the geologic media through which they travel, colloids can be quite mobile; it is this characteristic that raises concerns for groundwater contamination. Colloids in groundwater present a risk not only as the contaminant, but they are surfaces available to partitioning or sorption of contaminants from groundwater; the high mobility and surface area will enhance contaminant transport with groundwater flow [3].

2.3.1.4 Multiphase Flow

Because they are covered in other texts and for brevity, this chapter gives a brief overview of factors related to multiphase flow and excludes the related governing equations. NAPLs partition to different phases dependent on the equilibrium characteristics of each constituent (ideally descried by Raoult's law) such as the nonmiscible organic and aqueous phases. As mentioned earlier in this chapter, some NAPLs are quite volatile (i.e., VOCs); as volatile NAPLs flow in the subsurface, they will partition to the gas phase; this process is dependent on the characteristic distinctive vapor pressure of the compound, causing the NAPL to vaporize into the soil. Furthermore, depending on the characteristic solubility, some NAPLs will partially dissolve into groundwater and become mobile as a free-phase NAPL [51]. More details related to the dissolved NAPL phase are offered earlier in this chapter (i.e., Section 2.2.1.1.2). Partitioning of NAPL to the gas and liquid phases generally results in a layered contaminant plume, with vaporized components, parent NAPLs, and dissolved components (Figure 2.5). However, if we consider the partitioning of NAPLs to the solid grains, contamination in the saturated zone may be present in four different physical states: gas, sorbed to soil materials, dissolved in water, or immiscible liquid. Sorbed phase compounds may become free phase and be subject to advective transport dependent on changes on the chemical status of the groundwater (i.e., changes in pH and other physical parameters); the transport rate of NAPL in the mobile form will normally be different from the general groundwater velocity, but the main flow direction will be similar. As indicated in the above section, adsorption is relevant to solute transport as it alters solute transport through retardation. Not only that the solute concentration in the aqueous phase is decreasing, but the contaminant transport velocity will also be affected by retardation. Therefore, retardation plays a major role in the multiphase flow of NAPLs. The vaporized components of NAPLs have been observed to move outward in a radial fashion and generally will tend to migrate upward to the unsaturated zone and ground surface. This is largely independent of advective flow; in some cases, it even moves against the hydraulic gradient, beyond the point source, which may cause confusion in site characterization and remediation. The vaporized NAPL will eventually reach the surface again. The innate chemical and physical properties of the multiphase NAPLs along with the aquifer media will dictate the degree of retardation [51].

Darcy's law, originally formulated to describe one-dimensional and steady flow of freshwater in saturated, homogenous, and isotropic media has been revised to accommodate any significant change in the original conditions. The modified version accommodates multiphase flow and helps characterize much more complex flow models. It can be applied to three-dimensional, transient flow of multiple fluids in heterogeneous, anisotropic, and deformable porous media [51]. This approach is utterly important when studying harmful contaminants such as VOCs. These compounds are highly toxic and come in a variety of forms and phases. They can originate from improper disposal of numerous household items and easily seep into the ground. Examples include paints and paint thinners, lacquers, cleaning supplies, photographic solutions, glues, and adhesives [52]. Other sources of VOCs are hydrocarbon fuels such as gasoline and diesel. This type of contaminant is of particular importance in groundwater resource and remediation studies given their high toxicity even at very low concentrations and their ability to partition and travel to and as different phases including NAPLs, gases, and dissolved; these three phases have markedly different flow tendencies [26]. An in-depth understanding of VOCs and other contaminants that can partition to different physical states is essential for best and improved assessments of groundwater contamination extents and severities and selection of appropriate remediation approaches.

2.4 POLLUTION PREVENTION AND CONTROLS

The quality of groundwater to provide adequate drinking supplies, both in quantity and in quality, is becoming an issue of great importance all over the world. This will be an ongoing area of concern for decades to come, until more advanced and cost-effective water-purifying technology becomes available. One significant groundwater pollutant is saltwater intrusions; a concern for coastal communities as groundwater production from deeper sources has increased [8]. Water desalination practices are considered and in place in areas with high water sustainability risks. For instance, in Carlsbad, California, a plant is currently under construction and the total cost of construction is approximately $922 million, all of which is privately financed [53]. Funds should be allocated to help prevent and solve groundwater contamination issues and support new initiatives. The dilemma that is often faced, depending on location, is whether to put efforts into preventive action, or to just focus on treating contaminated water. Preventive action seems to be the best long-term option, but it can be difficult owing to the required implementation of official government-level decisions, legal obligations, and the bureaucracy that entails. It also does not solve the inherited contamination problem that already exists and that needs to be addressed [8]. It is necessary to resolve the existing problems as well as find ways to prevent further groundwater contamination.

Contaminants such as fertilizers, pesticides, radioactive waste, and petroleum wastes have been long recognized as groundwater pollutants all over the world. However, the widespread and growing use of caffeine, nicotine, pharmaceuticals, antibiotics, body care products, domestic cleaners, endocrine disruptors, and psychedelic drugs in developed and developing countries is a rising concern and more is to learn about the characteristics and behavior of these contaminants in the subsurface as well as risks associated with human and ecosystem health. We certainly know that many of these substances are toxic to humans. While it is recognized that these products are seeping into the groundwater at any given concentration, it remains uncertain how they will affect the groundwater resource in decades to come [8].

2.4.1 Best Management Practices

A large part of the United States relies on groundwater for domestic use, with some areas where groundwater is the only source of freshwater. Aside from domestic use, it is widely used for irrigation purposes and in industry. Groundwater is especially important in arid and semiarid regions such as the southwest where surface water resources have been already allocated and precipitation is limited. Aside from human use, groundwater plays an important role on ecosystem health as it can play an important role in the freshwater and nutrient budgets. There are numerous stressors that affect groundwater supplies, but among the most escalating ones are over-pumping (i.e., aquifer depletion) and contamination.

Best management practices (BMPs) are vital for best appropriation and protection of groundwater resources. In the United States, managing groundwater resources (including quality and quantity) is mostly left up to individual states to handle. However, laws enacted by the US Federal Government that states must abide by do exist. These include the Safe Drinking Water Act (amended 1996) and the Clean Water Act (1972), among others. There are also federal laws that regulate the financial aspects of groundwater management, such as the Farm Security and Rural Investment Act (2002), which gives federal money to rural communities to ensure safe drinking water [54]. Other federal laws such as CERCLA (1980) heavily regulate hazardous waste dumping and finance the cleanup of hazardous waste sites all over the country. Hazardous waste, indubitably, has proven to pose the highest contaminants risk to groundwater. In addition to federal laws, state regulations can be quite specific and inclusive, when it comes to groundwater protection. For example, the New Hampshire Department of Environmental Services (DES) has an extensive outline for the protection of all water, surface and subsurface, in the state. The outline includes rules and regulations, as well as strict definitions of certain terms in order to avoid ambiguities and misinterpretations. In relation to best management practices for groundwater protection, there are guidelines covering purpose, applicability, definitions, storage and transferring of regulated materials, release response information, and more [55]. On the other end of the spectrum is the state of Texas, where the predominant legal doctrine for groundwater is very simplistic. It is called the Rule of Capture. This states that landowners have vested property rights in the groundwater; the owners can pump as much water as they please, with disregard of the neighboring landowners' water supply. The Rule of Capture is unofficially known as the "law of the biggest pump." There are, of course, a few stipulations to the Rule of Capture. For example, a landowner cannot wastefully pump water, cause subsidence for a neighbor, or pump solely to spite a neighbor, and extract contaminated water. Also, the Texas Water Law forbids trespassing for water extraction [56].

Some of the most fundamental BMPs should be enforced on a grassroots level in order to be completely effective. The Environmental Fact Sheet developed by the New Hampshire DES in 2009 consists of just the sort of basic practices that should be ubiquitous for groundwater protection. The following is an example of rules issued by the New Hampshire State Legislature, applicable to other areas/states:

Storage

- Store regulated substances on an impervious surface.
- Secure storage areas against unauthorized entry.
- Label regulated containers clearly and visibly.
- Inspect storage areas weekly.
- Cover regulated containers in outside storage areas.
- Keep regulated containers that are stored outside more than 50 feet from surface water and storm drains, 75 feet from private wells, and up to 400 feet from public wells.
- Secondary containment is required for regulated containers stored outside, except for on-premise use heating fuel tanks, or aboveground or underground storage tanks otherwise regulated.

Handling

- Keep regulated containers closed and sealed.
- Place drip pans under spigots, valves, and pumps.
- Have spill control and containment equipment readily available in all work areas.
- Use funnels and drip pans when transferring regulated substances; perform transfers over impervious surface.

Release response information

- Post information on what to do in the event of a spill.

Floor drains and work sinks

- Cannot discharge into or onto the ground.

Many of these BMPs are aimed at industry, which is the major source of contamination. Unfortunately, illegal disposal of contaminants such as antifreeze, paint, motor oil, or gasoline occur with increased frequency from the general public and small businesses. Although regulations at the state or federal level may withhold the contamination problem at regional- or large-scale levels, it is simply not possible to regulate every single occurrence of groundwater pollution. A concerted effort to bring awareness to groundwater contamination on the smallest, most ordinary scale is necessary to infiltrate society from the ground up.

2.4.1.1 Regulatory Efforts

There have been numerous regulations enacted affecting groundwater safety on a state and federal level, especially since the 1970s. One of the most predominant laws passed in the United States regulating groundwater is the Safe Drinking Water Act (SDWA), originally passed in 1974 and amended four times since. The SDWA, enforced by the EPA, was enacted to ensure safe drinking water for all citizens and has numerous measures to do so. This includes the protection of public groundwater wells. Key aspects of the act include pollution prevention against known contaminants and the treatment of contaminated water for public use. The SDWA was initially intended to focus on treatment of groundwater after it was recovered through pumping, in order to bring it to drinking standards. The amendments in 1986 and 1996 added several provisions that focused on preventive measures. These were largely seen as improvements, as found by peer-reviewed scientific studies over the years. Many of the additions made in 1986 and 1996 included the regulation of more toxic contaminants than before. The 1996 amendment mandated the implementation of new technology for water treatment systems, as well as the provision of more funding for treatment facilities and new technology [57]. The Underground Injection Control Program (UIC) is part and parcel of the SDWA. The UIC is a federal mandate via the EPA (dating back to the late 70s) that regulates injection well sites with a set of standards in order to prevent ground contamination. This sort of contamination was not previously regulated and was rampant in the 1950s, as companies disposed of various toxic wastes by injection into deep aquifers [58].

The SDWA is not without controversy, however; in 2005, a third amendment was made to the act that explicitly excluded hydraulic fracturing, known as fracking, from its jurisdiction. This was done after an EPA study, conducted from 2000 to 2004, concluded that the practice of fracking and the chemicals therein used posed minimal threat of contamination to underground sources of drinking water [59]. This was largely seen as a flawed study and the conclusion came on the heels of a booming petroleum industry, which was heavily utilizing hydraulic fracturing. A *Washington Post* article suggests that the regulation of fracking will be out of the Federal Government's hands and will fall into the laps of state governments (*Washington Post* fracking article). A somewhat less controversial amendment was enacted in January 2014 and targeted the lead (Pb) content in drinking water; potential for leaching of Pb from pipes and other plumbing components into drinking water was considered [60].

2.4.1.2 Managing Environmental Impacts

The impact that groundwater contamination has on society is multifaceted. It affects the supply of drinking water, agricultural irrigation, industrial manufacturing, and the petroleum industry, among others. One aspect of groundwater pollution, which is a concern not only in the United States but all over the world, is contamination as a result of landfill leachate. There are a multitude of studies summarizing facts regarding groundwater contamination in the area of domestic waste disposal sites known as landfills. These studies, although site specific, offer much insight that can be applied to landfills elsewhere. Among potential landfill pollutants that seep into groundwater are the household disinfectant cleaners, aerosol cans, and acid from batteries. A detailed assessment of the affected area and the surroundings is the first step in managing environmental impacts related to groundwater contamination. Factors such as availability of resources, human health, and ecological risks from contaminated groundwater (or environmental impacts), as well as costs for protection/prevention measures, the expected beneficial outcomes in terms of uses and values generated through preventive measures, and sustainable development, need to be considered for implementation of best management approaches. In general, a network of monitoring wells is placed to observe and assess contaminant levels and spreading. Such investigations reveal that contaminants in the groundwater exceed the standards established by regulatory agencies for multiple parameters. In general, initial assessments show the presence of potentially harmful microorganisms, acidic pH levels, and concentrations of metal constituents (i.e., Iron [Fe], Pb, zinc [Zn], and chromium [Cr]) at levels conducive to toxicity.

In order to manage the impact of waste disposal sites on groundwater quality, certain measures must be implemented. Most of these are common practices and are outlined and enforced by various environmental agencies around the world, depending on the country. For instance, it is important that a highly impermeable layer underlies waste disposal sites to prevent leachate of harmful pollutants into groundwater. Protective layers are commonly made of a combination of clay and high-quality synthetic plastics such as high-density polyethylene or polypropylene. A downside of relying on these layers is that they are not completely impenetrable. Puncturing and breaking occur and allow leachate to seep down into the soils and groundwater [61]. New and innovative technologies and materials need to be continually developed that better serve this purpose. Another best practice for managing environmental impacts is to account for the hydrogeology and the demographics of the surrounding and construct landfills in areas with reduced groundwater vulnerability risk. One preventive step is sorting waste before disposal; this, although a small step, can have extremely high positive impacts on the environment by limiting seepage of toxic chemicals from sources such as refrigerant products, automotive parts, and cleaning supplies that should otherwise be disposed of properly [62].

Shale gas can contribute significantly to the world's energy demand. Hydraulic fracturing or fracking (Figure 2.8) [63] on horizontal drill lines developed over the last 15 years makes formerly inaccessible hydrocarbons economically available. Through fracking, large amounts of freshwater combined with sand and other substances (some toxic) are injected under high pressure into wells developed into deep stratigraphic layers composed of shale to create cracks large enough that would allow bubbles of trapped oil and natural gas to escape into the well for recovery. From 2000 to 2035, shale gas is predicted to rise from 1% to 46% of the total natural gas for the United States. A vast energy resource is available in the United States. While there is concern about environmental impacts to groundwater and air quality (i.e., methane contribution to deep and shallow aquifers) [64], there is a strong financial advantage to the application of fracking [65]. Replacing coal with natural gas would reduce total greenhouse gas emissions by 45% and health risks attributed to coal

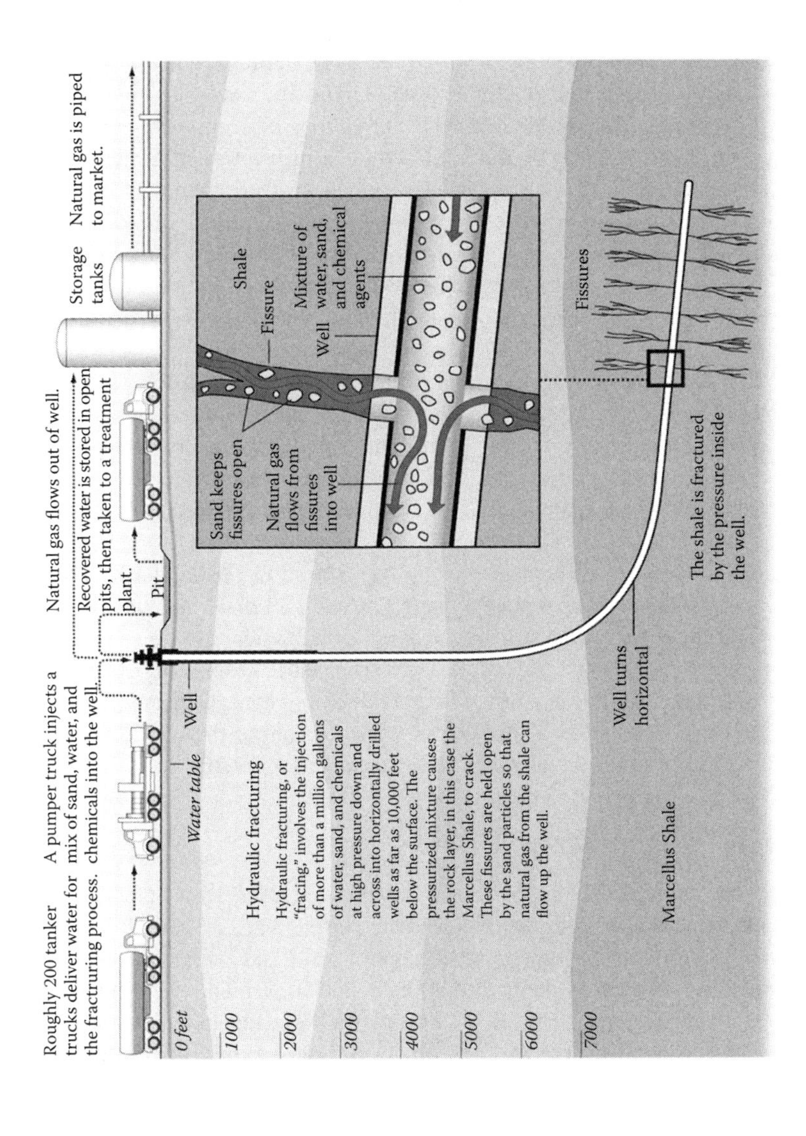

FIGURE 2.8 Schematic showing a typical hydraulic fracturing operation including surface equipment and services and subsurface fracking. (Graphic by Al Granberg. From ProPublica, What is hydraulic fracturing? [cited August 21, 2014]; Available from: http://www.propublica.org/special /hydraulic-fracturing-national.)

burning at industrial and electrical plants. On the other hand, methane seepage may result from casing failure in old wells or natural low level gas diffusion through geologic fractures in areas with active horizontal drilling and in regions with naturally occurring high gas pressures. It has been demonstrated that methane may be best related to topographic and geologic features, rather than shale-gas extraction [66].

Proper management of wastewater disposal is imperative as wastewater may contain some hydrocarbon constituents released from the oil or gas reservoirs, in addition to the original fracking solution [67]. Drilling companies often use open pits to store the toxic water; this may lead to seepage of contaminants into groundwater aquifers. For instance, in 2012 alone, 280 billion gallons of chemical-laced wastewater was produced as a result of fracking activities in the United States. In the state of New Mexico alone, such waste pits have caused more than 400 instances of contaminated groundwater to date. Wastewater is also injected into deep aquifers for storage; these wells can fail over time, leading to groundwater contamination. More recently, water treatment plants or water recycling facilities are used to make the wastewater suitable for reuse, or simply improving the quality of water for storage [68].

The revised Safe Drinking Water Act by the Energy Policy Act of 2005 exempts key aspects of hydraulic fracturing from rules that had previously regulated underground injection of fluids; Texas is the first state in the United States to require public disclosure of chemicals used in the fracking process [67]. On the other hand, to avoid or limit incidents of water resource contamination, the Natural Resources Defense Council recommends the following: wastewater testing, reuse of wastewater when appropriate, and banning the use of open-air waste pits, among many others [69]. With proper management of drilling processes along with regulations currently in place, it is unlikely that hydraulic fracturing will have adverse impacts on groundwater resources. However, since hydrocarbons may be naturally occurring from normal geologic processes, it is important to assess the baseline chemistry of groundwater in areas with active drilling. Periodic testing thereafter (typically once per year) to monitor the quality of water at active sites is recommended to detect and verify any changes that may occur [67].

2.4.2 Groundwater Remediation

As have been discussed, there are a host of problems and risks deeply embedded in groundwater contamination and prevention issues. Groundwater pollution is widespread and ongoing; once it occurs, remediation should and must be considered. Remediation in this context is defined as finding and implementing a remedy for contaminated groundwater. However, before any of these technologies and techniques can be implemented, there are some important steps to take to ensure proper and efficient remediation of groundwater. A systematic approach of assessment and remediation is necessary to effectively develop a strategic and feasible remediation plan. As stated before, the first step in remediation is a detailed site characterization. Every contamination site is different and therefore must be approached as such. This first step consists of data collection and assessment of contaminant type, concentration, and distribution; evaluation of site geology, hydrogeology, and chemistry are also included [70]. It is critical that site characterization is completed accurately to ensure accurate interpretation of data and avoid poorly designed operations and unnecessary spending. Risk assessments are conducted once the site has been thoroughly characterized; it is through this step that an evaluation of the true risk to the environment and human health exists. If the contaminated site poses minimal threat, remediation action may be postponed after later assessments. If a moderate or severe harmful risk is expected, the site is evaluated rigorously using guidelines set forth by the US EPA. This includes identifying hazards and toxicity and exposure risks [70]. After careful planning of site-specific details, the remediation process begins.

There are numerous operative remediation techniques that are widely used today. Increasing understanding of the different phases and locations in which contamination may occur at NAPL sites and their interactions has led to increasing use of several treatment technologies operating in parallel or in series. More frequently, treatment trains for NAPL include innovative remedies such as in situ thermal, chemical oxidation, and bioremediation, in addition to or as replacement to traditional remedies such as excavation and P&T. Remedy implementation often provides additional insight into the true nature and extent of subsurface contamination. The ongoing trend since the 1970s has been the use of bioremediation to clean up hydrocarbon contaminated sites. Bioremediation, in general, is the process by which microorganisms degrade pollutants through use and transformation resulting in a safer environment [71]. One of the more recent directions scientists have been going toward regarding NAPL remediation is Enhanced Aerobic Bioremediation. Through this process, organic pollutants are remediated by microorganisms enhanced by increasing the concentration of oxygen and other electron acceptors (Bioremediation Chlorinated Solvents viii); electron acceptors are often the limiting factor in the feasibility of this naturally occurring process. However, by supplementing the electron acceptor compounds, the process of remediation is expedited [72]. There are different methods of enhanced aerobic bioremediation, which include biosparging, bioventing, pure oxygen injection, hydrogen peroxide injection, and ozone injection. Through these methods, extra supply of oxygen is supplied and made available to pollutant-degrading bacteria. These technologies have proven to increase the rates of biodegradation of harmful petroleum-based contaminants by at least one order of magnitude or more, versus naturally occurring rates of degradation [73]. Recent advances in groundwater treatment techniques include the use of green chemistry technologies that have been developed and implemented for remediation of recalcitrant environmental contaminants, including NAPLs (i.e., VeruTEK Innovative Technologies). Plant-based and biodegradable surfactants and co-solvent mixtures, VeruSOL, with low concentrations of peroxide in an injection-and-extraction process facilitate removal of free-product and residual NAPL from the subsurface as shown in Figure 2.9 [74], which depicts the steps of the SEPR/S-ISCO implementation procedure.

P&T, air sparging, in situ flushing, and the use of permeable reactive barriers (PRBs) are among the most common techniques where natural attenuation or enhanced biodegradation is not practical [70]. The P&T method has been one of the most common methods of remediation historically. It is straightforward in nature. Water is pumped from the contaminated aquifer, treated on site, and then injected back into the ground on site (Figure 2.10) [75]. Air sparging, another method implemented for remediation of soil and groundwater, is particularly effective for remediating groundwater polluted with VOCs. Basically, air is injected below the

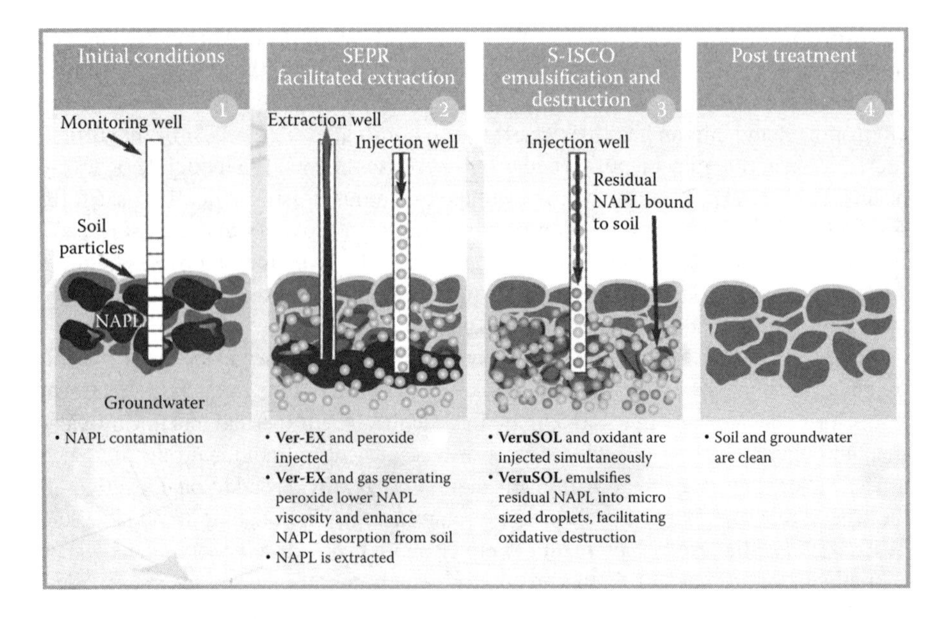

FIGURE 2.9 Steps of the SEPR/S-ISCO implementation procedure. (From VeruTEK Innovative Technologies, *VeruTEK's Surfactant-enhanced product recovery (SEPR™)* [cited August 26, 2014]; Available from: http://www.verutek.com/technologies/soil---groundwater-remediation/s-epr/.)

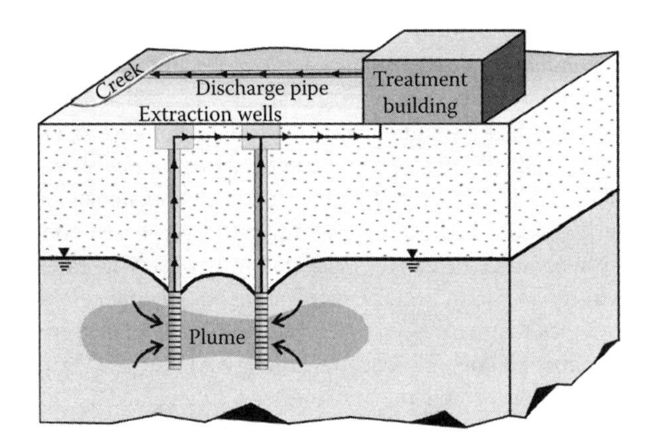

FIGURE 2.10 Simplified schematic of a pump and treat system. (From Stewart R., *Environmental Geoscience in the 21st Century*—An online textbook; groundwater remediation [cited August 21, 2014]; Available from: http://oceanworld.tamu.edu/resources/environment-book/groundwaterremediation.html.)

contaminant plume, and it naturally rises up above the saturated zone; through this process, contaminants are brought up to the vadose zone and extracted to the ground surface and treated on site (Figure 2.11) [76].

Flushing, another common method in groundwater remediation, is accomplished by injecting flushing liquids, usually clean water or surfactants, into the

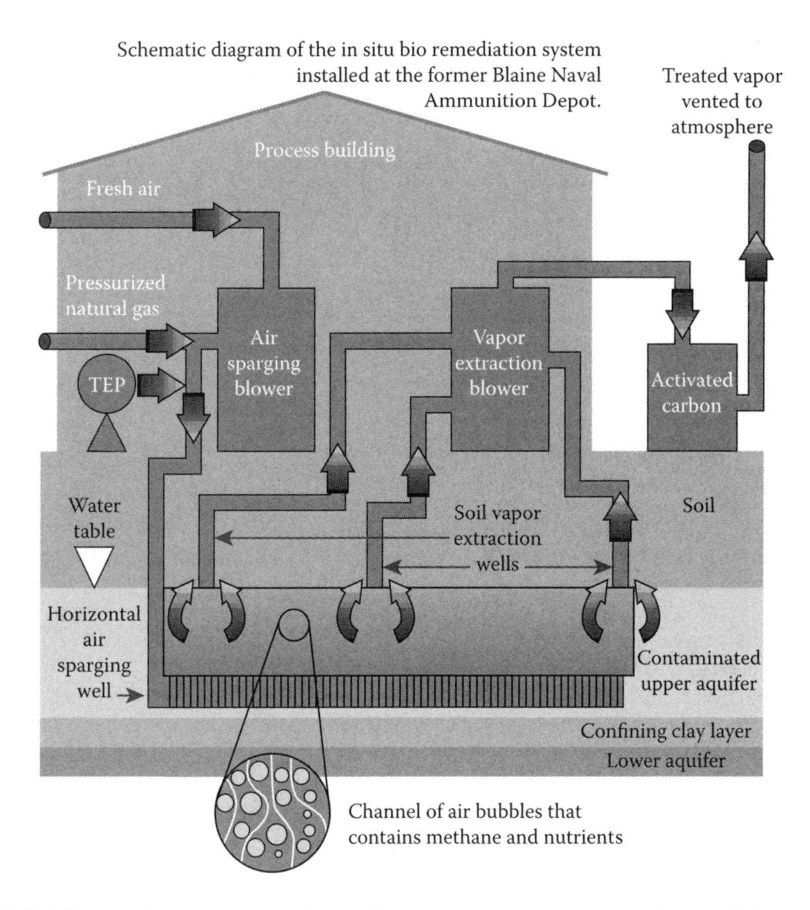

FIGURE 2.11 Schematic of an air sparging system used to remove volatile trichloroethylene from soil and aquifer. (Graphic provided by the U.S. Army Corps of Engineers. From North Carolina Division of Pollution Prevention and Environmental Assistance, *Initiatives online: Successful bioremediation recognized in Nebraska*, vol. 5, 1998 [cited August 21, 2014]; Available from: http://infohouse.p2ric.org/ref/14/0_initiatives/init/winter98/success.htm.)

contaminated groundwater zone. Contaminants in this case are transported downgradient with the groundwater flow, while reacting chemically with the host matrix and ambient groundwater. This process also involves pumping and treating the extracted contaminants on-site or off-site. Flushing has proven to be effective for treating DNAPLs; flushing liquids for DNAPL remediation are generally surfactants and alcohols. PRBs is a straightforward method through which a wall of reactive material is placed across (or perpendicular to) the groundwater flow path to intercept the contaminated groundwater (Figure 2.12) [75]. Contaminants react with the barrier and are degraded or removed as they pass through the wall. Thus, treated water emerges from the other side of the barrier [70].

Long-term studies indicate that it may take a minimum of 10 years for the reactive granular iron PRBs to exceed the P&T systems performances [77]. Also, it has been demonstrated that, in some cases, the PRBs affect the geochemical characteristics of

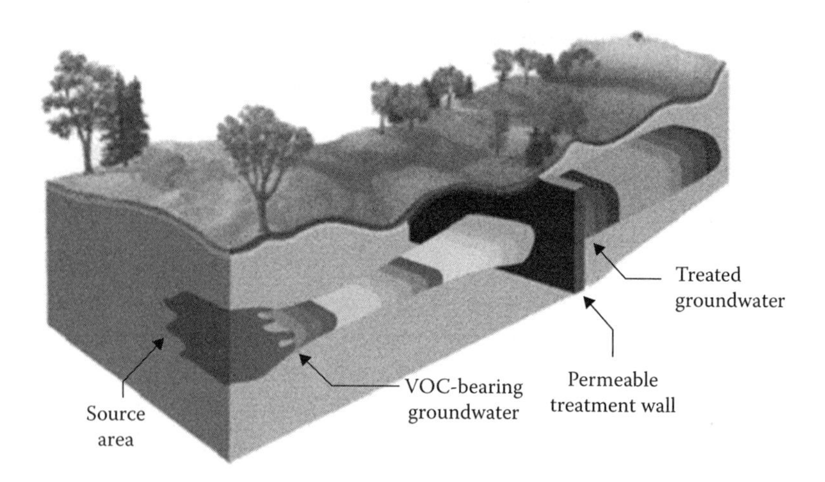

FIGURE 2.12 Schematic illustration of a PRB system. In this example, a barrier or trench backfilled with reactive material such as iron filings, activated carbon, or peat is used; the contaminant is absorbed and transformed as water from the aquifer passes through the barrier. This works only for relatively shallow aquifers. The barrier absorbs contaminants (in this image, VOC is volatile organic carbon), leaving treated water to flow downstream in the aquifer. (From Stewart R., *Environmental Geoscience in the 21st Century*—An online textbook; groundwater remediation [cited August 21, 2014]; Available from: http://oceanworld.tamu.edu/resources/environment-book/groundwaterremediation.html.)

the treated/remediated groundwater (i.e., downgradient of the RPB); higher pH and lower redox potential (EH) conditions were observed in wells located immediate downgradient of the barrier [78]. For this reason, a systematic approach to remediation is paramount in order to properly and efficiently implement cutting-edge technologies and techniques.

REFERENCES

1. Heath, R.C., *Basic Ground-Water Hydrology*, vol. 2220. 1983, US Geological Survey.
2. The COMET Program. *Basic Hydrologic Science Course Understanding the Hydrologic Cycle Section Five: Groundwater* [cited August 24, 2014]; Available from: http://www.goesr.gov/education/comet/hydro/basic/HydrologicCycle/print_version/05-groundwater.htm.
3. Fitts, C.R., *Groundwater Science*. 2002, Academic Press.
4. Buddermeier, R.W., and Schloss, J.A., *Groundwater Storage and Flow*. 2000 [cited August 24, 2014]; Available from: http://www.kgs.Ukans.edu/Hightplains/atlas//apgengw.htm.
5. NGWA Press Publication. *Ground Water Hydrology for Water Well Contractors*, Chapter 14. 1999 [cited July 14, 2014]; Available from: http://www.ngwa.org/fundamentals/hydrology/pages/unconfined-or-water-table-aquifers.aspx.
6. Kirsch, R., and Hinsby, K., *Aquifer vulnerability. Groundwater Resources in Buried Valleys—A Challenge for Geosciences*, pp. 149–155. 2006, GGA Institute: Hannover.
7. Costudio, E. *Trends in Groundwater Pollution: Loss of Groundwater Quality and Related Services. Groundwater Governance*, p. 74 [cited July 24, 2014]; Available

from: http://www.groundwatergovernance.org/fileadmin/user_upload/groundwatergovernance/docs/Thematic_papers/GWG_Thematic_Paper_1.pdf.
8. Liggett, J.E., and Talwar, S., Groundwater vulnerability assessments and integrated water resource management. *Streamline Watershed Management Bulletin*, 2009. **13**(1): 18–29.
9. Palmer, R.C., and Lewis, M.A., Assessment of groundwater vulnerability in England and Wales. In: Robins, N. S. (ed.). *Groundwater Pollution. Aquifer Recharge and Vulnerability*, vol. 130, pp. 191–198. 1998, Geological Society: London. Special Publication.
10. Walt, I.J., Pretorius, S.J., and Schoeman, C.B., The role of geohydrology in the determination of a spatial development framework in the Vredefort dome world heritage site. *Water*, 2010. **2**: 742–772.
11. Jiang, Y., Wu, Y., and Yuan, D., Human impacts on karst groundwater contamination deduced by coupled nitrogen with strontium isotopes in the Nandong underground river system in Yunan, China. *Environmental Science & Technology*, 2009. **43**(20): 7676–7683.
12. Walt, I.J., Pretorius, S.J., and Schoeman, C.B., The role of geohydrology in the determination of a spatial development framework in the Vredefort dome world heritage site. *Water*, 2010. **2**: 742–772.
13. Murgulet, D. and Tick, G.R., Characterization of flow dynamics and vulnerability in a coastal aquifer system. *Ground Water*, 2013. **51**(6): 893–903; doi: 10.1111/gwat.12020.
14. Mew, H.E., Hirth, D.K., Lewis, D.K., Daniels, R.B., and Keyworth, A.J., Methodology for compiling groundwater recharge maps in the Piedmont and Coastal Plain provinces of North Carolina. North Carolina Department of Environment and Natural Resources. *Ground Water Bulletin*, 2002. **25**: 68.
15. Foster, S.S.D., B.L., Morris, B., and Lawrence, A.R., Effects of urbanization on groundwater recharge. In: *Proceedings of the ICE International Conference on Groundwater Problems in Urban Areas*, pp. 43–63. 1994, London.
16. Chilton, P.J., Groundwater in the urban environment. Vol. 1: Problems, processes and management. In: *Proceedings of the 27th IAH Congress*, Nottingham, UK, p. 682. 1997, A.A. Balkema: Rotterdam.
17. Chilton, P.J., Groundwater in the urban environment. Vol. 2: Selected city profiles. *IAH Int. Contrib. Hydrogeol*, vol. 21, p. 342. 1999, A.A. Balkema: Rotterdam.
18. Seiler, K.-P., and Gat, J.R., (Eds.). Recharge under Different Climate Regimes. In: *Book Series: Groundwater Recharge from Runoff, Infiltration, and Percolation*, vol. 55. pp. 159–186. 2007, Springer: Netherlands.
19. Bedient, P.B., Rifai, H.S., and Newell, C.J., *Ground Water Contamination: Transport and Remediation*. 1994, Prentice-Hall International, Inc.
20. Kuntz, M. et al. *Background Paper for Final Beyond Waste Summary: Hazardous Waste History*. 2004, Department of Ecology: State of Washington.
21. US Environmental Protection Agency. *Basic Information|National Priorities List (NPL)*. 2012 [cited May 30, 2014]; Available from: http://www.epa.gov/superfund/sites /npl/.
22. Foster, S., Kemper, K., Tuinhof, A., Koundouri, P., Nann, M., and Garduño, H., Natural Groundwater Quality Hazards avoiding problems and formulating mitigation strategies; The GW•MATE Briefing Notes Series. In: *Sustainable Groundwater Management Concepts & Tools*. 2006, World Bank: Washington, D.C. [cited July 30, 2014]; Available from: http://www-wds.worldbank.org/external/default/WDSContentServer/WDSP/IB/2011/01/18/000334955_20110118041923/Rendered/PDF/321960BRI0REVI101public10BOX353822B.pdf.
23. Moujabber, M.E., Samra, B.B., Darwish, T., and Atallah, T., Comparison of different indicators for groundwater contamination by seawater intrusion on the Lebanese Coast. *Water Resources Management*, 2006. **20**: 161–180; doi: 10.1007/s11269-006-7376-4.

24. US Geological Survey, Water Science School. *Contaminants Found in Groundwater.* 2014 [cited May 30, 2014]; Available from: http://water.usgs.gov/edu/groundwater-contaminants.html.
25. Mayer, A.S., and Hassanizadeh, S.M., eds. *Soil and Groundwater Contamination: Nonaqueous Phase Liquids*, No. 17. 2005, American Geophysical Union.
26. Newell, C.J., Acree, S.D., Ross, R.R., and Huling, S.G., US EPA Groundwater issue: Light nonaqueous phase liquids, EPA/540/S-95/500 [cited August 14, 2014]; Available from: http://www.epa.gov/superfund/remedytech/tsp/download/lnapl.pdf.
27. Durnford, D.S., McWhorter, D.B., Miller, C.D., Swanson, A., Marinelli, F., Trantham, H.L., *DNAPL and LNAPL distributions in soils; experimental and modeling studies.* 1998, Colorado State University, AFRL-SR-BL-TR-98-0398 [cited August 14, 2014]; Available from: http://www.clu-in.org/download/contaminantfocus/dnapl/Chemistry_and_Behavior/dnapl_modeling_fate_and_transport.pdf.
28. Stewart, S., *Groundwater Remediation.* Environmental Geoscience: Environmental Science in the 21st Century—An Online Textbook [cited August 31, 2014]; Available from: http://oceanworld.tamu.edu/resources/environment-book/groundwaterremediation.html.
29. USGS Environmental Health—Toxic Substances. *Polynuclear Aromatic Hydrocarbons (PAHs)/Polycyclic Aromatic Hydrocarbons (PAHs).* 2014 [cited June 5, 2014]; Available from: http://toxics.usgs.gov/definitions/pah.html.
30. Thomas, H. III. PGBs—The rise and fall of an industrial miracle. *Natural Resources & Environment*, 2005. **19**(4): 15–19; Print.
31. Huling, S.G., and Weaver, J.W., *Dense Nonaqueous Phase Liquids*. EPA Groundwater Issue: EPA/540/4-91-002. 1991 [cited August 25, 2014]; Available from: http://www.epa.gov/superfund/remedytech/tsp/download/issue8.pdf.
32. USGS Environmental Health—Toxic Substances, DNAPL. *Definition Page.* 2014 [cited June 5, 2014]; Available from: http://toxics.usgs.gov/definitions/dnapl_def.html.
33. Sauer, T.C., and Costa, H.J., Fingerprinting of gasoline and coal tar NAPL volatile hydrocarbons dissolved in groundwater. *Environmental Forensics*, 2003. **4**(4): 319–329; doi: 10.1080/714044376.
34. Schmidt, T.C., Kleinert, P., Stengel, C., Goss, K.U., and Haderlein, S.B., Polar fuel constituents: Compound identification and equilibrium partitioning between non-aqueous phase liquids and water. *Environ. Sci. Technol.*, 2002. **36**(19): 4074–4080.
35. Lane W.F., and Loehr, R.C., Estimating the equilibrium aqueous concentrations of polynuclear aromatic hydrocarbons in complex mixtures. *Environ. Sci. Technol.*, 1992. **26**(5): 983–990.
36. Peters, C.A., and Luthy, R.G., Coal tar dissolution in water-miscible solvents: Experimental evaluation. *Environ. Sci. Technol.*, 1993. **27**(13): 2831–2843.
37. Mahjoub, B., Jayr, E., Bayard, R., and Gourdon, R., Phase partition of organic pollutants between coal tar and water under variable experimental conditions. *Water Resour.*, 2000. **34**(14): 3551–3560.
38. Cline, P.V., Delfino, J.J., and Rao, P.S.C., Partitioning of aromatic constituents into water from gasoline and other complex solvent mixtures. *Environ. Sci. Technol.*, 1991. **25**(5): 914–920.
39. Stout, S.A., Uhler, A.D., McCarthy, K.J., and Emsbo-Mattingly, S., Chemical fingerprinting of hydrocarbons. In: Murphy, B. L., and Morrison, R. D. (ed.). *Introduction to Environmental Forensics*, pp. 137–260. 2002, Academic Press: London.
40. Zheng, C., and Gorelick, S.M., Analysis of solute transport in flow fields influenced by preferential flowpaths at the decimeter scale. *Ground Water*, 2003. **41**(2): 142–155.
41. Troisi, S., and Vurro, M., Aspects of preferential flow paths in an apulian fissured aquifer using in-situ measurements. *Water Resources Management*, 1987. **1**(4): 305–312; doi: 10.1007/BF00421882.

42. EUGRIS, portal for soil and water management in Europe. *Contaminant Hydrology* [cited August 21, 2014]; Available from: http://www.eugris.info/FurtherDescription.asp?e=70&Ca=2&Cy=0&T=Contaminant%20hydrology.
43. Focazio, M.J., Reilly, T.E., Rupert, M.G., and Helsel, D.R., *Assessing Ground-Water Vulnerability to Contamination: Providing Scientifically Defensible Information for Decision Makers*, USGS Circular 1224. 2013 [cited August 10, 2014]; Available from: http://pubs.usgs.gov/circ/2002/circ1224/pdf/circ1224_ver1.01.pdf.
44. Artiola, J., Pepper, I.L., and Brusseau, M.L., *Environmental Monitoring and Characterization*, Ed 1, p. 410. 2004, Elsevier Science. ISBN-13: 9780080491271.
45. Weinberger, G., and Rosenthal, E., Reconstruction of natural groundwater flow paths in the multiple aquifer system of the northern Negev (Israel), based on lithological and structural evidence. *Hydrogeology Journal*, 1998. **6**(3): 421–440.
46. Fetter, C.W., *Contaminant Hydrogeology*, p. 500. 1999, Prentice Hall: New Jersey.
47. Bear, J., *Dynamics of Fluids in Porous Media*. 1972, American Elsevier: New York.
48. Elango, L., Hydrogeochemical reactions in aquifers and its identification by geochemical modelling. *Journal of Applied Hydrology*, 2006. **21**(3): 35–44.
49. US Environmental Protection Agency. *North Belmont PCE Site Remedial Investigation North Belmont, Gaston County, NC*; SESD Project No. 96S-058/June. 1997 [cited August 2, 2014]; Available from: http://www.epa.gov/region4/sesd/reports/1999-0219/nbsect5.pdf.
50. Metzge, M.R., Groundwater contamination: Sources & prevention. Water quality products. *Water Quality Products Magazine*. 2003 [cited May 30, 2014]; Available from: http://www.wqpmag.com/groundwater-contamination-sources-prevention.
51. Hinchee, R.E., and Reisinger, H. J., A practical application of multiphase transport theory to ground water contamination problems. *Groundwater Monitoring & Remediation*, 1987. **7**(1): 84–92.
52. US Environmental Protection Agency. *An Introduction to Indoor Air Quality: Volatile Organic Compounds (VOCs)*. 2011 [cited August 23, 2014]; Available from: http://www.epa.gov/iaq/voc.html.
53. Water-Technology. *Carlsbad Desalination Project, San Diego, California, United States of America*. Carlsbad Desalination Project, San Diego, California. 2012 [cited June 20, 2014]; Available from: http://www.water-technology.net/projects/carlsbaddesalination/.
54. Job, C.A., *Groundwater Economics*. 2009, CRC Press. ISBN 9781439809006.
55. New Hampshire Department of Environmental Services. *Best Management Practices for Groundwater Protection*. 2013. Env-Ws 401-405 (WSPC).
56. Dowell, T., *Texas Water: Basics of Groundwater Law*. Texas Agriculture Law. Texas A&M AgriLife Extension Service. 2013 [cited June 28, 2014]; Available from: http://agrilife.org/texasaglaw/2013/10/22/texas-water-basics-of-groundwater-law/.
57. U.S. Environmental Protection Agency. *Safe Drinking Water Act (SDWA)*. 2009 [cited June 28, 2014]; Available from: http://water.epa.gov/lawsregs/rulesregs/sdwa/index.cfm.
58. U.S. Environmental Protection Agency. *History of the UIC Program—Injection Well Time Line*. 2012 [cited July 4, 2014]; Available from: http://water.epa.gov/type/groundwater/uic/history.cfm.
59. Cupas, A.C., Not-So-Safe Drinking Water Act: Why We Must Regulate Hydraulic Fracturing at the Federal Level. *The. Wm. & Mary Envtl. L. & Pol'y Rev.*, 2008. **33**: 605.
60. Laughlin, J., *Water Industry Plans for Transition to Lead-Free Rules*. WaterWorld Industries newsletter. 2014 [cited July 4, 2014]; Available from: http://www.waterworld.com/articles/print/volume-29/issue-1/editorial-features/industry-plans-transition-lead-free-rules.html.
61. Reddy, D.V., and Butul, B., *A comprehensive literature review of liner failures and longevity*. Florida Center for Solid and Hazardous Waste Management. 1999 [cited July 20,

2014]; Available from: http://www.hinkleycenter.org/pubs/538-a-comprehensive-literature-review-of-liner-failures-and-longevity.pdf.
62. Akinbile, C.O., and Yusoff, M.S., Environmental impact of leachate pollution on groundwater supplies in Akure, Nigeria. *International Journal of Environmental Science and Development*, 2011. **2**(1): 81–89.
63. ProPublica. What is hydraulic fracturing? [cited August 21, 2014]; Available from: http://www.propublica.org/special/hydraulic-fracturing-national.
64. Osborn, S.G., Vengosh, A., Warner, N.R., and Jackson, R.B., Methane contamination of drinking water accompanying gas-well drilling and hydraulic fracturing. *PNAS* early edition, 2011. **108**(20): 8172–8176 [cited August 18, 2014]; Available from: https://nicholas.duke.edu/cgc/pnas2011.pdf.
65. Pierce, Jr, R.J., Natural gas fracking addresses all of our major problems. *Geo. Wash. J. Energy & Envtl. L.*, 2013. **4**: 22.
66. Molofsky, L.J., Connor, J.A., Wylie, A.S., Wagner, T., and Farhat, S.K., Evaluation of Methane Sources in Groundwater in Northeastern Pennsylvania. *Groundwater*, 2013. **51**: 333–349.
67. Uhlman, K., Boellstorff, D.E., McFarland, M.L., and Smith, J.W., Facts about fracking… and your drinking water well [cited July 30, 2014]; Available from: http://region8water.colostate.edu/PDFs/facts-about-fracking_TXAM.pdf.
68. Ridlington, E., and Rumpler, J., *Fracking by the Numbers: Key Impacts of Dirty Drilling at the State and National Level*, p. 46. 2013, Environment America Research & Policy Center.
69. Hammer, R., Van Briesen, J., and Levine, L., In *Fracking's Wake: New Rules Are Needed to Protect Our Health and Environment from Contaminated Wastewater*, p. 11. 2012, Natural Resources Defense Council.
70. Reddy, K.R., *Physical and Chemical Groundwater Remediation Technologies. Overexploitation and Contamination of Shared Groundwater Resources*, pp. 257–274. 2008, Springer: Netherlands.
71. US Environmental Protection Agency. *Engineered Approaches to In Situ Bioremediation of Chlorinated Solvents*, p. 133. Fundamentals and Field Applications.
72. Chapelle, F.H., Bioremediation of petroleum hydrocarbon-contaminated ground water: The perspectives of history and hydrology. *Groundwater*, 1999. **37**(1): 122–132.
73. US Environmental Protection Agency, OSWER, Office of Underground Storage Tanks. Enhanced Aerobic Bioremediation. *How to Evaluate Alternative Cleanup Technologies for Underground Storage Tank Sites—A Guide for Corrective Action Plan Reviewers, Chapter 12, Enhanced Aerobic Bioremediation*. 2004 [cited August 20, 2014]; Available from: http://www.epa.gov/oust/pubs/tums.htm.
74. VeruTEK Innovative Technologies. *VeruTEK's Surfactant-enhanced product recovery (SEPR™)* [cited August 26, 2014]; Available from: http://www.verutek.com/technologies/soil---groundwater-remediation/s-epr/.
75. Stewart, R., *Environmental Geoscience in the 21st Century*—An online textbook; groundwater remediation [cited August 21, 2014]; Available from: http://oceanworld.tamu.edu/resources/environment-book/groundwaterremediation.html.
76. North Carolina Division of Pollution Prevention and Environmental Assistance. *Initiatives online: Successful bioremediation recognized in Nebraska*, vol. 5. 1998 [cited August 21, 2014]; Available from: http://infohouse.p2ric.org/ref/14/0_initiatives/init/winter98/success.htm.
77. Higgins, M.R., and Olson, T.M., Life-cycle case study comparison of permeable reactive barrier versus pump-and-treat remediation. *Environ. Sci. Technol.*, 2009. **43**(77): 9432–9438.
78. Wilkin, R.T., Acree, S., Ross, R., Lee, T., Puls, R., and Woods, L. Fifteen-year assessment of a permeable reactive barrier for treatment of chromate and trichloroethylene in groundwater. *Science of the Total Environment*, 2014. **468–469**: 186–194.

3 Groundwater Protection and Remediation

Mohamed K. Mostafa, Jude O. Ighere, Ramesh C. Chawla, and Robert W. Peters

CONTENTS

3.1 INTRODUCTION

When rain falls to the ground, it takes several routes to different destinations. Some of it flows along the land surface to streams or lakes, some is used by plants, some evaporates and returns to the atmosphere, and some seeps underground into pores between sand, clay, and rock formations called aquifers. This subsurface water is termed *groundwater* (U.S. Environmental Protection Agency [EPA] 2012a).

Many communities obtain their drinking water from aquifers. Water suppliers drill wells through soil and rock into aquifers to reach the groundwater and supply the public with drinking water. Many homes also have their own private wells drilled on their property to tap this supply. Unfortunately, groundwater can become contaminated by human activity. These chemicals can enter the soil and rock, polluting the aquifer and eventually the well (U.S. EPA 2012a).

Groundwater is an essential component of the hydrologic cycle. It can be defined as subsurface water that occurs below the water table and is located in a fully saturated zone (Charbeneau 2006). Groundwater is an important natural resource with high social and economic benefits. The use of groundwater should be properly planned to ensure its sustainable use (Naftz et al. 2002). It is also very important to understand the groundwater systems' behavior in order to prevent a continuous decline in the water table and the progressive depletion of the source (Page 1987). Groundwater is considered safe and a readily available source of water for agricultural, domestic, and industrial use. Figure 3.1 indicates the amount of groundwater withdrawals in the United States in 2005 by various sectors. Primary sectors include agriculture, municipalities, and industry. The agricultural sector is clearly the largest groundwater consumer, accounting for approximately 68% of the total groundwater withdrawals (Bedient et al. 1997). In 2005, groundwater provided 23% of the total freshwater used in the United States, while the remaining 77% was derived from surface waters (Bedient et al. 1997).

Groundwater is exposed to pollution from different sources, such as landfills, septic systems, and hazardous waste sites (Jain et al. 1993). Groundwater protection

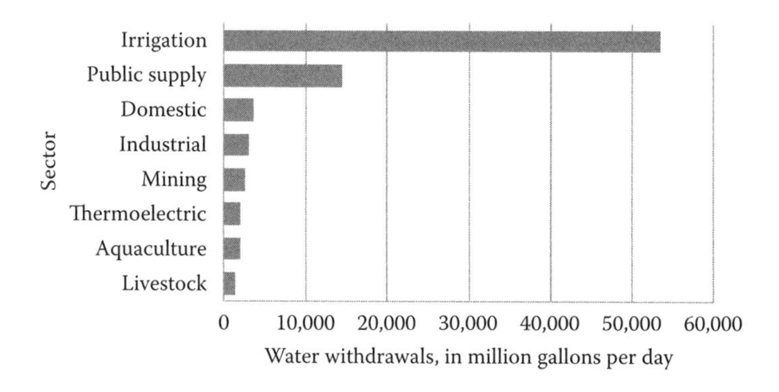

FIGURE 3.1 Total groundwater usage in the United States, 2005. (Adapted from USGS [2005], "Groundwater Use in the United States," Available online at: http://ga.water.usgs.gov/edu/wugw.html, Reston, Virginia.)

and remediation have emerged as critical issues (Naftz et al. 2002). There is also increasing awareness of the importance of protecting groundwater for people, as it represents an important source of water supply for human consumption, agriculture, and industry (Charbeneau 2006). This chapter reviews groundwater production plans and the potential remediation technologies.

3.2 SOURCES OF GROUNDWATER CONTAMINATION

Subsurface contamination is the presence of undesirable materials in the groundwater or soil under a site (Bedient et al. 1997; Rail 2000). Different means have been used in the past to eliminate wastes, such as storage in the ground, placement in streams, and burning. These disposal methods were considered the best technologies available at that time because of their cost-effectiveness (Jain et al. 1993). The use of these disposal methods led to the contamination of soil and groundwater and caused a serious threat to both human health and the environment (Rail 2000). Remedial technologies must be applied if harmful contaminants are detected at the site (Bedient et al. 1997). The characteristics of existing remedial technologies should be considered carefully, because improper implementation can worsen the site contamination.

During the World War II era, a variety of chemical wastes were generated from wartime products that required use of polymers, chlorinated solvents, metal finishing, paints, plastics, and wood preservatives (Bedient et al. 1997). In 1972, a tannery industry located in Woburn, Massachusetts, dumped toxic chemicals on the ground, which resulted in groundwater contamination with chlorinated chemicals (Bedient et al. 1997; Rail 2000). The investigations showed that the dumped waste contaminated two drinking water wells installed in a small community near the contaminated site (Rail 2000). Since the 1990s, a variety of chemicals and hazardous wastes increased significantly as a result of the expansion in the production of steel, iron, petroleum, lead batteries, and other industrial practices (Bedient et al. 1997).

The Palmerton Zinc Pile Superfund site is located in a valley near the confluence of the Lehigh River and Aquashicola Creek in Carbon County, Pennsylvania. This site was polluted by the smelting of primary ore zinc sulfide, which resulted in shallow groundwater and soil contamination (Bedient et al. 1997). High concentrations of cadmium and zinc in the soil were responsible for the spread in pollution and the vegetation damage in the surrounding areas (Blue Mountain and Stoney Ridge) (Rail 2000). After a thorough site examination, it was estimated that the removal of pollutants would take approximately 45 years and cost more than $4 billion. The initial assessment amount was estimated at $9 million for the first phase of the treatment strategy on 850 acres (Bedient et al. 1997).

The Love Canal hazardous waste site is located in Niagara Falls, New York. This site was polluted by chemical wastes, which caused serious environmental impacts on nearby residents (Rail 2000). The investigations showed that approximately 20,000 metric tons of chemical waste were dumped at this site. The federal government and New York State spent approximately $140 million to clean up and remediate the site, and relocate the residents (Bedient et al. 1997). Other sites that received national attention during the 1980s include the dioxin-contaminated sites in

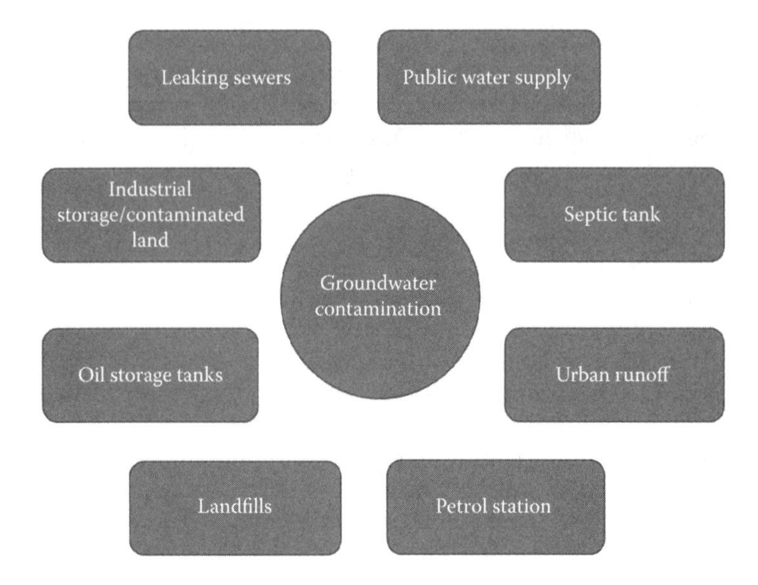

FIGURE 3.2 Schematic diagram showing major sources of groundwater contamination. (Adapted from USGS [2015a], "Groundwater Quality," Available online at: http://water.usgs.gov/edu/earthgwquality.html, Reston, Virginia.)

Missouri; the Motco and Brio chemical waste sites in Texas; the Stringfellow Acid Pits site in Riverside County, California; the Valley of the Drums site in Kentucky; and the Vertac site in Arkansas (Bedient et al. 1997).

This section describes major sources of contamination that threaten groundwater quality. Figure 3.2 shows major sources of groundwater pollution, which include underground storage tanks, septic tanks, surface impoundments, and landfills.

Table 3.1 lists sources of groundwater contamination. The major sources include underground storage tanks, landfills, septic tanks, agricultural activities, abandoned hazardous waste sites, injection wells, and land application (Jain et al. 1993; Rail 2000). The highest priority rankings were given to underground storage tanks, septic tanks, agricultural activity, municipal landfills, surface impoundments, and abandoned hazardous waste sites (Bedient et al. 1997). Figure 3.3 shows the ranking of each pollution source according to the presence in states and territories as reported to Congress in 1990.

Table 3.2 provides a list of the most common contaminants found in groundwater. These compounds can be divided into a number of categories: halogenated aliphatics (e.g., trichloroethylene); polychlorinated biphenyls (PCBs); alcohols; ketones; fuels and derivatives; benzene, toluene, ethyl benzene, and xylene (BTEX); and polycylic aromatic hydrocarbons (PAHs) (Bedient et al. 1997). The fuel contaminants emerged after the installation of the underground storage tanks caused a serious groundwater problem (Rail 2000). Over the years, the other organic compounds have been discharged to the environment in several ways, especially after World War II (Bedient et al. 1997). Figure 3.4 shows the ranking of each contaminant according to the presence in states and territories as reported to Congress in 1990. Volatile

TABLE 3.1
Sources of Groundwater Contamination

Category I—Sources designed to discharge substances	Category II—Sources designed to store, treat, or dispose of substances; discharge through unplanned release
Subsurface percolation (e.g., septic tanks and cesspools)	Landfills
Injection wells	Open dumps
Land application	Surface impoundments
	Waste tailings
	Waste piles
	Materials stockpiles
	Aboveground storage tanks
	Underground storage tanks
	Radioactive disposal sites
Category III—Sources designed to retain substances during transport or transmission	**Category IV**—Sources discharging as a consequence of other planned activities
Pipelines	Irrigation practices
Materials transport and transfer	Pesticide applications
	Fertilizer applications
	Animal feeding operations
	De-icing salts applications
	Urban runoff
	Percolation of atmospheric pollutants
	Mining and mine drainage
Category V—Sources providing conduit or inducing discharge through altered flow patterns	**Category VI**—Naturally occurring sources whose discharge is created or exacerbated by human activity
Production wells	Groundwater–surface water interactions
Other wells (nonwaste)	Natural leaching
Construction excavation	Saltwater intrusion/brackish water upcoming

Source: Adapted from the Office of Technology Assessment (1984). *Protecting the Nation's Groundwater from Contamination*, Washington, D.C.

organic compounds (VOCs) are a class of pollutants that may pose danger to human health (Rail 2000). Long-term exposure to high levels of VOCs increases the risk of cancer, liver and kidney damage, and serious damage to the central nervous system. Short-term exposure to high levels of VOCs can cause headaches, dizziness, nausea, worsening of asthma symptoms, and irritation of throat, nose, and eyes (Rail 2000).

3.2.1 Underground Storage Tanks

Underground storage tanks are located below ground surface and are used mainly to store liquid fuels, oils, industrial chemicals, hazardous chemicals, and solvents (Rail 2000). In 1984, the Office of Technology Assessment estimated the number of underground storage tanks, both still in use and those long abandoned, at approximately

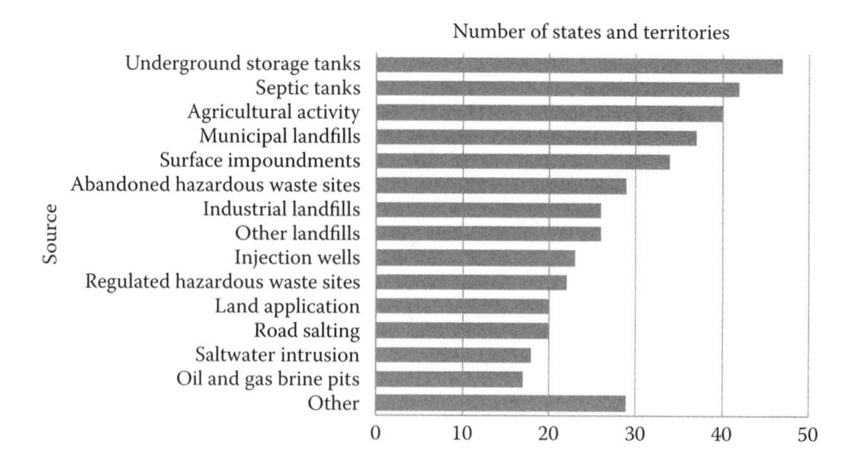

FIGURE 3.3 The ranking of each pollution source according to the presence in states and territories. (Adapted from U.S. EPA [2000b], *Groundwater Quality*, National Water Quality Inventory, 1998 Report to Congress, Ground Water and Drinking Water Chapters. Available online at: http://water.epa.gov/type/drink/protection/upload/2006_08_28_sourcewater_pubs_guide_nwiq98305b_gwqchap.pdf, Washington, D.C.)

TABLE 3.2
The Common Organic Compounds Found in Groundwater

	Groundwater Contaminant	
Acetone	1,2-dichloroethane	Toluene
Benzene	Di-*n*-butyl phthalate	1,2-*trans*-dichloroethane
Bis-(2-ethylhexyl)phthalate	Ethyl benzene	1,1,1-trichloroethane
Chlorobenzene	Methylene chloride	Trichloroethene
Chloroethane	Naphthalene	Vinyl chloride
Chloroform	Phenol	Xylene
1,1-Dichloroethane	Tetrachloroethene	

Source: Adapted from USGS (2015b). "Contaminants Found in Groundwater," Available online at: http://water.usgs.gov/edu/groundwater-contaminants.html, Reston, Virginia.

2.5 million (Bedient et al. 1997). An EPA survey study conducted in 1990 found that 47 states reported the occurrence of groundwater pollution as a result of the faulty underground tanks. These tanks can leak as a result of external or internal corrosion (Rail 2000). The leaks occur through cracks or holes in the tank or in associated valves and pipes. An EPA survey found that 280,000 motor fuel storage tanks leaked out of 800,000 tanks (Bedient et al. 1997).

A US Coast Guard Air Station, located in Traverse City, Michigan, experienced a leakage of jet fuel from underground storage tank. A spill of fuel and aviation gas resulted in a wide area of pollution approximately 500 feet wide and 1 mile long, which polluted nearly 100 municipal water wells (Bedient et al. 1997).

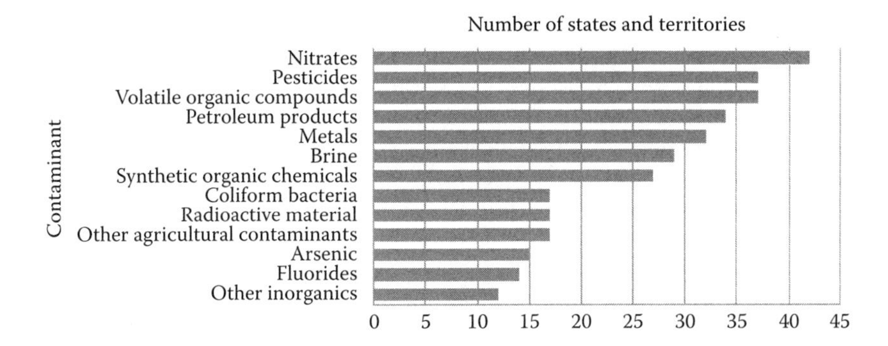

FIGURE 3.4 The ranking of each contaminant according to the presence in states and territories. (Adapted from U.S. Department of the Interior [1993], *National Desalting and Water Treatment Needs Survey*, Bureau of Reclamation, Water Treatment Technology Program Report No. 2. Available online at: https://www.usbr.gov/research/AWT/reportpdfs/report002.pdf, Denver, CO.)

3.2.2 Landfills

An EPA survey study found that between 24,000 and 36,000 abandoned or closed landfills, and approximately 2395 open dumps, exist in the United States (Bedient et al. 1997). In addition, there are approximately 75,000 industrial landfills. According to a survey study conducted by EPA, hazardous wastes were detected in 12,000 to 18,000 municipal landfills (Bedient et al. 1997). Materials placed in these landfills include trash, garbage, debris, incinerator ash, sludge, and hazardous substances (Geophysics Research Forum and Geophysics Study Committee 1984). The disposal process includes three steps: filling the landfill with solid wastes and liquid, compacting with a bulldozer, and then covering the top surface with a layer of soil. During rainfall, the water level increases inside the landfill, which in turn leads to leakage of organic and inorganic contaminants into the groundwater (Geophysics Research Forum and Geophysics Study Committee 1984). Recently, leak prevention systems have been used in landfills to monitor and control the leakage.

3.2.3 Septic Systems

A septic system is designed mainly to dispose of household waste and it is used in areas with no connection to sewage networks. Septic systems consist of two major parts, the septic tank and the leach field (Bedient et al. 1997). A physical process is used in the tank to separate the inflow into wastewater, sludge, and scum. Sludge accumulates at the bottom of the tank and the scum floats on top of the wastewater. The septic tank must be pumped out when the sludge and scum levels begin to reach the tank's normal storage capacity (Patrick et al. 1983).

The failure of such systems can lead to flooding of the wastes out of the tank, causing odors and exposing people to viruses and pathogenic bacteria (Patrick et al. 1983). Domestic septic systems can contaminate groundwater with organic and inorganic compounds from chemical oxygen demand, biological oxygen demand, phosphorus, nitrates, fecal coliform bacteria, nitrites, total dissolved solids, total suspended solids, and ammonia (Patrick et al. 1983). Industrial and commercial septic systems represent a greater risk to groundwater than the domestic systems, where these systems receive hazardous chemical waste. Chemical contaminants include nitrates and heavy metals such as copper, zinc, and lead. Synthetic organic chemicals such as perchloroethylene (PCE), trichloroethylene (TCE), benzene, and chloroform can also be discharged into industrial septic systems (Bedient et al. 1997). In the past, many small businesses including restaurants, laboratories, service stations, hardware stores, and dry cleaners have contaminated soil and groundwater through commercial septic systems (Patrick et al. 1983).

3.2.4 Agricultural Wastes

Groundwater has been contaminated by pesticides in more than 35 states. Pesticides are used for many purposes, including fungicides, insecticides, weed control, and defoliants (Rail 2000). They are used on lawns and gardens, roadsides, agricultural fields, golf courses, roadsides, parks, home foundations, and in wood products. A report issued by the United States Geological Survey (USGS) indicates that 50% of the sampled wells contained residues of one or more pesticides (Bedient et al. 1997).

The use of fertilizers in agriculture leads to increased nutrient concentrations in the subsurface (Rail 2000). Typically, fertilizers contain nitrogen, potassium, and phosphorus. Nitrogen is very mobile and easily leaches into groundwater, while phosphorus is immobile in the soil and does not constitute a major threat to groundwater (Rail 2000). A survey conducted by USGS found that 6% of the groundwater samples exceeded the 10 mg/L maximum limit for drinking water specified by the US EPA (Bedient et al. 1997). Nitrates represent a major threat to groundwater according to the National Water Quality Inventory report published in 1988 (U.S. EPA 2000a).

3.2.5 Waste Disposal Wells

Waste disposal wells are used to dispose of municipal sewage, oil waste, liquid hazardous waste, agricultural and urban runoff, brine, and mining waste into the subsurface (Jorgensen 1989). In the United States, millions of tons of hazardous and nonhazardous wastes are dumped every year directly into the subsurface through injection wells (Jorgensen 1989). This waste has contaminated the aquifer system in more than 20 states, and it mainly comes from the petroleum, aerospace, chemical, metals, minerals, and wood-preserving industries (Bedient et al. 1997). Faulty well construction or poor design can cause groundwater contamination. The waste disposal wells that constitute the greatest threat to the aquifer system include septic systems wells, agricultural wells, deep injection wells for hazardous waste, and brine injection wells (Jorgensen 1989).

3.2.6 Land Application and Mining

Land application technique is basically a treatment and disposal method, and is often called land farming and land treatment (Liu and Bela 2000). This technique involves spreading wastewater and sludge generated by industrial operations, livestock farms, and public treatment works below the surface of the ground. Industrial operations include textile manufacturing, extracting oil and gas, paper and pulp manufacturing, and tanning. Wastewater is applied to the soil through a spray irrigation system, while the sludge is applied to land as a fertilizer (Bedient et al. 1997). Wastes from oil refining operations are also applied to land to be decomposed by soil microbes (Liu and Bela 2000). Contamination occurs when toxic chemicals, pathogens, nitrogen, and heavy metals leach into the groundwater. Wastewater and sludge should receive adequate treatment, and the depth to groundwater should be properly considered to prevent groundwater contamination (Liu and Bela 2000). More than 20 states have reported that land application constitutes a major threat to groundwater (Bedient et al. 1997).

Mining operations also constitute a serious threat to the quantity and quality of nearby groundwater, where acids are likely to leak from the mines causing severe groundwater contamination (Liu and Bela 2000). Vast areas of land in the United States have been mined for uranium, copper, coal, and other minerals. Active and inactive mines as well as abandoned mines are considered serious sources of contamination (Bedient et al. 1997).

3.2.7 Surface Impoundments

Surface impoundments serve as temporary storage or disposal sites for nonhazardous and hazardous wastes (Liu and Bela 2000). These impoundments vary in size from a few acres to several thousand acres and are often called ponds, pits, or lagoons (Bedient et al. 1997). Surface impoundments are often used by wastewater treatment plants for settling solids, chemical treatment, and biological oxidation (Liu and Bela 2000). These impoundments are also used by farms and animal feedlots and by several industries including chemical, mining, oil, and paper.

Investigations found that the leakage may occur from the surface impoundments, which in turn leads to the spread of contamination in the subsurface (Liu and Bela 2000). The Rocky Mountain Arsenal site is located in Commerce City, Colorado. This site discharged pesticides and nerve gas into unlined ponds from 1942 until 1956, resulting in a groundwater plume extending more than 8 miles long. The estimated cleanup cost was approximately $1 billion (Bedient et al. 1997). An EPA survey study conducted in 1982 found that more than 180,000 waste impoundments exist in the United States including 65,688 pits for oil and gas, 37,000 municipal, 27,912 industrial, 25,000 mining, and 19,400 agricultural (Bedient et al. 1997).

3.2.8 Radioactive Contaminants

The expansion of radioactive isotope production since World War II has led to increasing concern about their health and environmental effects (Bedient et al. 1997). Radioactive waste is extremely hazardous because it contains radioactive materials (Aral and Stewart 2011). Large amounts of radioactive wastes are produced every year in the United States by the nuclear weapons industries. The radioactive wastes are generated mainly from uranium mining and milling, power plant operation, and fuel fabrication and reprocessing (Patrick et al. 1983). Radioactive wastes must be properly disposed in well-designed sites to prevent the migration to groundwater (Aral and Stewart 2011). The disposal of uranium mill tailings and civilian radioactive wastes is licensed under the Nuclear Regulatory Commission regulations (Bedient et al. 1997).

3.2.9 Military Sources of Contamination

According to a survey study conducted by the Citizens Clearinghouse for Hazardous Waste, the US military industries produce more than 1 billion pounds of hazardous waste per year and are considered the largest producer of hazardous waste in the country (Bedient et al. 1997; Mittal 2005). Many waste sites belonging to the US Air Force were listed on the EPA National Priorities List (NPL) of Superfund cleanup sites.

Air Force Plant 44 is located in Tucson, Arizona. Leakages of TCE and chromium to groundwater were detected (Bedient et al. 1997; Mittal 2005). The operations at the site include missiles manufacturing and aircraft repair and painting. The spill of TCE and chromium resulted in a wide area of pollution that reached approximately half a mile in width and 6 miles in length and polluted many water wells in Tucson (Bedient et al. 1997). The contaminated wells remained out of service for many years. Large pump-and-treat systems were used to treat groundwater located more than 100 feet below the ground surface at a rate of 5000 gallons per minute (Bedient et al. 1997; Mittal 2005).

3.2.10 Industrial Waste Site

Large areas in abandoned industrial sites are leaking organic contaminants into groundwater (Geophysics Research Forum and Geophysics Study Committee 1984). The main sources of pollution include drum storage areas, leaking industrial sewer lines, process areas for wastes and chemicals, brine and liquid waste injection wells, surface pits, and unlined landfills for liquid and solid wastes (Bedient et al. 1997). The most common organic contaminants found in groundwater from abandoned industrial sites include toluene, benzene, ethyl benzene, xylenes, TCE, PCE, and 1,2-dichloroethane (Geophysics Research Forum and Geophysics Study Committee 1984; Bedient et al. 1997).

3.3 TRANSPORT OF CONTAMINANTS THROUGH THE SOIL ZONE TO THE WATER TABLE

Transmission mechanisms of waste through the subsurface to reach groundwater include vertical migration through the vadose zone (unsaturated zone), lateral migration in the phreatic zone (saturated zone), and adsorption to soils (Charbeneau 2006). During the migration phases, the pollutant fate is controlled by physical, chemical, and biological processes (Page 1987). The physical processes include dispersion, advection, diffusion, and capillarity, and the abiotic and biotic processes include degradation, bioaccumulation, immobilization, volatilization, and retardation (Charbeneau 2006).

Numerical and analytical models have been used to simulate the subsurface transport of contaminants. Analytical models are computationally much more efficient than numerical models (Page 1987). In addition, analytical models require less specific data and simple assumptions. This section addresses three transport models, which include column experiments, chemical plumes, and chemical spills (Charbeneau 2006).

3.3.1 Column Experiments

The column experiment is the most popular method used to estimate the dispersion coefficient through a porous media (Bedient et al. 1997; Charbeneau et al. 1992). A tracer is commonly used to displace entrapped air. The effluent concentration can then be evaluated as a function of time. The following approximate equation can be used to analyze column experiment data (Charbeneau et al. 1992):

$$\frac{c}{c_o} \approx \frac{1}{2} erfc\left(\frac{x - vt}{\sqrt{4Dt}}\right), \tag{3.1}$$

where c/c_o = relative concentration, *erfc* () = complementary error function, D = combined dispersion and diffusion coefficient, x = distance from contaminant source, t = time since start of transport, and v = contaminant transport velocity.

The retardation processes may slow contaminant movement; hence, an equivalent form of Equation 3.1 can be expressed as (Charbeneau et al. 1992)

$$\frac{c}{c_o} \approx \frac{1}{2} erfc\left(\frac{x - \frac{vt}{R}}{\sqrt{\frac{4Dt}{R}}}\right), \tag{3.2}$$

where R = retardation factor and $vt/R = L$ = distance of migration of the contaminated mass.

3.3.2 Spill Model in Two and Three Dimensions

The release of contaminants to the subsurface can occur at a slow rate over a long period or at a fast rate over a short period (Bedient et al. 1997). This model assumes an instantaneous release of contaminants to the subsurface (Charbeneau et al. 1992). If a volume V_o containing contaminant at a concentration c_o is released to the subsurface over a short period, the contaminant is subsequently transported horizontally and vertically across the thickness of the aquifer (Bedient et al. 1997). If the flow has a constant velocity v in the x direction, the three-dimensional concentration is given by (Charbeneau et al. 1992)

$$c(x,y,z,t)=\frac{2c_oV_oe^{-\lambda t}\exp\left(-\frac{\left(x-\frac{vt}{R}\right)^2}{\frac{4D_{xx}t}{R}}-\frac{y^2}{\frac{4D_{yy}t}{R}}-\frac{z^2}{\frac{4D_{zz}t}{R}}\right)}{nR\sqrt{\left(\frac{4\pi D_{xx}t}{R}\right)\left(\frac{4\pi D_{yy}t}{R}\right)\left(\frac{4\pi D_{zz}t}{R}\right)}},\quad L<\frac{0.036vb^2}{D_{zz}}. \tag{3.3}$$

According to Equation 3.3, the maximum concentration occurs at the location $y = z = 0$, $x = vt/R$, and is given by (Charbeneau et al. 1992)

$$c_{\max}=\frac{2c_oV_oe^{-\lambda t}}{nR\sqrt{\left(\frac{4\pi D_{xx}t}{R}\right)\left(\frac{4\pi D_{yy}t}{R}\right)\left(\frac{4\pi D_{zz}t}{R}\right)}}. \tag{3.4}$$

Away from the source of pollution, the contaminant spreads over the full thickness of the aquifer (Bedient et al. 1997). In this case, the contaminant transport can be estimated by a two-dimensional model. Equation 3.3 can be modified to (Charbeneau et al. 1992)

$$c(x,y,t)=\frac{c_oV_oe^{-\lambda t}\exp\left(-\frac{\left(x-\frac{vt}{R}\right)^2}{\frac{4D_{xx}t}{R}}-\frac{y^2}{\frac{4D_{yy}t}{R}}\right)}{4\pi nbt\sqrt{D_{xx}D_{yy}}},\quad L>\frac{0.8vb^2}{D_{zz}} \tag{3.5}$$

where b = thickness of the aquifer.

3.3.3 Contaminant Plume Model

If a contaminant is released from a source at a constant rate for a long period, a steady-state contaminant plume will be developed (Bedient et al. 1997). The steady-state plume model can predict the maximum possible concentration in a certain location (Charbeneau et al. 1992). This model assumes that the contaminant is well mixed over the aquifer thickness and the release occurs at a centralized point at a rate $\dot{m}$ (Bedient et al. 1997). The concentration distribution is given by (Charbeneau et al. 1992)

$$c(x,y) = \frac{\dot{m}\exp\left(-\frac{\lambda Rx}{v}\right)}{nb\sqrt{4\pi v D_{yy} x}} \exp\left(-\frac{y^2}{\frac{4D_{yy}x}{v}}\right), \tag{3.6}$$

where $\dot{m}$ = release rate (in kilograms per day).

3.4 GROUNDWATER MONITORING

There are four tests commonly applied at a site to identify the percentage of contamination. These tests include corrosivity, ignitability, reactivity, and toxicity (Canter et al. 1988; Fouillac et al. 2009). Corrosive wastes include acids or bases (pH ≤ 2 or pH ≥ 12.5) that are able to corrode metal containers, such as drums, storage tanks, and barrels (Canter et al. 1988; LaGrega et al. 1994). The corrosivity toward steel is the most common method to determine corrosivity. Ignitable wastes have a flash point less than 140°F and therefore are flammable automatically. Typical examples include used solvents and waste oils. The Pensky–Martens Closed Cup is the most common method to determine ignitability. Reactive wastes are unstable and can react violently with water or air (Canter et al. 1988; Fouillac et al. 2009). They can cause toxic fumes and explosions. There is no available method to determine reactivity. Typical examples include explosives and lithium–sulfur batteries (Canter et al. 1988; LaGrega et al. 1994). Toxic wastes are harmful when ingested and may pollute groundwater as a result of disposing of wastes in the landfill (Fouillac et al. 2009). Typical examples include lead and mercury. The Toxicity Characteristic Leaching Procedure, SW-846 Method 1311 (U.S. EPA 2014), is the most common method to determine toxicity and the concentrations of contaminants that may be unsafe to human health.

3.5 ENVIRONMENTAL LAWS, EXECUTIVE ORDERS, AND REGULATIONS

A number of laws for protecting the environment and public health were enacted by the US Congress and signed into law by various presidents over the past several decades. In addition, executive orders were issued by the presidents to further help protect the environment and public health. Some of these laws are summarized below (U.S. EPA 2015).

The National Environmental Policy Act (NEPA) was established in 1970 to foster and promote the general welfare, to create and maintain conditions under which man and nature can exist in productive harmony, and to fulfill the social, economic, and other requirements of present and future generations of Americans.

The Occupational Safety and Health Act was enacted in 1970 to "assure safe and healthful working conditions for working men and women by setting and enforcing standards and by providing training, outreach, education and assistance."

The Clean Air Act was signed by President Richard Nixon on December 31, 1970, to foster the growth of a strong American economy and industry while improving human health and the environment. Clean Air Act Amendments were passed in 1990.

The Federal Water Pollution Control Act was first enacted in 1948, but took on its modern form when completely rewritten in 1972 entitled the Federal Water Pollution Control Act Amendments of 1972. Major changes have subsequently been introduced via amendatory legislation including the Clean Water Act of 1977 and the Water Quality Act of 1987. The Clean Water Act does not directly address groundwater contamination. Groundwater protection provisions are included in the Safe Drinking Water Act (SDWA), the Resource Conservation and Recovery Act (RCRA), and the Superfund act.

The SDWA, Title XIV of the Public Health Service Act, is the key federal law for protecting public water supplies from harmful contaminants. First enacted in 1974 and substantially amended in 1986 and 1996, the act is administered through programs that establish standards and treatment requirements for public water supplies, control underground injection of wastes, finance infrastructure projects, and protect sources of drinking water. The RCRA is the primary law governing the disposal of solid and hazardous waste. Congress passed RCRA on October 21, 1976, to address the increasing problems the nation faced from our growing volume of municipal and industrial waste. RCRA was amended and strengthened by Congress in November 1984 with the passing of the Federal Hazardous and Solid Waste Amendments (HSWA). These amendments to RCRA required phasing out land disposal of hazardous waste. RCRA has been amended on two occasions since HSWA: (i) the Federal Facility Compliance Act of 1992 (strengthened enforcement of RCRA at federal facilities) and (ii) the Land Disposal Program Flexibility Act of 1996 (provided regulatory flexibility for land disposal of certain wastes). RCRA regulations aim to protect public health, conduct proper waste management, reduce the quantity of waste, control leaking underground tanks that contain liquid hazardous material, and conserve natural resources (American Planning Association 2006). A voluntary remediation program has been offered by many states in the United States, which aims to enhance and encourage the remediation of contaminated sites (LaGrega et al. 1994). RCRA focuses only on active and future facilities and does not address abandoned or historical sites, which are managed under the Comprehensive Environmental Response, Compensation, and Liability Act (CERCLA)—commonly known as Superfund. In 1980, the US Congress enacted the Superfund program, to provide funding for the treatment of hazardous waste sites (American Planning Association 2006; LaGrega et al. 1994). The US EPA is responsible for administering the Superfund program. The main objective of this program is to reduce risks to public health by eliminating harmful contaminants from the contaminated sites. The NPL identifies the most serious hazardous waste sites across the United States. These sites require extensive remediation and have a priority for funding from the Superfund program (LaGrega et al. 1994).

The Superfund Amendments and Reauthorization Act (SARA) amended the CERCLA on October 17, 1986. SARA reflected EPA's experience in administering

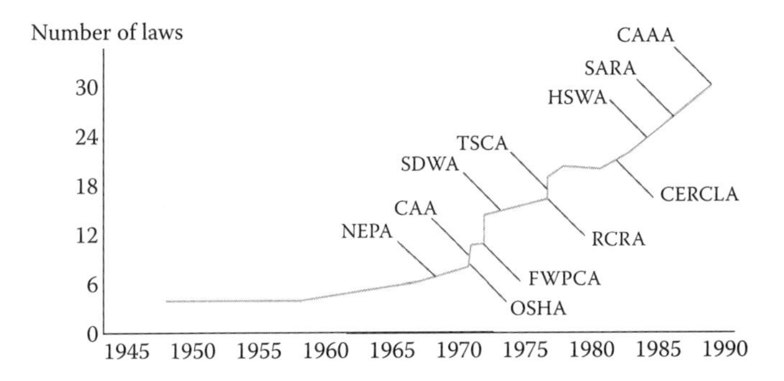

FIGURE 3.5 Growth of environmental laws in the United States. (Adapted from U.S. EPA [2002], *Profile of the Organic Chemical Industry*, EPA Office of Compliance Sector Notebook Project, Available online at: http://archive.epa.gov/sectors/web/pdf/organic-2.pdf, Washington, D.C.)

the complex Superfund program during its first 6 years and made several important changes and additions to the program. SARA

- Stressed the importance of permanent remedies and innovative treatment technologies in cleaning up hazardous waste sites
- Required Superfund actions to consider the standards and requirements found in other state and federal environmental laws and regulations
- Provided new enforcement authorities and settlement tools
- Increased state involvement in every phase of the Superfund program
- Increased the focus on human health problems posed by hazardous waste sites
- Encouraged greater citizen participation in making decisions on how sites should be cleaned up
- Increased the size of the trust fund to $8.5 billion

SARA also required EPA to revise the Hazard Ranking System to ensure that it accurately assessed the relative degree of risk to human health and the environment posed by uncontrolled hazardous waste sites that may be placed on the NPL. The Toxic Substances Control Act of 1976 provides EPA with authority to require reporting, record-keeping and testing requirements, and restrictions relating to chemical substances or mixtures. Figure 3.5 shows the growth of environmental laws in the United States.

3.6 GROUNDWATER PROTECTION PLANNING

Pollution prevention is an essential tool for protecting the groundwater from contamination. The pollution prevention program includes identifying and estimating all sources of pollution, managing chemicals to reduce risk, and waste minimization. There are several contaminated sites in the United States and worldwide that constitute a threat to the public's health and the environment. These sites must be examined carefully to identify the potential pollution (LaGrega et al. 1994). The examination includes defining the site's hydrology, geology, and contamination; communities near these sites; and potential releases to the environment (Canter et al. 1988). A risk assessment is the next step after classifying the site, and such assessment includes identifying quantity and type of pollutants that exist in the site and choosing a suitable remedial action.

The CERCLA aims to clean up the contaminated sites by conducting 10 remedial action stages for each site, as shown in Figure 3.6 (LaGrega et al. 1994): (1) site discovery, (2) preliminary assessment, (3) site inspection and hazard ranking analysis, (4) national priorities determination, (5) feasibility study and remedial investigation, (6) remedy selection and record of decision, (7) remedial design, (8) remedial action, (9) operation and maintenance, and (10) NPL delisting.

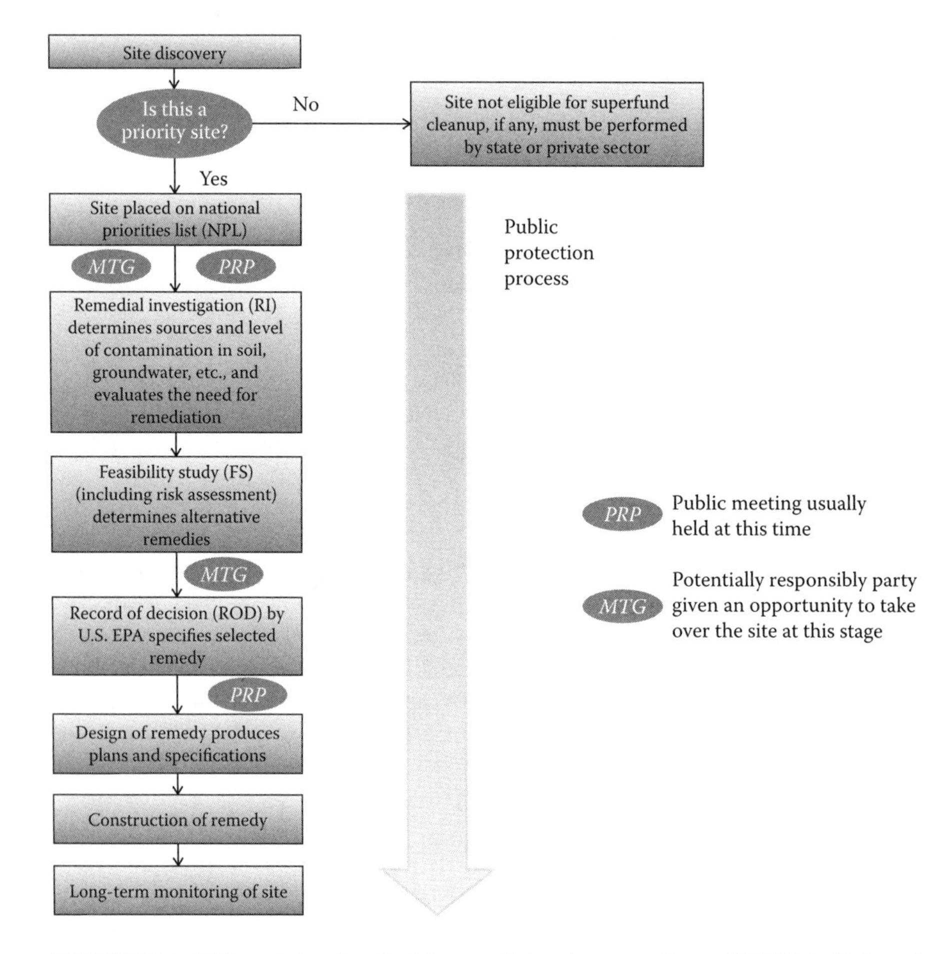

FIGURE 3.6 Different steps involved in remedial action according to CERCLA. (Adapted from U.S. Department of Energy [1994], *A Comparison of the RCRA Corrective Action and CERCLA Remedial Action Processes*, Office of Environmental Guidance. Available online at: http://www.qe3c.com/dqo/project/level5/rcracomp.pdf, Washington, D.C.)

Operators or owners can enter into an agreement with the state Department of Environmental Management (DEM) to clean up and remediate contaminated property (LaGrega et al. 1994). A Certificate of Completion will be issued by the DEM to the sites that have been cleaned up correctly. There are some criteria that must be available for the sites to share in a state's voluntary remediation programs: (1) sites that are not listed on the NPL, (2) sites that do not have environmental problems, (3) sites that do not pose a risk to human health, (4) sites that do not affect the drinking water supplies, and (5) sites that are not under discussion by the RCRA (LaGrega et al. 1994).

Laws and regulations require a methodology for the evaluation and cleanup of contaminated sites in order to facilitate the cleanup process and prevent any delays. Three aspects should be taken into consideration: (1) site description, (2) risk evaluation, and (3) choice of an effective treatment measure (LaGrega et al. 1994). Remedial actions must be implemented at sites containing hazardous wastes. Remedial actions include building barriers at the site, in situ drilling and disposal in a landfill, and on-site or ex situ treatment (Reddy 2004). The choice of remedial action also depends on the cost of the cleanup, end use of the site, health and safety aspects, and environmental liability (Canter et al. 1988). The cost of remediation depends on the applicable regulations and the site conditions. Levels of treatment or cleanup are selected according to the end use of the site. Federal regulations provide protection for employees working in hazard waste sites and thus require strict safety procedures. The party responsible for contamination must pay for the remediation. Four classes have been identified by the CERCLA (LaGrega et al. 1994): (1) the present owner or operator of the contaminated site, (2) the person responsible for the site at the time of disposal of hazardous contaminants, (3) anyone who is responsible for treating or disposing of hazardous contaminants at the site, and (4) anyone who is responsible for transporting hazardous substances to the site (LaGrega et al. 1994).

3.7 GROUNDWATER REMEDIATION TECHNOLOGIES

This section provides an overview of the physicochemical and biological treatment techniques used to treat contaminated groundwater (Bhandari et al. 2007). The physicochemical processes involve changing the structure of molecules and atoms, and their interactions. Biological treatment technology uses microorganisms to consume organic contaminants (Bhandari et al. 2007).

3.7.1 AIR STRIPPING

The air stripping process enhances the transformation of contaminants from a liquid phase into a gas phase. Air stripping can be performed by using stripping towers or stripping basins. Stripping towers include tray towers, packed towers, and spray systems; stripping basins consist of mechanical aeration or diffused aeration (Bedient et al. 1997; Canter et al. 1988; Hyman and DuPont 2001; LaGrega et al. 1994). Air stripping is a cost-effective method for removing VOCs from contaminated groundwater; it is most effective for VOC concentrations less than 200 mg/L. The removal efficiency of VOCs can exceed 99.9%, if the system is designed correctly (Bhandari et al. 2007; Canter et al. 1988). Air stripping techniques are effective in removing volatile and semivolatile compounds, but are not effective in removing nonvolatile compounds. The contaminated water stream is pumped to the top of the stripper and distributed uniformly over the packing through a distributor (Bedient et al. 1997; Canter et al. 1988; LaGrega et al. 1994). Contaminated water flows downward by gravity through the packing materials. The packing can be a structure piled in the column, or it can be composed of different pieces (Hyman and DuPont 2001). The packing materials must provide appropriate transfer surface to facilitate the transition of volatile compounds from the liquid phase to the air phase. The packing materials are often in plastic forms that have a high surface-to-volume ratio (Bedient et al. 1997; Bhandari et al. 2007; Canter et al. 1988). The air stream is blown into the bottom of the column and flows upward, making contact with the contaminated water (Bhandari et al. 2007; LaGrega et al. 1994). Finally, the treated groundwater stream leaves the tower at the bottom, while the air stream leaves the tower at the top. The main problem is that the stripping towers foul because of the growth of bacteria, algae, or fungi; iron oxidation; and fine particulates in the water (Canter et al. 1988; Hyman and DuPont 2001; LaGrega et al. 1994). The process is illustrated in Figure 3.7.

Removal efficiencies could reach 99% for towers that have 15 to 20 feet of conventional packing. Removal efficiencies can be improved by changing the configuration of the packing material, heating the contaminated water, or adding a second air stripper in series with the first stripper (FRTR 2000).

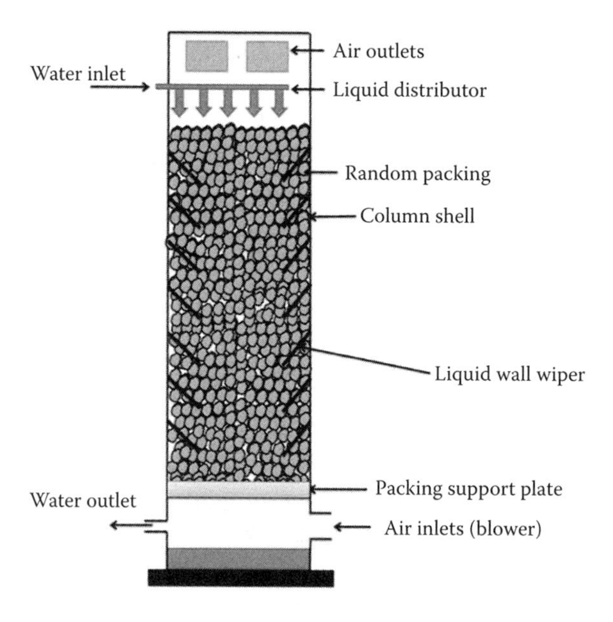

FIGURE 3.7 Packed tower air stripper. (Adapted from U.S. EPA [2012b], *A Citizen's Guide to Air Stripping*, Office of Solid Waste and Emergency Response, Washington, D.C.)

3.7.2 Carbon Adsorption

Sorption is a process by which one substance holds another in a different phase. Activated carbon is the most widely used adsorbent in environmental applications (Canter et al. 1988; LaGrega et al. 1994; Nyer 1993). The adsorption characteristics of the carbon depend on the nature of the used raw materials and processing techniques. Activated carbon adsorption is a sophisticated technique that is used most often to remove a wide range of organic and inorganic compounds from groundwater (Bhandari et al. 2007; LaGrega et al. 1994; Nyer 1992). Activated carbon is available in both granular and powdered forms. Powdered activated carbon is mainly used in biological treatment systems, while granular activated carbon is used in groundwater remediation (Canter et al. 1988; LaGrega et al. 1994; Nyer 1993).

Carbon adsorption systems consist of a series of continuous flow columns. A plenum plate is used to hold the carbon in place. The contaminated water stream is pumped to the top of the adsorption column to make contact with the activated carbon. Finally, the water stream leaves the last column through a drain system at the bottom (Bhandari et al. 2007; Nyer 1993). The final column in the system is a polishing unit. The process is illustrated in Figure 3.8. Backwashing carbon is a very important step to avoid buildup of excessive head loss resulting from the accumulation of solid particles. Down-flow beds must also be backwashed regularly to flush away accumulated solids. The adsorption capacity of the carbon is not unlimited; hence, the spent carbon should be regenerated several times to maintain the same adsorption capacity (Canter et al. 1988; LaGrega et al. 1994; Nyer 1993).

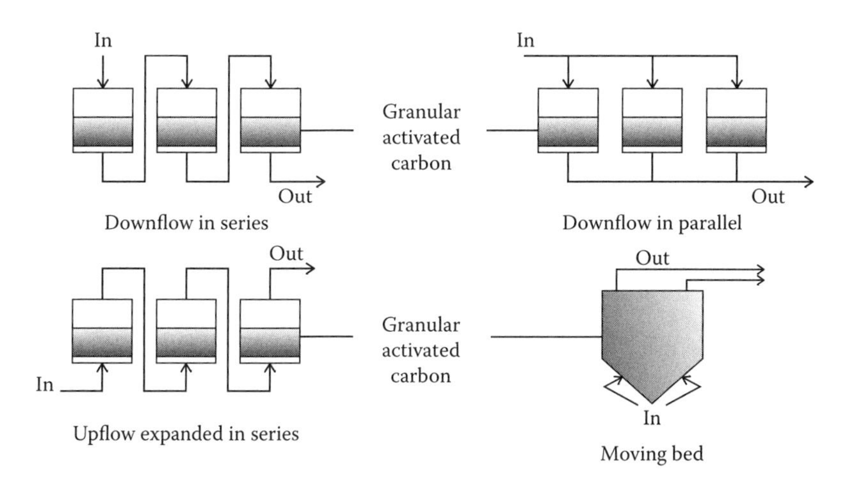

FIGURE 3.8 Activated carbon adsorption systems. (Adapted from U.S. EPA [2000c], *Granular Activated Carbon Absorption and Regeneration*, Wastewater Technology Fact Sheet, Office of Water. Available online at: http://water.epa.gov/scitech/wastetech/upload/2003_05_15_mtb_carbon_absorption.pdf, Washington, D.C.)

3.7.3 Steam Stripping

Steam stripping technology is used to remove VOCs and sometimes semi-VOCs from contaminated groundwater (LaGrega et al. 1994). This process is capable of reducing VOC concentrations to extremely low levels (Bhandari et al. 2007; LaGrega et al. 1994; Reddy and Claudio 2009). Removal efficiencies can reach 99% (Celenza 2000). Both air and steam strippers are based on the transfer of organic compounds from the liquid phase to the vapor phase. Using an air stripper is better than using a steam stripper, if the contaminated groundwater has a high concentration of organic compounds. In this case, the steam stripper will require more complex design techniques. Another difference is that the steam stripper requires a much higher temperature than an air stripper to operate efficiently (Bhandari et al. 2007; LaGrega et al. 1994).

The steam stripper can operate at atmospheric pressure or under vacuum. In the atmospheric pressure steam stripper, the contaminated water stream enters the system at the feed point and flows downward by gravity through the stripping section of the column (LaGrega et al. 1994; Reddy and Claudio 2009). The steam is blown into the bottom of the column and flows upward, making contact with the contaminated water. The operating temperature of the column ranges from 215°F to 220°F (above the boiling point of water). The elevated temperature causes the VOCs in the water to transfer from the liquid phase to the vapor phase. The vapor stream exits the top of the column and condenses back to a liquid in the overhead condenser (Bhandari et al. 2007; Reddy and Claudio 2009). Chlorinated hydrocarbons can be separated from the liquid (condensed water) and recovered separately. An activated carbon adsorption unit is used to treat any uncondensed vapor before venting to the atmosphere (LaGrega et al. 1994; Reddy and Claudio 2009). Finally, the stripped water flows from the bottom of the column via a polishing unit. The process is illustrated in Figure 3.9.

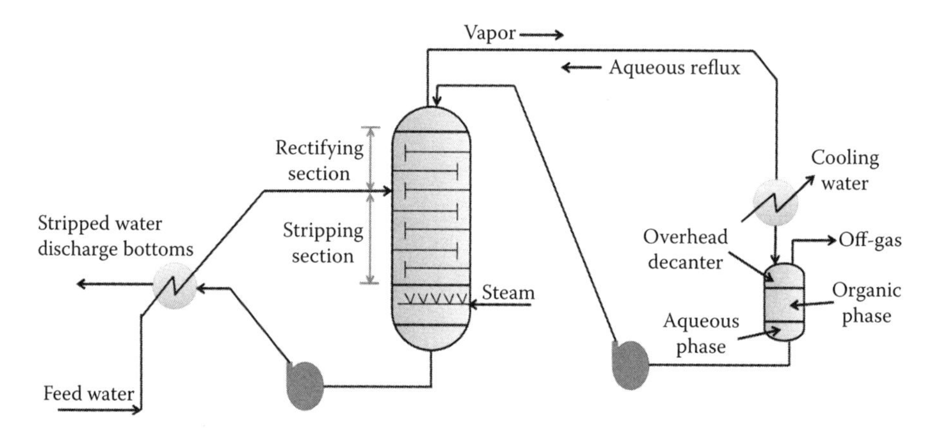

FIGURE 3.9 Atmospheric pressure steam stripping column. (Adapted from U.S. EPA [1988], *Industrial Wastewater Steam Stripper Performance*, Office of Air Quality Planning and Standards. Available online at: http://nepis.epa.gov/Exe/ZyPDF.cgi/9100EQYG.PDF?Dockey=9100EQYG.PDF, Research Triangle Park, NC.)

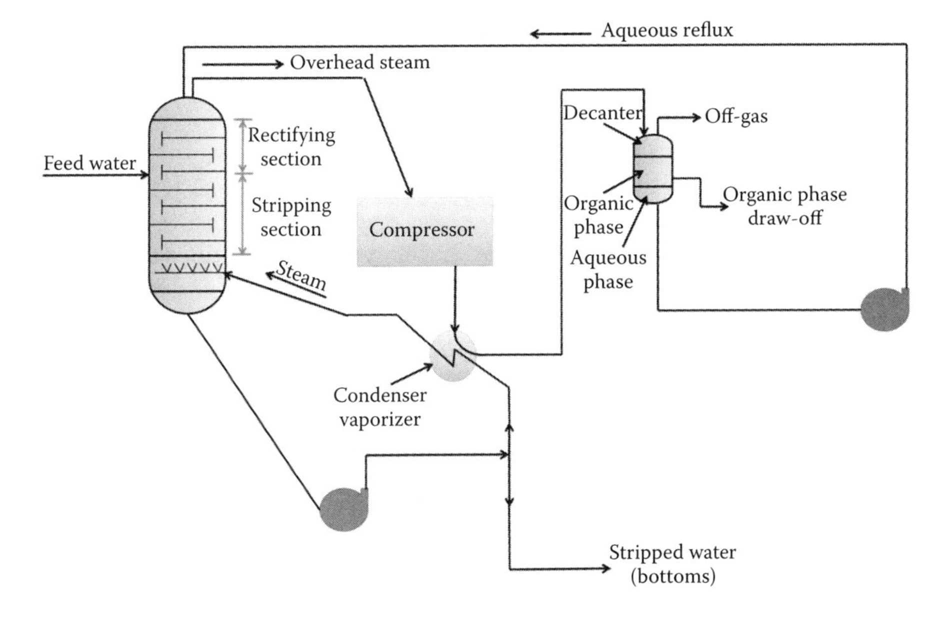

FIGURE 3.10 Vacuum steam stripper. (Adapted from U.S. EPA [1988], *Industrial Wastewater Steam Stripper Performance*, Office of Air Quality Planning and Standards. Available online at: http://nepis.epa.gov/Exe/ZyPDF.cgi/9100EQYG.PDF?Dockey=9100EQYG.PDF, Research Triangle Park, NC.)

A vacuum stripper has two sections for stripping and rectification, and has many similarities to an atmospheric steam stripper (Bhandari et al. 2007; LaGrega et al. 1994). However, the atmospheric stripping column operates at a higher temperature than the vacuum stripping column, as a result of reducing the pressure inside the vacuum stripping column. The operating temperature of the vacuum stripping column ranges from 140°F to 180°F (LaGrega et al. 1994; Reddy and Claudio 2009). The process is illustrated in Figure 3.10.

3.7.4 Chemical Oxidation

Chemical oxidation is an emerging technology for in situ remediation and is most commonly applied at EPA hazardous waste sites (Bhandari et al. 2007; LaGrega et al. 1994). This technology can destroy a wide range of organic contaminants, including VOCs, mercaptans, and phenols, and inorganic contaminants such as cyanide and thiocyanate (Siegrist et al. 2011). The most common oxidizing agents used for hazardous waste treatment include chlorine, ozone, and hydrogen peroxide. Ultraviolet radiation is often used with hydrogen peroxide or ozone to accelerate the oxidation of VOCs (Bhandari et al. 2007; Hyman and DuPont 2001; LaGrega et al. 1994).

In situ chemical oxidation is generally conducted in plug flow reactors or completely mixed tanks (LaGrega et al. 1994; Nyer 1992; Siegrist et al. 2011). The oxidation tank is divided into two sections. One section contains contaminated water while the other contains the treated water. The oxidizing agent is either dosed directly into the oxidation tank or added to the contaminated water just before entering the oxidation tank (Hyman and DuPont 2001; LaGrega et al. 1994; Siegrist et al. 2011). Complete mixing between the oxidizing agent and the contaminants for a short period is necessary to help reduce the chemical dosage needed to obtain a certain target effluent concentration (Bhandari et al. 2007; LaGrega et al. 1994; Nyer 1992).

This technology is not efficient to use if organic compounds other than those of concern are present in high concentrations (LaGrega et al. 1994; Siegrist et al. 2011). For example, during the oxidation of cyanide, the presence of a high concentration of other organic compounds will require adding large amounts of oxidizing agents (Hyman and DuPont 2001; Siegrist et al. 2011). Additionally, reactions involving some organic compounds (e.g., hydrocarbons) and oxidizing agents (e.g., chlorine) may lead to the production of toxic substances rather than destruction of the organic compounds (Bhandari et al. 2007; LaGrega et al. 1994; Nyer 1992).

3.7.5 Biological Treatment

Biological treatment uses living organisms, including algae, bacteria, fungi, and protozoa, to degrade organic compounds (Canter et al. 1988; Hyman and DuPont 2001). The living organisms are biologically distinct and also differ in characteristics. All organic chemicals contained in groundwater can be effectively removed, if proper microbial communities are maintained (Bhandari et al. 2007; Canter et al. 1988). Bioreactors are used to treat contaminated groundwater in the presence of microorganisms through suspended or attached growth systems. In suspended growth systems, contaminant degradation takes place in activated sludge units (Bhandari et al. 2007). Aerators are used in activated sludge units to provide the oxygen required by microorganisms to live and reproduce by feeding on organic compounds. In attached growth systems, the microorganisms grow on the filter media (e.g., gravel, rocks, sand, and plastic) (Canter et al. 1988; Hyman and DuPont 2001). A rotating distributor located above the unit is used to uniformly distribute the wastewater. The microorganisms break down organic compounds and remove pollutants from the groundwater. The most common application for attached growth systems is the trickling filter (Bhandari et al. 2007; Hyman and DuPont 2001).

3.7.6 Air Sparging (Biosparging)

Air sparging involves injecting air through a network of injection wells into the saturated zone in order to volatilize contaminants from the groundwater (Bhandari et al. 2007; LaGrega et al. 1994). Application of this technique helps increase the oxygen concentration in the subsurface, which, in turn, leads to enhanced biodegradation of contaminants in saturated and unsaturated soils (LaGrega et al. 1994; Nyer 1998). Air sparging technology is used to remove VOCs and semi-VOCs from contaminated groundwater, in addition to removing soil contaminants including toluene, benzene, ethylbenzene, xylene, and chlorinated solvents (LaGrega et al. 1994; Nyer 1998). A soil vapor extraction system is commonly used to extract volatile substances migrating into the unsaturated zone (Bhandari et al. 2007; LaGrega et al. 1994).

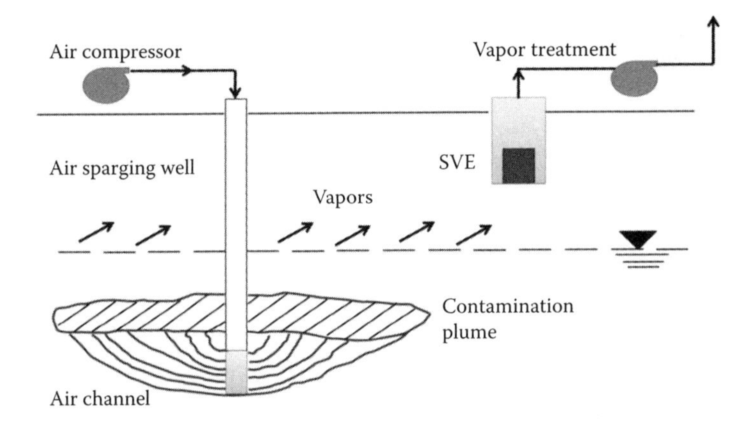

FIGURE 3.11 Air sparging process flow diagram with soil vapor extraction. (Adapted from Grindstaff [1998], "Bioremediation of chlorinated solvent contaminated groundwater," U.S. EPA Technology Innovation Office, Washington, D.C. Available online at: http://nepis.epa.gov/Exe/ZyPDF.cgi/P1003FI0.PDF?Dockey=P1003FI0.PDF.)

Air sparging techniques are better to use at sites with homogeneous and highly permeable soils to promote the contact between the media being treated and sparged air (Bhandari et al. 2007). Air sparging can also be a cost-effective method if applied in sites with large saturated thicknesses (LaGrega et al. 1994). For example, if the saturated thickness is small, the number of injection wells required to cover the contaminated area may become prohibitively expensive (LaGrega et al. 1994; Nyer 1998). The process is illustrated in Figure 3.11.

3.7.7 Passive Treatment Walls

Permeable treatment walls, also called permeable barriers, are installed across the migration path of the contaminant plume (Naftz et al. 2002). The plume moves passively through the wall and the contaminants interact with a catalyst located in the porous media of the wall (Morrison and Spangler 1993). Permeable barriers can be classified into sorption barriers, precipitation barriers, and degradation barriers (Naftz et al. 2002). Sorption barriers adsorb contaminants from groundwater to the barrier surface. Precipitation barriers react with contaminants to produce insoluble products (Bhandari et al. 2007; Morrison and Spangler 1993). Degradation barriers break down contaminants into harmless by-products. These walls may contain chelating agents to immobilize metals, oxygen, and nutrients for microorganisms to promote biological treatment, and metal-based catalysts to help degrade VOCs.

Successful application of this technology requires knowing the nature of subsurface geology, the properties of contaminants, and the theory of groundwater flux. These barriers are suitable to use in shallow aquifers and are designed to operate for several years without an external energy source and with minimal maintenance (Bhandari et al. 2007; Morrison and Spangler 1993). These barriers are effective in remediating groundwater, which contains VOCs and semi-VOCs, including PCE, 1,2-dichloroethane, TCE, trichlorotrifluoroethane (Freon 113), toluene, benzene, ethylbenzene, xylene, chloroform, carbon tetrachloride, and vinyl chloride; nitrate; and metals such as uranium, arsenic, lead, copper, chromium, radium, phosphate, molybdenum, nickel, zinc, cadmium, and radium (Morrison and Spangler 1993).

REFERENCES

American Planning Association, (2006). *Planning and Urban Design Standards*, John Wiley & Sons, Inc. Publication, Hoboken, New Jersey.

Aral, M.M., and W.T. Stewart, (2011). *Groundwater Quantity and Quality Management*, American Society of Civil Engineers Press, Reston, Virginia.

Bhandari, A., Y.S. Rao, C. Pasclae, K.O. Say, R.D. Tyagi, and M.C. Irene, (2007). *Remediation Technologies for Soils and Groundwater*, American Society of Civil Engineers, Reston, Virginia.

Bedient, P.B., H.S. Rifal, and C.J. Newell, (1997). *Ground Water Contamination: Transport and Remediation*, Prentice Hall, Inc., Upper Saddle River, New Jersey.

Canter, L.W., R.C. Knox, and D.M. Fairchild, (1988). *Ground Water Quality Protection*, Lewis Publishers, Inc., Chelsea, Michigan.

Celenza, G., (2000). *Industrial Waste Treatment Processes Engineering: Specialized Treatment Systems*, Technomic Publishing Company, Inc., Lancaster, Pennsylvania.

Charbeneau, R.J., P.B. Bedient, and R.C. Loehr, (1992). *Groundwater Remediation*, Technomic Publishing Company, Inc., Lancaster, Pennsylvania.

Charbeneau, R.J., (2006). *Groundwater Hydraulics and Pollutant Transport*, Waveland Press, Inc., Long Grove, Illinois.

Federal Remediation Technologies Roundtable (FRTR), (2000). *Air Stripping, Remediation Technologies Screening Matrix and Reference Guide*. Available online at: http://www.frtr.gov/matrix2/section4/4-46.html.

Fouillac, A.M., G. Johannes, and W. Rob, (2009). *Groundwater Monitoring*, John Wiley & Sons, Inc., West Sussex, UK.

Geophysics Research Forum and Geophysics Study Committee, (1984). *Groundwater Contamination: Studies in Geophysics*, National Academy Press, Washington, D.C.

Grindstaff, M., (1998). "Bioremediation of Chlorinated Solvent Contaminated Groundwater," U.S. EPA Technology Innovation Office, Washington, D.C. Available online at: http://nepis.epa.gov/Exe/ZyPDF.cgi/P1003FI0.PDF?Dockey=P1003FI0.PDF.

Hyman, M., and R.R. DuPont, (2001). *Groundwater and Soil Remediation Process Design and Cost Estimating of Proven Technologies*, American Society of Civil Engineers Press, Reston, Virginia.

Jain, R.K., L.V. Urban, G.S. Stacey, and H.E. Balbach, (1993). *Environmental Assessment*, Elsevier Science, Orlando, Florida.

Jorgensen, P., (1989). *The Poisoned Well: New Strategies for Groundwater Protection*, Sierra Club Legal Defense Fund, Washington, D.C.

LaGrega, M.D., P.L. Buckingham, and J.C. Evans, (1994). *Hazardous Waste Management*, McGraw Hill, Inc., New York.

Liu, D.H.F. and G.L. Bela, (2000). *Groundwater and Surface Water Pollution*, CRC Press LLC, Boca Raton, Florida.

Mittal, A.K., (2005). "Groundwater Contamination: DOD Uses and Develops a Range of Remediation Technologies to Clean Up Military Sites," United States Government Accountability Office, Report to Congressional Committees, Washington, D.C.

Morrison, S.J., and R.R. Spangler, (1993). "Chemical Barriers for Controlling Groundwater Contamination," *American Institute of Chemical Engineers Journal*, 12 (3): 175–181.

Naftz, D.L., S.J. Morrison, J.A. Davis, and C.C. Fuller, (2002). *Handbook of Groundwater Remediation Using Permeable Reactive Barriers: Applications to Radionuclides, Trace Metals, and Nutrients*, Elsevier Science, Orlando, Florida.

Nyer, E.K., (1992). *Groundwater Treatment Technology*, 2nd edition, Van Nostrand Reinhold, New York.

Nyer, E.K., (1993). *Practical Techniques for Groundwater and Soil Remediation*, CRC Press LLC, Boca Raton, Florida.

Nyer, E.K., (1998). *Groundwater and Soil Remediation: Practical Methods and Strategies*, Sleeping Bear Press, Volume II, Chelsea, Michigan.

Office of Technology Assessment (OTA), (1984). *Protecting the Nation's Groundwater from Contamination*, Washington, D.C.

Page, W.L., (1987). *Planning for Groundwater Protection*, McGraw Hill, Inc., New York.

Patrick, R., F. Emily, and Q. John, (1983). *Groundwater Contamination in the United States*, 2nd edition, University of Pennsylvania, Philadelphia, Pennsylvania.

Rail, C.D., (2000). *Groundwater Contamination: Sources and Hydrology*, Volume I, CRC Press LLC, Boca Raton, Florida.

Reddy, K.R., (2004). "Subsurface Contaminant Remediation: Regulations and Case Studies," Department of Civil and Materials Engineering, University of Illinois at Chicago, Chicago.

Reddy, K.R. and C. Claudio, (2009). *Electrochemical Remediation Technologies for Polluted Soils, Sediments and Groundwater*, John Wiley and Sons, Inc., Hoboken, New Jersey.

Siegrist, R.L., C. Michelle, and J.S. Thomas, (2011). *In Situ Chemical Oxidation for Groundwater Remediation*, Springer Publisher, New York.

U.S. Department of Energy, (1994). *A Comparison of the RCRA Corrective Action and CERCLA Remedial Action Processes*, Office of Environmental Guidance. Available online at: http://www.qe3c.com/dqo/project/level5/rcracomp.pdf, Washington, D.C.

U.S. Department of the Interior, (1993). *National Desalting and Water Treatment Needs Survey*, Bureau of Reclamation, Water Treatment Technology Program Report No. 2. Available online at: https://www.usbr.gov/research/AWT/reportpdfs/report002.pdf, Denver, CO.

U.S. EPA, (1988). *Industrial Wastewater Steam Stripper Performance*, Office of Air Quality Planning and Standards. Available online at: http://nepis.epa.gov/Exe/ZyPDF.cgi/9100EQYG.PDF?Dockey=9100EQYG.PDF, Research Triangle Park, NC.

U.S. EPA, (2000a). *National Water Quality Inventory, 1998 Report to Congress, Ground Water and Drinking Water Chapters*, Report to Congress. Available online at: http://water.epa.gov/type/drink/protection/upload/2006_08_28_sourcewater_pubs_guide_nwiq98305b_gwqchap.pdf, Washington, D.C.

U.S. EPA, (2000b). *Groundwater Quality*, National Water Quality Inventory, 1998 Report to Congress, Ground Water and Drinking Water Chapters. Available online at: http://water.epa.gov/type/drink/protection/upload/2006_08_28_sourcewater_pubs_guide_nwiq98305b_gwqchap.pdf, Washington, D.C.

U.S. EPA, (2000c). *Granular Activated Carbon Absorption and Regeneration*, Wastewater Technology Fact Sheet, Office of Water. Available online at: http://water.epa.gov/scitech/wastetech/upload/2003_05_15_mtb_carbon_absorption.pdf, Washington, D.C.

U.S. EPA, (2002). *Profile of the Organic Chemical Industry*, EPA Office of Compliance Sector Notebook Project, Available online at: http://archive.epa.gov/sectors/web/pdf/organic-2.pdf, Washington, D.C.

U.S. EPA, (2012a). *Groundwater*, U.S. Environmental Protection Agency, Available online at: http://water.epa.gov/type/groundwater/, Washington, D.C.

U.S. EPA, (2012b). *A Citizen's Guide to Air Stripping*, Office of Solid Waste and Emergency Response, Washington, D.C.

U.S. EPA, (2014). *SW-846 Methods*, Available online at: http://www.epa.gov/osw/hazard/testmethods/sw846/, Washington, D.C.

U.S. EPA, (2015). *Laws and Executive Orders*, Available online at: http://www2.epa.gov/laws-regulations/laws-and-executive-orders#majorlaws, Washington, D.C.

United States Geological Survey (USGS), (2005). "Groundwater Use in the United States," Available online at: http://ga.water.usgs.gov/edu/wugw.html, Reston, Virginia.

United States Geological Survey (USGS), (2015a). "Groundwater Quality," Available online at: http://water.usgs.gov/edu/earthgwquality.html, Reston, Virginia.

United States Geological Survey (USGS), (2015b). "Contaminants Found in Groundwater," Available online at: http://water.usgs.gov/edu/groundwater-contaminants.html, Reston, Virginia.

4 GIS, GPS, and Satellite Data

Kevin Urbanczyk

CONTENTS

4.1 INTRODUCTION

Geographical Information Systems (GIS) are utilized extensively in assessing water occurrence on and in the Earth. This technology allows users to make spatial assessments of data. It has the capability to bring multiple types of data into the same analysis scenario. Data for a GIS can come from multiple sources. One primary source of spatial information for field water science is the Global Positioning System (GPS). This is a system that uses multiple navigational satellites to allow a user to precisely locate a point on the Earth. Satellite data are another type of data frequently used by water scientists. Satellite data are remotely sensed data types that focus on spatially and temporally variable features such as land cover types.

4.2 GIS

A GIS is a computerized data management system. It is designed to assemble, store, manipulate, analyze, and display geographically referenced information (Environmental Systems Research Institute [ESRI] 2015a; United States Geological Survey [USGS] 2007). The GIS works in a georeferenced environment. Provided that the spatial data used are properly prepared in either a geographic or a projected coordinate system, the multiple layers (data types) that are displayed in the GIS will line up properly on the screen and be available for spatial analysis. This allows for multiple data types to be analyzed simultaneously. A GIS for water science will typically include basic base map information such as aerial photography, topography, roads, and rivers. Project-specific data might include water or soil sample locations, locations of toxic spills or site elevation models collected by Lidar technology. A user can obtain data from a GPS unit in the field, or new data can be created on screen by digitizing features on an aerial photograph. Historic maps can be scanned and georeferenced to a coordinate system and therefore be incorporated into a GIS.

Understanding the concept of georeferenced data is important for the proper use of a GIS. The GIS layers are models of real-world features. The real world is a spherical surface while the GIS and analog equivalent paper maps are essentially flat surfaces. Spherical coordinates used on a globe are referred to as latitude (which changes in a north–south direction and ranges from 0° to 90° north and 0° to 90° south from the equator) and longitude (which changes in an east–west direction and ranges from 0° to 180° east and 0° to 180° west from the prime meridian). Displaying spherical coordinates in a GIS can be done using a type of coordinate system referred to as "geographic." In a geographic coordinate system, latitude is displayed on a Cartesian y axis and longitude is displayed on a Cartesian x axis. The units for a geographic coordinate system are still spherical (degrees, minutes, seconds). This is an intuitive and simple way to display map data, but it has severe limitations because of the increasing distortion of the longitude to the north and south of the equator. Lines of latitude on a globe are all parallel to each other. This means that the actual distance represented by a degree of latitude on a sphere is the same regardless of the actual latitude (actually, a degree of latitude is just slightly more than 100 km). This distance is the same at the equator as it is at the poles. On the other hand, lines of longitude are all great circles that pass through the north and the south poles. This means that they are not parallel to each other and that the actual distance on a sphere represented by 1° of longitude is not constant. A degree of longitude represents the same ~100-km distance as a degree of latitude only at the equator. The distance represented by a degree of longitude declines systematically with distance north or south of the equator ultimately becoming zero at the North or the South Pole. Since a planar Cartesian coordinate system is rectangular, a map displaying geographic coordinates has increasingly severe distortion with distance north or south of the equator.

A solution to this distortion problem that is available in most GIS packages is the use of a projected coordinate system. "Projecting" is an attempt to display the curved surface of the Earth on a flat surface. The original map projections involved light sources inside wire frame Earth models of latitude and longitude and the projection of the shadow of the wire frame onto a flat surface (Demers 2005). An example of three projection types is included in Figure 4.1.

Cylindrical projections convert the lines of latitude and longitude onto a rectangular plane with the lines mutually perpendicular. Standard cylindrical projections such as shown in Figure 4.1 suffer from a similar distortion as geographic projections. Conic projections place the lines on a conic surface. Lines of latitude

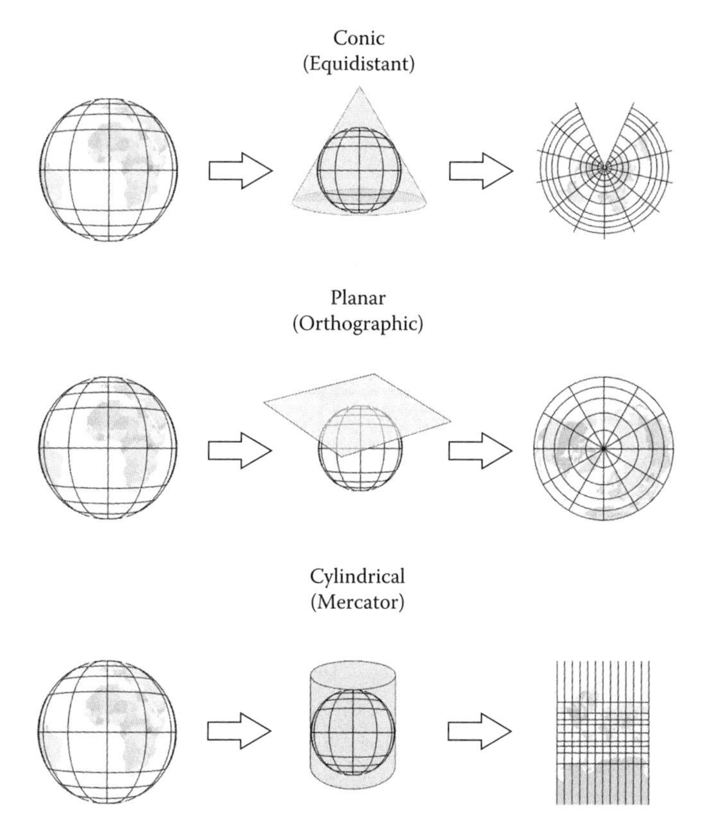

FIGURE 4.1 Map projection examples. (Adapted from Geosphere 2015, Mapping and Projections: Map projection methods, http://www.geog.ucsb.edu/~dylan/mtpe/geosphere/topics/map/map1.html#proj [accessed May 2015].)

are curved and lines of longitude radiate away from the top or the bottom of the map. Planar projections convert the spherical lines to a plane that is shown in Figure 4.1 as being located above the sphere. Lines of latitude are now circles and lines of longitude radiate away from the center of the map.

There are many varieties of projections (Snyder and Stewart 1988), but since they are all flat representations of a curved surface, they all suffer from some type of distortion. Distortion occurs in various forms. Typically, one or more of the following distortions will be inherent in a map projection: shape, area, distance, and direction (Price 2013). A cartographer or GIS analyst will choose a projection that causes the least distortion of the property of interest.

A modern GIS can simultaneously work with multiple types of coordinate systems provided that the coordinate system for the data is properly identified. One way to recognize projected data is to determine the units for the Cartesian display. If the units for a GIS data set are some version of degrees/minutes/seconds, then the coordinate system for that data set is geographic and there is no projection. If the units for the data set are some version of distance such as feet or meters, then the data set is of a projected type. Viewing the metadata for a data set should reveal the projection/coordinate system details.

The most widely used map projection for field water science applications is the Universal Transverse Mercator (UTM) system (Bernhardsen 2002). This system was established in 1947 by the US Department of Defense. It is a cylindrical, Mercator-type projection similar to that shown in Figure 4.1, but it is transverse (Figure 4.2) and cuts through the sphere ("ellipsoid" is the technically correct term) at two locations. The UTM system is made up of 60 zones, each 6° of longitude wide. The only difference between each zone is the longitude of the central meridian that divides the zone in half.

The intersection of the conceptual cylinder of the projection and the ellipsoid is located at a distance of 180 km either side of the central meridian. The distortion inherent in the projection is minimal along these lines 180 km either side of the central meridian, and it increases either way east or west from that line. Coordinates in the UTM system have units of meters. The *X* coordinate is referred to as an "easting" and is relative to an arbitrary "false easting" of 500,000 m along the line of longitude that is the central meridian (determined to eliminate negative values for a coordinate). A UTM value of >500,000 meters East (mE) indicates a location in the eastern portion of the zone. Since lines of longitude converge to the north or south, the actual width in meters of a UTM zone decreases with distance from the equator. The *Y* coordinate in the UTM system is referred to as a "northing." For a position in the northern hemisphere, it is defined as the number of meters north of the equator. Since there are 60 identical UTM zones on the Earth, it is required that a UTM location include the zone designation. Figure 4.3 shows the UTM zones for the United States.

The design of the UTM zone is such that the distortion reaches a maximum at the center of the zone. This distortion is best described in terms of distance measured by a traditional survey instrument such as a laser total station. If a surveyor was located 180 km east or west of the central meridian in any zone, and was on the ellipsoid (most areas on the Earth are above the ellipsoid), and he or she measured a distance

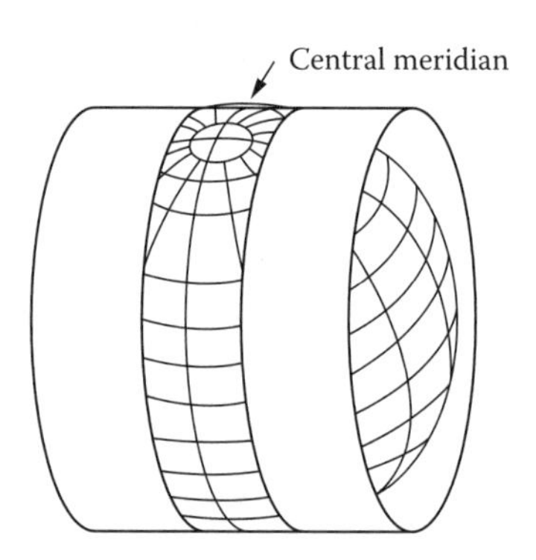

FIGURE 4.2 A depiction of a UTM projection. Note that the projection cylinder intersects the ellipsoid at two locations symmetrical about the central meridian. (From UNSTATS 2012, Plane Rectangular Coordinate Systems, http://unstats.un.org/UNSD/geoinfo/UNGEGN/docs/_data_ICAcourses/_HtmlModules/_Selfstudy/S06/S06_03.html [accessed May 2015].)

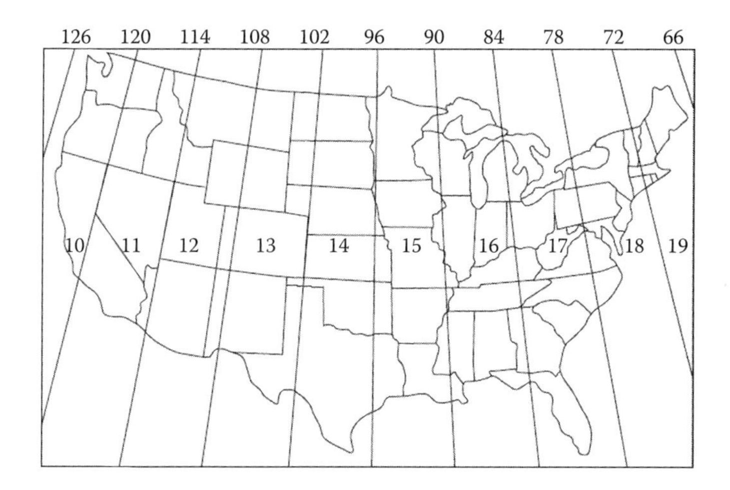

FIGURE 4.3 A map of the UTM zones for the United States. Each zone is 6° wide. (From USGS 2001, The Universal Transverse Mercator (UTM) Grid, USGS Fact Sheet 077-01, http://pubs.usgs.gov/fs/2001/0077/report.pdf [accessed May 2015].)

on the ground of 100 m ("ground" distance), this distance would be identical to the distance between the two points in the UTM system ("grid" distance). If the surveyor measured the same distance, only at the center of the zone, he or she would find that the UTM map projection grid distance is 99.6 m (the scale factor for the UTM system is 0.9996; Limp and Barnes 2014). The difference of 0.4 m is much larger than the precisions of modern Real-Time Kinematic (RTK) GPS or Total Station survey devices (~centimeters). The opposite relationship would occur for an area outside of the 180-km lines. Understanding this difference is extremely important when working with high-precision total station laser survey equipment and combining data with the GPS technology described later in this chapter. Figure 4.4 illustrates the difference between an ellipsoid distance (ground distance) and a projected UTM map distance (grid distance). Elevation above the ellipsoid adds to the distortion factor and a combined factor is reported by the National Geodetic Survey (the scale factor × the elevation factor = the combined factor).

An example scenario might be a geomorphology study of a sandbar adjacent to a river. A team of researchers have both Total Station and RTK GPS survey instruments. The Total Station would be collecting data using ground distances while the RTK would most likely be collecting data in projected grid distances. In order to get both data sets merged into a GIS, the Total Station distances would have to be corrected by the UTM scale factor at the location of the survey.

Being a map, a GIS always incorporates a component of scale. Scale is nearly infinitely variable in a GIS. Scale is expressed as a ratio of distance on a screen or map to the same distance in the real world. A "small-scale" map is a map that covers a large area with relatively little detail. A "large-scale" map is a map that covers a small area with a relatively large amount of detail. An example would be the standard paper topographic quadrangle map available from the USGS. One series of these maps is referred to as the 7.5-minute quadrangle map. These maps cover 7.5 minutes

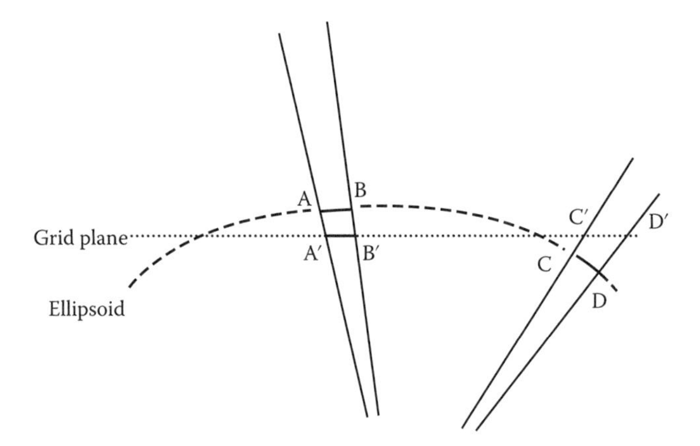

FIGURE 4.4 A representation of the grid versus ground distance issue in the UTM system. The UTM grid is projected onto a planar surface (the "Grid Plane" on the diagram, referred to as the grid). Points A and B are located outside the intersection of the ellipsoid and the grid; therefore, a ground distance measured on the ellipsoid (AB on the diagram) would be longer than the projected distance on the grid (A′B′). The opposite relationship holds for points C and D and distances CD and C′D′. Elevation above the ellipsoid adds to the scale factor required to relate ground to grid distance. (Adapted from Limp, F. and Barnes, A. 2014, *Advances in Archaeological Practice: A Journal of the Society for American Archaeology*, How-To Series, May 2014, 138–143.)

of latitude and 7.5 minutes of longitude. They are intended to be printed at a scale of 1:24,000. Therefore, 1 cm on a printed map would be equivalent to 24,000 cm on the real world (240 m). These maps are sometimes referred to as "24K" maps. The printed 7.5-minute quadrangle map will extend ~60 cm (~23 inches) in the north–south direction (excluding the map collar area). The width will be variable depending on the latitude of the location being displayed. When a digital version of a 7.5-minute quadrangle map (referred to as a Digital Raster Graphic) is displayed in a GIS, the operator must be careful to not attempt to display or use the map at a scale far from the intended scale. In the USGS series of topographic maps, the 7.5-minute (1:24,000) map is large scale compare to the 1:100,000 and 1:250,000 series topographic maps. The 1:24,000 maps are useful as a base map for detailed site studies, while the 1:100,000 topographic maps are useful as base maps for large areas such as multiple counties and the 1:250,000 or smaller scale maps are useful as base maps for projects that are displaying areas as large as a state or region.

4.2.1 Available Packages

The most widely used GIS package is the ArcGIS family of GIS software made available by the ESRI. This software package has evolved from an initial attempt in 1969 to analyze geographic information to help land planners and resource managers make environmental decisions (ESRI 2015b). The first ESRI software package available was ArcInfo (1982), which moved to the desktop environment with pcArcInfo in 1986. The available GUI version was referred to as ArcView (1991). The latest version is referred to as ArcGIS 10.3. The ESRI package includes the standard ArcGIS map display that is used for most viewing, analysis, and map printing; ArcCatalog that assists with dealing with file maintenance and organizing GIS content; ArcScene for displaying three-dimensional data; and ArcGlobe for displaying small-scale data on a globe. The software is available in three levels: Arc View, Arc Editor, and ArcInfo (in increasing order of functionality and cost). ESRI also has a web-based GIS referred to as ArcGIS online.

ESRI's primary competitors include GE Energy, Intergraph, MapInfo, and AutoDesk (GISWIKI 2006).

4.2.2 Standard GIS Data Types

GIS data fall into two distinct categories: vector based and raster based. Each of these two data types has advantages and disadvantages and most natural earth/water features can be effectively modeled by one or the other, or both of these types.

Vector data are stored as the *XY* coordinates of vertices. The coordinates will be either in a geographic or in a projected coordinate system. A "point" layer is composed of one or more individual *XY* coordinates for the points. A "line" layer is composed of line segments that have *XY* coordinates of individual vertices that define the line (at least two are required for a line). A "polygon" layer is composed of individual vertices that define a closed loop (at least three are required). Water sample locations will likely be stored as a point layer, a river or a road would likely be stored as a line layer, and county boundary would likely be stored as a polygon layer.

The ESRI ArcGIS package stores points, lines, and polygons in several different formats. In this discussion, the point, line, or polygon layers are referred to as "feature classes." A feature class has both geometry (the location of the vertices) and a table (attributes about the features). A "coverage" is a type of storage format that can contain multiple feature classes (points, lines, or polygons). It was the standard storage type for the original ArcInfo GIS package but has more recently fallen out of favor. A "shapefile" is a single feature class that is limited to a point, line, or polygon. A "geodatabase" is the latest version utilized by ESRI. A geodatabase can contain multiple feature classes similar to the coverage, and it can also group feature classes into feature data sets. The coverage and the geodatabase storage types have more functionality in terms of topology and behaviors and are necessary for more robust GIS analysis. The shapefile type of storage is simple and easy to use and, as a result, remains a viable data storage format.

Raster data are composed of rows and columns of individual raster cells. Raster data have a resolution that is the cell size in the *X* and *Y* directions, and they have a fixed extent that is defined by the number of rows and columns. A standard digital photograph is an example of raster data. It has a resolution and each grid cell has a value that is the color signal for that pixel. These are commonly stacked in groups of three individual rasters that are merged to form a display image (red, green, blue) on a computer monitor. A "tiff" type file is an example of this. Other specific raster storage types include the ESRI grid format that stores raster data in a relatively complex folder and subfolder scheme and the ERDAS *.img format.

TABLE 4.1
An Example of an Attribute Table for a Water Sample Database

Sample Locations

Reach	mE	mN	SC_uScm	Temp_C	pH	GP5_Rcvr
Outlaw flats	705,856.2	3,347,963.5	730	18.1	7.84	Trimble GeoXT08
Hot springs canyon	681,312.2	3,294,420.8	1380	34.3	6.7	Trimble GeoXT08
Lower canyons	723,065.4	3,359,315.8	727	20.7	8.28	Trimble GeoXT08
Lower canyons	739,555.6	3,370,759	438	14.8	7.81	Trimble GeoXT08

Note: The data in the table were collected with a GPS unit with a data dictionary. The data dictionary on the GPS allows for the field parameters to be input as the point is collected. This functionality prepopulates the attribute table before it is exported into a GIS.

No GIS data type is perfect for storing or modeling all natural phenomena. One can, though, categorize natural features into groups that are better stored in one format or the other. Vector data have the advantage of generally taking less file storage space because only the *XY* location of the vertices are necessary while a raster has a fixed extent regardless of whether the feature of interest covers that full extent. The vector format is preferred for discrete features such as sample locations (point), roads (line), or area of interest boundaries (polygons). The raster format is preferred for continuously varying features such as elevation (digital elevation models [DEMs] are used for this), precipitation, or satellite or aerial imagery.

Both vector and raster data types have the additional feature that they can store multiple attributes about the features that they are modeling. For example, a water sample location feature class (point type) can have additional fields in the attribute table that could include data such as water temperature, pH, dissolved oxygen, and specific conductivity (Table 4.1). Raster data that are discrete, meaning that they have a limited number of possible values such as land cover type, can have multiple attributes in the attribute table. For example, an area may have 100 rows and 100 columns, but there might be only five land cover types. The attribute table for this raster would have only five summary records and would record the total number of grid cells per land cover type and any other attributes about the cover type that might be necessary. A continuous raster such as a DEM that has 100 rows and 100 columns could have up to 10,000 unique values (especially if the data are stored as floating point values). The attribute table for a raster such as this is not viewable because it is much too large and would be a meaningless display of all grid cell values.

4.3 GPS

The GPS is a United States–owned and –operated system that provides users with positioning, navigation, and timing services (GPS.GOV 2015). For the purposes of water science, the system allows users to determine geographic locations of sample locations and other field site location information. These data can then be imported into a GIS for real-time or postprocessed analysis.

The GPS system includes three segments: the space segment, the control segment, and the user segment. The space and control segments are operated by the US Air Force. The user segment includes a wide variety of government, military, and civilian applications.

The Space segment includes a constellation of satellites that transmit radio signals to users. The system includes 31 operational satellites. The satellites fly in a medium Earth orbit at an altitude of 20,200 km and each circles the Earth twice daily. The orbit details are designed such that at any time a user can view at least four satellites from any place on Earth (GPS.GOV 2015).

The control segment is made up of multiple ground facilities that track and control the GPS satellites. These facilities are strategically located around the globe. This segment includes a master control station in Colorado, an alternate master control station in California, 12 command and control ground antennas used to communicate with the GPS satellites, and 16 monitoring sites used to track the GPS satellites.

The GPS user segment includes any type of equipment that can receive and utilize the GPS signal. This includes military and civilian equipment. The civilian community includes a large and diverse user group. This includes applications in surveying, commercial transportation, in-car navigation, farming, recreation, basic navigation, and resource management (FAA 2014). Multiple augmentation systems are available that improve the GPS system. The Nationwide Differential GPS System is a ground-based system that improves the GPS accuracy for users on US land and waterways. It is operated by the US Coast Guard and the Department of Transportation. The system can improve the GPS system from a standard 4- to 20-m accuracy to better than 10-cm accuracy (FHWA 2003). The Wide Area Augmentation System (WAAS) is a satellite-based augmentation system operated by the Federal Aviation Administration. It is designed for aviation use but is widely available for other uses across the United States. The WAAS can improve GPS accuracy to better than 3 m (Garmin 2015). The Continuously Operating Reference Stations (CORS) are coordinated by the National Oceanic and Atmospheric Administration (NOAA). This system manages a network of continuously operating stations and is a cooperative endeavor with government, academic, and private organizations involved (NGS 2015a). Data from the CORS are made available via a web-based Online Positioning User Service (OPUS; NGS 2015b). This free service makes available postprocessed GPS data at the centimeter or better level of accuracy depending on the GPS hardware and collection conditions. Table 4.2 is an example of an OPUS solution. Note that the precision estimates for the latitude, longitude, and height vary from 0.004 to 0.015 m.

Other countries besides the United States are involved in satellite location technology. The US version of the GPS is referred to as NAVSTAR. Russia operates the only other global system (GLONASS). Other countries are in the development phase for more future navigation systems. The term *Global Navigation Satellite System* (GNSS) is used to describe a positioning system that utilizes all available global systems. Currently, this includes only the US NAVSTAR and the Russian GLONASS, but the combination of the two makes available 20–30 satellites for high-precision location data collection (Wikipedia 2015).

TABLE 4.2
An Example of an OPUS Solution Report

NGS OPUS SOLUTION REPORT

All computed coordinate accuracies are listed as peak-to-peak values.
For additional information: http://www.ngs.noaa.gov/OPUS/about.jsp#accuracy

SOFTWARE: page 5	1209.04 master 52.pl 022814		START: 2014/10/08	17:25:00
EPHEMERIS: igr18133.eph [rapid]			STOP: 2014/10/08	21:11:00
NAV FILE: brdc2810.14n			OBS USED: 7330/7801	94%
ANT NAME: TRMR6	NONE		# FIXED AMB: 37/40	93%
ARP HEIGHT: 1.550			OVERALL RMS: 0.015 (m)	

REF FRAME:	NAD_83(2011)(EPOCH: 2010.0000)		IGS08 (EPOCH:2014.7693)	
X:	−1,242,953.934 (m)	0.006 (m)	−1,242,954.713 (m)	0.006 (m)
Y:	−5,427,591.179 (m)	0.003 (m)	−5,427,589.734 (m)	0.003 (m)
Z:	3,101,450.807 (m)	0.006 (m)	3,101,450.621 (m)	0.006 (m)
LAT:	29 16 54.68917	0.004 (m)	29 16 54.70351	0.004 (m)
E LON:	257 6 4.70827	0.005 (m)	257 6 4.66820	0.005 (m)
W LON:	102 53 55.29173	0.005 (m)	102 53 55.33180	0.005 (m)
EL HGT:	538.355 (m)	0.005 (m)	537.188 (m)	0.005 (m)
ORTHO HGT:	559.644 (m)	0.015 (m)	[NAVD88 (Computed using GEOID12A)]	

(*Continued*)

TABLE 4.2 (CONTINUED)
An Example of an OPUS Solution Report

	NGS OPUS SOLUTION REPORT	
	UTM COORDINATES UTM (Zone 13)	STATE PLANE COORDINATES SPC (4204 TXSC)
Northing (Y) [meters]	3,241,045.116	4,166,848.033
Easting (X) [meters]	704,133.579	221,272.660
Convergence [degrees]	1.02811900	−1.91001850
Point scale	1.00011422	0.99986367
Combined factor	1.00002966	0.99977913
US NATIONAL GRID DESIGNATOR: 13RGN0413341045(NAD 83)		

Source: NGS 2015b, OPUS: Online Positioning User Service, http://www.ngs.noaa.gov/OPUS/ (accessed May 2015).

Note: UTM coordinates indicate that the location is in the eastern portion of zone 13 (the UTM Easting is >500,000) and is more than 180 km east of the central meridian (500,000 mE). This is consistent with the point scale factor being >1. If the position had been less than 180 km from the center of the zone, the point scale factor would be <1. A distance measured on the ground at this location would be shorter than the same distance measured in the projected UTM grid.

4.3.1 How GPS Works

The GPS constellation is made up of 31 satellites that are placed in six different orbits around the Earth. Each satellite circles the planet twice per day. Each satellite broadcasts radio signals that include location information, status, and a precise time. A GPS device that receives the signals can determine an exact distance from the satellite based on the arrival time of the signal. If the device receives information from at least four satellites, it can determine its geographic position via a process of triangulation (GPS.GOV 2015).

4.3.2 Providers of GPS Systems

For professional field science applications, the primary vendors of GPS equipment include Trimble, Leica, and Topcon. Each of these vendors makes available hardware and software for the three grades of GPS systems: Consumer grade, Map grade, and Survey grade. Consumer grade systems are those that are available over the counter at many stores and are designed for low cost and ease of use. A WAAS-enabled consumer grade system can achieve accuracies of better than 3 m. Survey grade systems are more expensive and utilized by professional surveyors and others that need precise locations either real-time or postprocessed. The most common survey grade GPS instrument is the RTK system. This technology requires the simultaneous operation of a GPS base station positioned on a known location and a GPS rover instrument collecting data at various locations of interest. The GPS base can be part of the CORS system or a local base location can be established by a user. In order to obtain local, high-precision, georeferenced location information, the user must occupy a location of interest with the RTK GPS base station for at least 4 h and then transmit the collected information via the web to the OPUS system, which will calculate the base station location (NGS 2015b). The user can then use the base station location combined with a rover GPS instrument to determine accurate locations of sites of interest in real time. RTK is expensive and cumbersome, but for many field water science applications, it provides a better solution than traditional surveying methods such as a level or a total station. Map grade GPS systems occupy the market position between the Consumer grade and the Survey grade. With the incremental increase in the number of satellites in the GNSS system and improvements in technology, the boundaries between the grades are becoming increasingly vague.

4.4 SATELLITE DATA

Non-GPS satellite data are used extensively in water science applications. Information from these satellites falls under the category of "remotely sensed" data. Remote sensing is a technique where the scientist utilizes data acquired by a device that is not in contact with the area being studied (Lillesand and Kiefer 2000). Most of the content described in this section refers to satellite systems, but remote sensing is not limited to that. Aerial photography and Lidar data acquisition will also be briefly described.

Since the first satellite, Sputnik 1, was launched by the Soviet Union in 1957, satellite technology has evolved to include communications, GPS/GNSS navigation and location, weather studies, Earth observation, and space stations (Wikipedia 2015). Earth satellite orbits are classified according to altitude. A Low Earth Orbit (LEO) has altitudes ranging from 0 to 2000 km, a Medium Earth Orbit (MEO) has altitudes ranging from 2000 to 35,786 km, a Geosynchronous Orbit (GEO) has an altitude of 35,786 km, and a High Earth Orbit (HEO) has an altitude >35,786 km (NASA 2015). Orbital velocity varies as a function of orbital height. The altitude of 35,786 km has a velocity that is coincident with the orbit of the Earth. This altitude is required for geostationary orbit. A sun-synchronous orbit is one that allows the satellite to observe the Earth at the same local sun time each day (Campbell 2007).

Communication satellites such as the Iridium satellites are placed into LEO near the Earth. These are used for satellite phone applications and can be used for voice and data. Navigation satellites include the GPS and GLONASS systems described above. They are used for various government, military, and civilian location applications. Weather satellites are operated by agencies such as the NOAA. The NOAA geostationary satellite group provides weather-related imagery and NASA's Earth observing satellite group include multiple missions for various Earth observation purposes, including the Landsat program (NASA 2015).

The Earth observing group of satellites is the most useful for water science studies and will be discussed here. They all utilize a passive observation technique. They acquire data from the electromagnetic spectrum that reaches the satellite. This starts as energy from the sun passes through the Earth's atmosphere, interacts with land cover on the Earth, then is reflected off of or emitted from the land cover material, passes back through the Earth's atmosphere, and is detected by the satellite. The sensors on the satellites are designed to collect this radiation and transfer it to ground-based stations.

Multispectral systems such as the Landsat group of satellites collect data in discreet bands, or ranges, of the electromagnetic spectrum. Examples of multispectral systems are SPOT 1 HRV, Landsat MSS, and Lansdsat TM. The Landsat series 4 and 5 satellites utilized a Thematic Mapper (TM) and a Multispectral Scanner (MSS) to collect data in the visible and infrared regions in four and seven channels, respectively (Table 4.3). The ground cell resolution of the Landsat 4 and 5 series satellites is 30 m.

TABLE 4.3
Landsat Mission Dates

Satellite	Launch	Decommissioned	Sensors
Landsat 1	July 23, 1972	January 6, 1978	MSS/RBV
Landsat 2	January 22, 1975	July 27, 1983	MSS/RBV
Landsat 3	March 5, 1978	September 7, 1983	MSS/RBV
Landsat 4	July 16, 1982	June 15, 2001	MSS/TM
Landsat 5	March 1, 1993	2013	MSS/TM
Landsat 6	October 5, 1993	Did not achieve orbit	ETM
Landsat 7	April 15, 1999	Operational	ETM+
Landsat 8	February 11, 2013	Operational	OLI/TIRS

Source: USGS 2013, Landsat—A Global Land-Imaging Mission, Fact Sheet 2012-3072, revised May 2013, http://pubs.usgs.gov/fs/2012/3072/fs2012-3072.pdf (accessed May 2015).

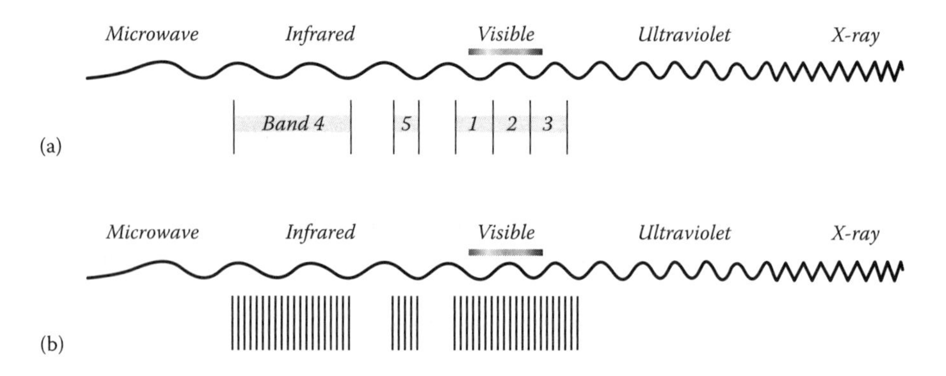

FIGURE 4.5 A comparison of multispectral versus hyperspectral data collectors. (a) An example of multispectral data. A collector of this type would record visible data in bands 1, 2, and 3, a wide band of infrared (4), and a near-infrared band (5). There are therefore only five discrete data components. A hyperspectral data collector (b) would collect much more narrow bands of data across the region of interest. Not to scale. (From Gisgeography 2015, Multispectral vs Hyperspectral Imagery Explained, http://gisgeography.com/multispectral-vs-hyperspectral-imagery-explained/ [accessed May 2015].)

Hyperspectral remote sensing utilizes many narrowly defined spectral channels. The hyperspectral sensors can provide 200 or more channels, each only 10 nm wide. Figure 4.5 shows a comparison of multispectral and hyperspectral bandwidths and locations on the electromagnetic spectrum.

The Airborne Visible/Infrared Imaging Spectrophotometer is an example of a hyperspectral spectrometer that is operated from an aircraft.

4.4.1 The Landsat Program

The Landsat program was designed to continuously collect moderate resolution remotely sensed data for the entire Earth over a long period. It is a joint program between the USGS and NASA and it collects data to support government, commercial, industrial, civilian, military, and educational entities worldwide (Landsat 2015). The program began in 1972 with the launch of Landsat 1 (Table 4.3). This was followed by launches of Landsat 2–8 (Landsat 6 did not reach orbit). The currently operating versions are Landsat 7 and 8. Landsat 5 was launched in 1984 and exceeded its 3-year design life. It was decommissioned in 2013.

The return beam vidicon (RBV) sensor on Landsat 1–3 produced high-resolution television-like images of the Earth's surface (Campbell 2007). It included three spectral channels (green, red, and near infrared). The MSS subsystem used a flat oscillating mirror to scan from west to east to produce a ground coverage of 185 km perpendicular to the satellite track. The MSS sensor collected data in four bands (green, red, and two infrared regions) with a ground cell resolution of approximately 79 m × 57 m. The Landsat TM was included on missions 4 and 5. It included seven bands: 1, blue-green (0.45 to 0.52 μm); 2, green (0.52 to 0.60 μm); 3, red (0.63 to 0.69 μm); 4, near infrared (0.76 to 0.90 μm); 5, mid infrared (1.55 to 1.75 μm); 6, far infrared (10.4 to 12.5 μm); and 7, mid infrared (2.08 to 2.35 μm). The resolution of bands 1–5 and 7 is 30 m and the resolution of band 6 is 120 m. The ETM+ scanner that is included on Landsat 7 is similar to the TM sensor with the following exceptions. The far-infrared band 6 resolution decreased to 60 m and a "panchromatic" band was added, which was located in a wide band from 0.52 to 0.90 um. This single band covers the visible range of the electromagnetic spectrum and has a ground cell resolution of 15 m. The Operational Land Imager (OLI) on the Landsat 8 satellite collects data in nine bands (Table 4.4). The arrangement is similar to the ETM+ scanner with the addition of narrow band added as band 1 for coastal zone observations, band 9 for cirrus cloud observations, and the Thermal Infrared Sensor

TABLE 4.4
The Landsat 8 OLI and TIRS Band Designations

Spectral Bands	Wavelength (μm)	Resolution (m)	Use
Band l—coastal/ aerosol	0.43–0.45	30	Increased coastal zone observations
Band 2—blue	0.45–0.51	30	Bathymetric mapping; distinguishes soil from vegetation; deciduous from coniferous vegetation
Band 3—green	0.53–059	30	Emphasizes peak vegetation, which is useful for assessing plant vigor
Band 4—red	0.64–0.67	30	Emphasizes vegetation slopes
Band 5—near IR	0.85–0.88	30	Emphasizes vegetation boundary between land and water, and landforms
Band 6—SWIR 1	1.57–1.65	30	Used in detecting plant drought stress and delineating burnt areas and fire-affected vegetation, and is also sensitive to the thermal radiation emitted by intense fires; can be used to detect active fires, especially during nighttime when the background interference from SWIR in reflected sunlight is absent
Band 7—SWIR-l	2.11–2.29	30	Used in detecting drought stress, burnt and fire-affected areas, and can be used to detect active fires, especially at nighttime
Band 8—panchromatic	0.50–0.68	15	Useful in "sharpening" multispectral images
Band 9—cirrus	1.36–1.38	30	Useful in detecting cirrus clouds
Band 10—TIRS 1	10.60–11.19	100	Useful for mapping thermal differences in water currents, monitoring fires and other night studies, and estimating soil moisture
Band 11—TIRS 2	11 50–12.51	100	Same as band 10

Source: USGS 2013, Landsat—A Global Land-Imaging Mission, Fact Sheet 2012-3072, revised May 2013, http://pubs.usgs.gov/fs/2012/3072/fs2012-3072.pdf (accessed May 2015).

Note: Instrument-specific relative spectral response functions may be viewed and compared using the Spectral Viewer tool (http://landsat.usgs.gov/tools_spectralViewer.php).

(TIRS) bands 10 and 11. Table 4.4 includes uses for the specific Landsat bands for Earth observations.

4.5 LIDAR

Lidar refers to "light detection and ranging" (Campbell 2007). It is a remote sensing technique that is different from the previously covered satellite techniques in that it utilizes an active as opposed to a passive data collection sensor. Lidar is a laser scanner. It uses a very narrow band of the electromagnetic spectrum transmitted as a narrow beam over long distances with only slight divergence (laser). An imaging Lidar transmits thousands of laser pulses per second in a swath across an area being imaged. The reflected pulses are detected as "returns" and the timing of the return is converted to a distance. Each return is a point in a large "point cloud" of data produced by the instrument. If the location of the scanner and several ground control points are known, the point cloud represents multiple georeferenced points on the three-dimensional surface that was scanned and can be imported into a GIS for analyses and comparison to other georeferenced data.

Lidar scanners can be mounted on airplanes or operated in a ground-based arrangement. They have become a standard technique for producing high-resolution DEMs. The Lidar technology has limited application in heavily vegetated areas and in areas with water. Multiple returns can partially resolve tree canopy versus bare ground in a vegetated area.

4.6 REMOTE SENSING

Remote sensing is a term used to refer to the science of studying remote objects from a distance. It utilizes modern sensors, data-processing equipment, communication, and space and airborne vehicles for the purpose of creating surveys of the Earth's surface (Campbell 2007; National Academy of Sciences 1970). A concise definition is that it is the "...practice of deriving information about the Earth's land and water surfaces using images acquired from an overhead perspective, by employing electromagnetic radiation in one or more regions of the electromagnetic spectrum, reflected or emitted from the Earth's surface" (Campbell 2007, p. 6).

Passive satellite-based remote sensing requires an understanding of electromagnetic radiation and how energy from the sun travels through the Earth's atmosphere, interacts with Earth materials and is reflected or r-emitted from the surface, travels back through the Earth's atmosphere, and is ultimately collected by a sensor on the satellite. The process includes photographic sensors, digital data storage, and image processing. Translating the images into useful information requires image interpretation skills (Campbell 2007). Classification is the task of assigning features on an image to distinct classes based on appearance. Enumeration is the task of counting recognizable features on an image. Measurement involves distances, heights, and volumes and also image brightness. Delineation is the task of outlining regions that are distinct in terms of tones or textures. All of these tasks are the job of the remote sensing specialist.

Active Lidar remote sensing requires an understanding of laser generation, laser travel and reflection, timing of reflections, and laser detection. The large volume of point data created by the Lidar system requires significant computer processing capability and an understanding of the conversion of the point cloud to a triangular irregular network (TIN) or a DEM. The TIN or DEM produced is analyzed in a GIS.

4.7 EXAMPLES

4.7.1 Land Use Land Cover

The combination of satellite imagery and GIS is exemplified by land use and land cover determination techniques (Sabins 1997). Land use describes how land is used such as for agriculture, industry, or municipal purposes. Land cover refers to the types of materials that are on the surface. Examples include vegetation, bare rock, or buildings. Remotely sensed data are used to develop land use land cover classification maps. Remote sensing can image large areas quickly, with reasonable resolution and perspective, and can be repeated over time.

The land use and land cover classification system includes three levels (Anderson et al. 1976). Level I is the coarsest scheme and includes the following general cover types: 100 Urban or built-up, 200 Agriculture, 300 Rangeland, 400 Forest Land, 500 Water, 600 Wetlands, 700 Barren Land, 800 Tundra, and 900 Perennial snow or ice. Each of these is further divided into Levels II and III. For example, the Level I Wetlands (600) is further divided into 610 Vegetated wetlands forested, 620 Vegetated wetlands nonforested, and 630 nonvegetated wetlands. The Level II Vegetated wetlands (610) is further divided into 611 Evergreen, 612 Deciduous, and 613 Mangrove.

Soulard et al. (2014) produced a report that describes land cover trends for selected areas in the conterminous United States for the period 1973 to 2000. The data include five dates (1973, 1980, 1986, 1992, and 2000) of land use and land cover data in sample blocks from 84 ecological regions. The maps were classified using Landsat MSS, TM, and ETM+ imagery and a modified Anderson level 1 classification scheme. This scheme consisted of 11 general LULC classes that included two transitional disturbance classes: "mechanically disturbed," which denotes human induced disturbance, and "nonmechanically disturbed," which denotes natural disturbances such as fire or insect infestation events (Table 4.5). The completed maps have the Level 1 classification scheme color coded. Figure 4.6 shows an example of the applied land cover trends data set for an area in the Ouachita Mountains Ecoregion. This set of maps shows forest cutting and regrowth patterns from 1973 to 2000. The area designated "mechanically disturbed" (magenta color) increases from 1973 to 1992 as a result of forest cutting (Soulard et al. 2014).

4.7.2 Geomorphic Change Detection

Recent work on the geomorphology of sand and gravel bars on the Rio Grande in Big Bend National Park represents an example of the combination of GIS, GPS, Lidar, and Total Station survey technology to an applied environmental problem

TABLE 4.5
The LULC Classification System Used for the Land Cover Trends Data Set

Class	Description
1. Water	Areas persistently covered with water, such as streams, canals, lakes, reservoirs, bays, or oceans.
2. Developed/urban	Areas of intensive use with much of the land covered with structures (e.g., high-density residential, commercial, industrial, transportation, mining, confined livestock operations), or less-intensive uses where the land-cover matrix includes both vegetation and structures (e.g., low-density residential, recreational facilities, cemeteries, etc.), including any land functionally attached to the urban or built-up activity.
3. Mechanically disturbed[a]	Land in an altered and often nonvegetated state that is in transition from one cover type to another because of disturbances by mechanical means. Mechanical disturbances include forest clear-cutting, earthmoving, scraping, chaining, reservoir drawdown, and other similar human-induced changes.
4. Barren	Land composed of natural occurrences of soils, sand, or rocks where less than 10% of the area is vegetated.
5. Mining	Areas with extractive mining activities that have a significant surface expression. This includes (to the extent that these features can be detected) mining buildings, quarry pits, overburden, leach, evaporative, tailing, or other related components.
6. Forests/woodlands	Tree-covered land where the tree-cover density is greater than 10%. Cleared forest land (i.e., clear-cut logging) will be mapped according to current cover (e.g., disturbed or transitional, shrubland/grassland).
7. Grassland/shrubland	Land predominantly covered with grasses, forbs, or shrubs. The vegetated cover must comprise at least 10% of the area.
8. Agriculture	Cropland or pastureland in either a vegetated or nonvegetated state used for the production of food and fiber. Forest plantations are considered as forests or woodlands regardless of the use of the wood products.
9. Wetland	Lands where water saturation is the determining factor in soil characteristics, vegetation types, and animal communities. Wetlands are composed of water and vegetative cover.
10. Nonmechanically disturbed[a]	Land in an altered and often nonvegetated state that is in transition from one cover type to another because of disturbances by nonmechanical means. Nonmechanical disturbances are caused by wind, floods, fire, insects, and other similar phenomenon.
11. Ice/snow	Land where the accumulation of snow and ice does not completely melt during the summer period.

Source: USGS 2013, Landsat—A Global Land-Imaging Mission, Fact Sheet 2012-3072, revised May 2013, http://pubs.usgs.gov/fs/2012/3072/fs2012-3072.pdf (accessed May 2015).

Note: The classification system for the Land Cover Trends Data Set consisted of 11 general LULC classes and is a modified version of the Anderson Level I classification system (Anderson et al. 1976).

[a] Indicates class included to capture anthropogenic or natural disturbance events.

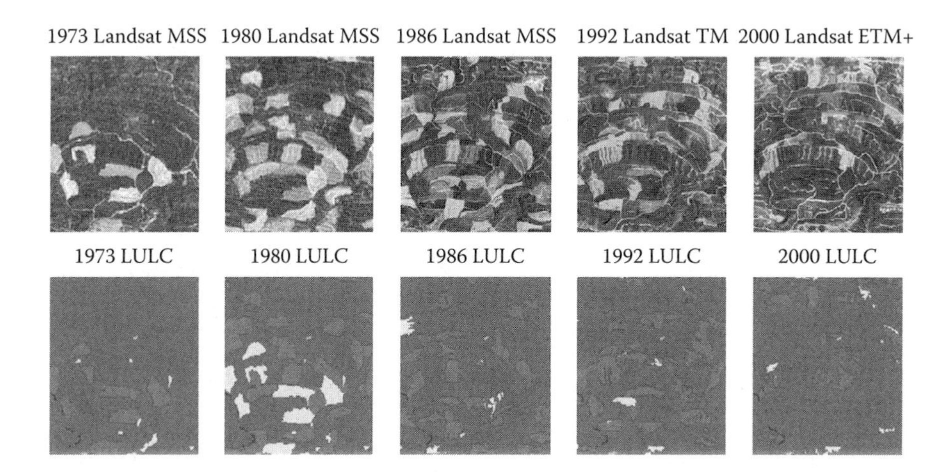

FIGURE 4.6 An example in the Ouachita Mountains Ecoregion of the five dates of the applied land cover trends dataset shown in greyscale. The level I classified maps are shown on the bottom row. (From Soulard, C.E. et al. 2014, Land Cover Trends Dataset, 1973–2000, U.S. Geological Survey, Data Series 844.)

(Urbanczyk and Bennett 2012). The project is designed to track geomorphic change in a river system that is in a sediment surplus condition. The Rio Grande has a long history of anthropogenic manipulation (Dean and Schmidt 2011). Over the past century, the channel has been transformed from a wide meandering channel with little vegetation dominated by periodic large flash floods to narrow confined channel choked with predominantly invasive vegetation. The geomorphic changes are driven by declining flows that are the result of upstream diversion and drought conditions. Smaller periodic floods now deposit the abundant sediment and the invasive plants cover and stabilize this degrading condition. The channel capacity and the aquatic habitat have suffered as a result of the buildup and stabilization of the sediment (Figure 4.7). Recent restoration efforts coordinated by multiple agencies are attempting to eradicate some of the problematic invasive species (salt cedar and giant river cane) in the hopes that this will liberate sediment during the periodic high flow conditions that still occur. The sandbar survey work described here is designed to monitor changes that might result from the removal of giant river cane in the vicinity of Boquillas canyon. The work described is located at the entrance to Boquillas canyon in Big Bend National Park.

The work started in 2004 with the collection of three cross sections as part of a reconnaissance trip through the canyon to consider solutions to the sediment problem. Subsequently, a large reset event occurred in 2008 that stripped much of the sediment and vegetation. Topographic surveys using Total Station laser survey techniques combined with RTK GPS techniques were then completed in the fall of 2011 and 2013. An aerial Lidar survey of the area was completed in the summer of 2012. The 2011, 2012, and 2013 data sets can all be processed into three-dimensional models and analyzed using the Geomorphic Change Detection (GCD) process described by Wheaton et al. (2010). This process compares three-dimensional surfaces (DEMs)

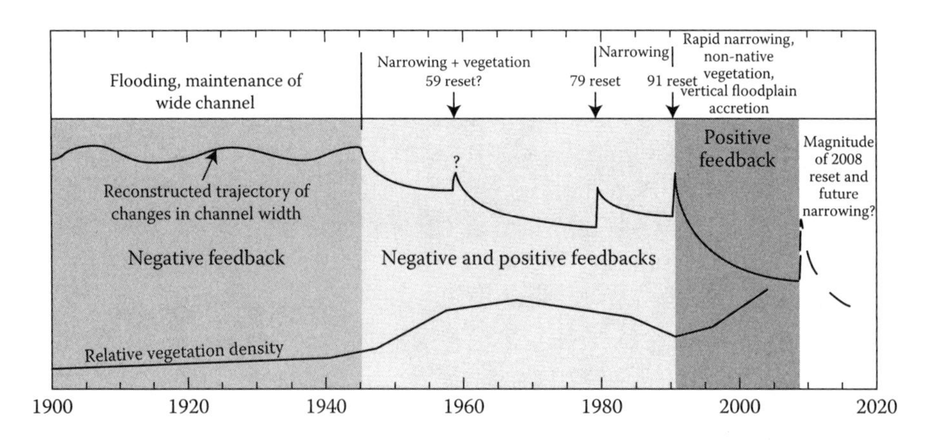

FIGURE 4.7 The history of channel narrowing on the Rio Grande in the Big Bend Region of Texas. (From Dean, D.J. and Schmidt, J.C. 2011, *Geomorphology*, 126(3–4), 333–349.)

from different periods and calculates a DEM of Difference (DOD). This DOD shows where the surface elevation has increased or decreased and calculates a total volumetric change. The technique also accounts for the uncertainties in the individual DEMs and propagates the uncertainty error through to the resultant DOD in the form of thresholds that exclude undetectable changes. Urbanczyk and Bennett (2012) describe a method to extrapolate volumetric change using the 2004 cross-section data set so that nearly a decade of geomorphic change can be determined for this location.

The area of interest is shown on Figure 4.8. Figure 4.9 shows the aerial photographs of the entrance bar in 2004, 2008, and 2010. To complete the GCD analysis, DEMs for the 2011 and 2013 (field data collected by Total Station and RTK GPS) needed to be constructed. This involved delineating a polygon area of interest that is covered by the data for each campaign. The field point data were processed in UTM coordinates and merged into a single feature class in a GIS. After this, they were converted to a TIN and then converted to a DEM. The Lidar data were already processed into a DEM, so at this point, the data were ready for the GCD analysis. The technique of Wheaton et al. (2010) was used for the analysis and the uncertainties in the RTK, Total Station, and Lidar data sets were incorporated into the threshold component of the analysis. Figures 4.10 and 4.11 show the DODs superimposed on the DEM from the 2012 Lidar data. Compare this to the aerial photos on Figure 4.9. On Figure 4.9, the river flows from left to right and the analyzed area is the bar on the left of the river (above the river in the images). The area analyzed includes only the bar on the left side of the river and not the river itself because there was little topographic data in the river channel itself. The reason that the DEM is visible through the DOD is that areas of small elevation change did not exceed the uncertainty threshold (the statistics suggest that the differences are too small to claim to be real). The 2011 to 2012 changes (Figure 4.10) were fairly small decreases visible at the edges and small changes along the bar in the center of the map. Figure 4.11

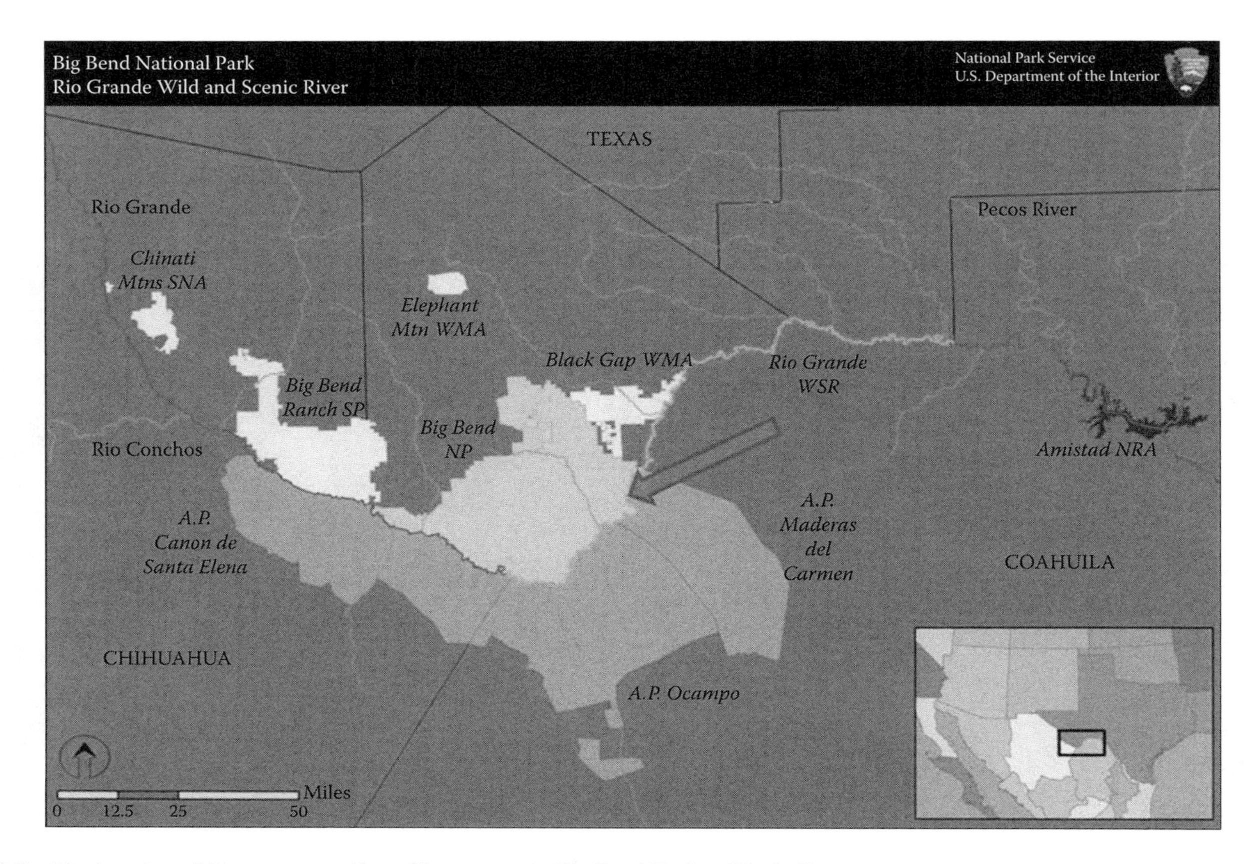

FIGURE 4.8 The location of the entrance to Boquillas canyon in Big Bend National Park, Texas.

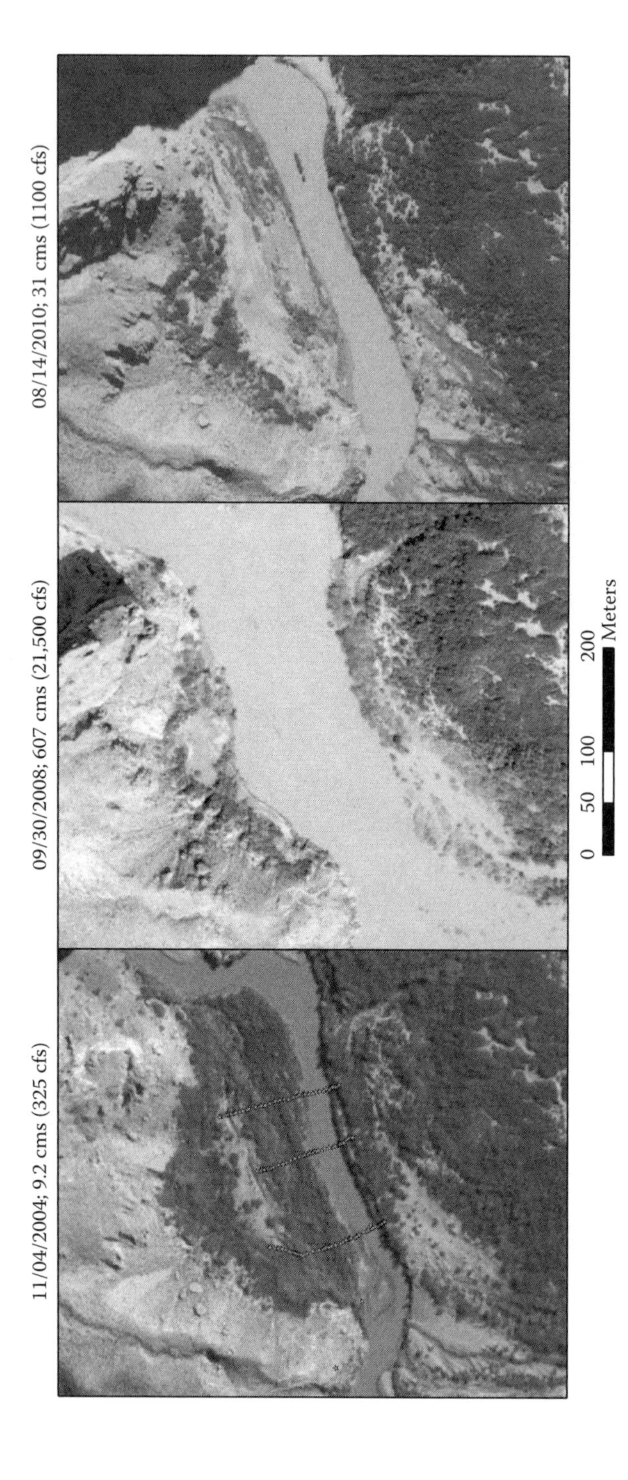

FIGURE 4.9 Aerial photographs of the entrance bar in 2004, 2008, and 2010. The location of the three cross sections surveyed in 2004 is shown on the left image.

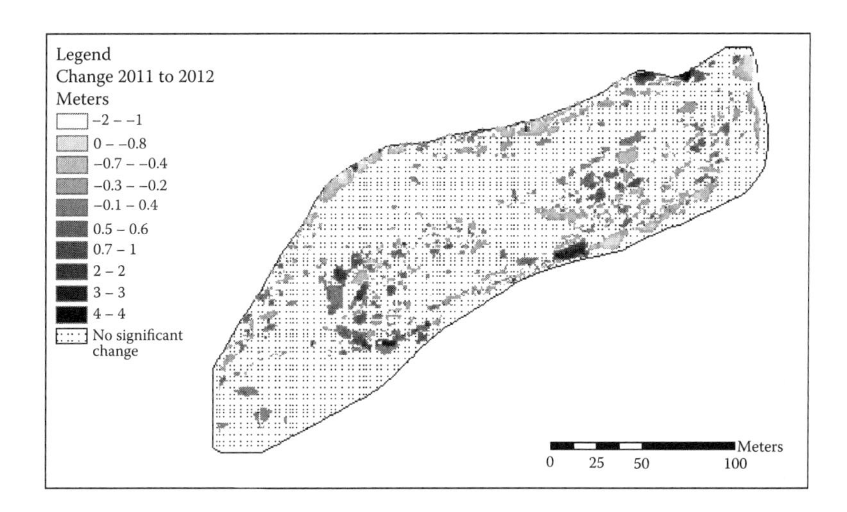

FIGURE 4.10 DOD for the 2011 to 2012 period.

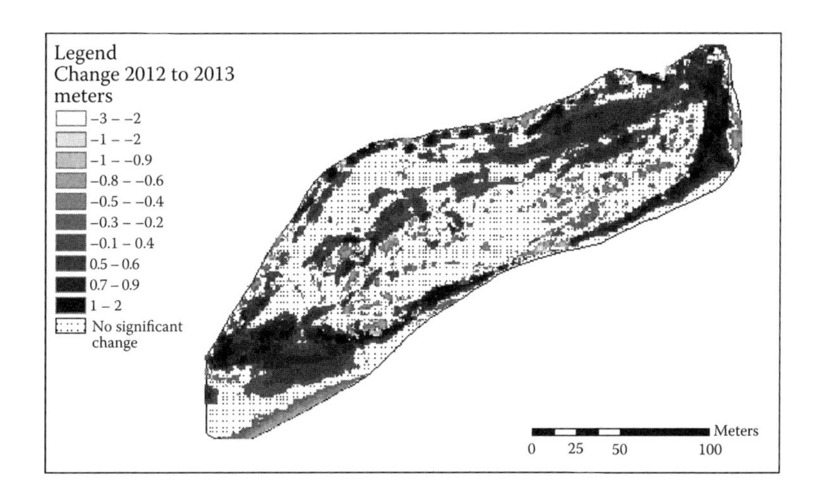

FIGURE 4.11 DOD for the 2012 to 2013 period. Note the significant addition of sediment particularly along the lower left and upper right edges and in the channel at the top of the bar.

shows the 2012 to 2013 DOD. This map shows significantly more change, mostly in the form of increases in elevation attributed to sediment deposition along the edges and in the channel at the back (top) part of the bar. This aggradation is consistent with small flood events that occurred in the period between the two data sets.

Table 4.6 includes summary information for all data for the Boquillas canyon entrance bar. The largest obvious change occurred during the 2004 to 2011 period (volumetric decrease estimated to be ~40,000 m^3). This was caused by the large channel resetting event that occurred in the fall of 2008 (these occur on a ~20-year cycle). The 2011 to 2012 period showed little change because of the general low flow

TABLE 4.6
Summary Data for the Three Periods in the Study

Volumetric	2004 to 2011	2011 to 2012	2012 to 2013
Total volume of erosion (m^3)		1161.15	1118.93
Total volume of deposition (m^3)		1260.15	5600.36
Total volume of difference (m^3)		2421.30	6719.79
Total net volume difference (m^3)	−40,517.00	99	4481.93

of the river during that period with balanced erosion or deposition. The 2012 to 2013 period shows the problematic aggradation that occurs sporadically between the large reset events. This is the aggradation that (with the associated growth of invasive species) gradually contributes to the narrowing of the channel.

REFERENCES

Anderson, J.R., Hardy, E.T., Roach, J.T. and Witmer, R.E., 1976, A land use and land cover classification system for use with remote sensor data: U.S. Geological Survey Professional Paper 964.

Bernhardsen, T., 2002, *Geographic Information Systems, An Introduction*, John Wiley & Sons, Inc., 428 pp.

Campbell, J.B., 2007, *Introduction to Remote Sensing*, Fourth Edition, The Guilford Press, 626 pp.

Dean, D.J. and Schmidt, J.C., 2011, The role of feedback mechanisms in historic channel changes of the lower Rio Grande in the Big Bend region, *Geomorphology*, **126**(3–4), 333–349.

Demers, M., 2005, *Fundamentals of Geographic Information Systems*, Third Edition, John Wiley & Sons, Inc., 468 pp.

ESRI, 2015a, What is GIS?, http://www.esri.com/what-is-gis (accessed May 2015).

ESRI, 2015b, History, in About ESRI, http://www.esri.com/about-esri/history (accessed May 2015).

FAA, 2014, Satellite Navigation–GPS–User Segment, http://www.faa.gov/about/office_org/headquarters_offices/ato/service_units/techops/navservices/gnss/gps/usersegments/ (accessed May 2015).

FHWA, 2003, Fact Sheet, 2003, High Accuracy-Nationwide Global Positioning System Program Fact Sheet, http://www.fhwa.dot.gov/publications/research/operations/03039/ (accessed May 2015).

Garmin, 2015, What is WAAS?, http://www8.garmin.com/aboutGPS/waas.html (accessed May 2015).

GISGeography, 2015, Multispectral vs Hyperspectral Imagery Explained, http://gisgeography.com/multispectral-vs-hyperspectral-imagery-explained/ (accessed May 2015).

Geosphere, 2015, Mapping and Projections: Map projection methods, http://www.geog.ucsb.edu/~dylan/mtpe/geosphere/topics/map/map1.html#proj (accessed May 2015).

GISWIKI, 2006, Environmental Systems Research Institute, http://giswiki.org/wiki/Environmental_Systems_Research_Institute (accessed May 2015).

GPS.GOV, 2015, The Global Positioning System, http://www.gps.gov/systems/gps/ (accessed May 2015).

Landsat, 2015, Landsat Missions, http://landsat.usgs.gov/index.php (accessed May 2015).

Lillesand, T.M. and Kiefer, R.W., 2000, *Remote Sensing and Image Interpretation*, Fourth Edition, John Wiley & Sons, Inc., 724 pp.

Limp, F. and Barnes, A., 2014, Solving the grid-to-ground problem when using high precision GNSS in archaeological mapping, *Advances in Archaeological Practice: A Journal of the Society for American Archaeology*, How-To Series, May 2014, 138–143.

NASA, 2015, Earth Observatory: Three Classes of Orbit, http://earthobservatory.nasa.gov/Features/OrbitsCatalog/page2.php (accessed May 2015).

National Academy of Sciences, 1970, Remote Sensing with Special Reference to Agriculture and Forestry, Washington, DC: National Academy of Sciences, 424 pp.

NGS, 2015a, Continuously Operating Reference Station (CORS), http://www.ngs.noaa.gov/CORS/ (accessed May 2015).

NGS, 2015b, OPUS: Online Positioning User Service, http://www.ngs.noaa.gov/OPUS/ (accessed May 2015).

Price, M., 2013, *Mastering ArgGiIS*, Sixth Edition, McGraw Hill, 611 pp.

Sabins, F.F., 1997, *Remote Sensing, Principles and Interpretation*, W.H. Freeman and Company, 494 pp.

Snyder, J. P. and Steward, H., eds., 1988, Bibliography of map projections, USGS Bulletin 1856.

Soulard, C.E., Acevedo, W., Auch, R.F., Sohl, T.L., Drummond, M.A., Sleeter, B.M., Sorenson, D.G., Kambly, S., Wilson, T.S., Taylor, J.L., Sayler, K.L., Stier, M.P., Barnes, C.A., Methven, S.C., Loveland, T.R., Headley, R. and Brooks, M.S., 2014, Land Cover Trends Dataset, 1973–2000, U.S. Geological Survey, Data Series 844.

UNSTATS, 2012, Plane Rectangular Coordinate Systems, http://unstats.un.org/UNSD/geoinfo/UNGEGN/docs/_data_ICAcourses/_HtmlModules/_Selfstudy/S06/S06_03.html (accessed May 2015).

Urbanczyk, K.M. and Bennett, J.B., 2012, Geomorphology of Sand Bars in Boquillas Canyon, Big Bend National Park, Geological Society of America Abstracts with Programs, March 2012, Vol. 44, No. 1, p. 2.

USGS, 2001, The Universal Transverse Mercator (UTM) Grid, USGS Fact Sheet 077-01, http://pubs.usgs.gov/fs/2001/0077/report.pdf (accessed May 2015).

USGS, 2007, What is a GIS? http://webgis.wr.usgs.gov/globalgis/tutorials/what_is_gis.htm (accessed May 2015).

USGS, 2013, Landsat—A Global Land-Imaging Mission, Fact Sheet 2012-3072, revised May 2013, http://pubs.usgs.gov/fs/2012/3072/fs2012-3072.pdf (accessed May, 2015).

Wheaton, J.M., Brasington, J., Darby, S.E., and Sear, D.A., 2010, Accounting for uncertainty in DEMs from repeat topographic surveys: Improved sediment budgets, *Earth Surface Processes and Landforms*, **35**, 136–156.

Wikipedia, 2015, Satellite navigation, http://en.wikipedia.org/wiki/Satellite_navigation (accessed May 2015).

5 Nanotechnology Applications

Changseok Han, Bangxing Ren, Mallikarjuna N. Nadagouda, George Em. Romanos, Polycarpos Falaras, Teik Thye Lim, Virender K. Sharma, Natalie Johnson, Pilar Fernández-Ibáñez, John Anthony Byrne, Hyeok Choi, Rachel Fagan, Declan E. McCormack, Suresh C. Pillai, Cen Zhao, Kevin O'Shea, and Dionysios D. Dionysiou

CONTENTS

5.1 INTRODUCTION

Water is an essential element to all life (i.e., plants, animals, and humans), and access to clean and safe drinking water is critical to human health. Recently, the World Health Organization (WHO) reported that more than 700 million people currently do not have sources of clean and safe drinking water (WHO 2014). The problem regarding the scarcity of clean water is especially prevalent in developing countries, where there is an urgent need for sustainable treatment processes and infrastructure for clean and safe drinking water. Developed countries also need to improve water treatment systems as a result of contamination caused by anthropogenic activities. Contaminants of emerging concern such as cyanotoxins, pesticides, industrial synthetic compounds, pharmaceuticals, hormones, and personal care products, as well as the effluent of wastewater treatment plants, have been found in drinking water sources (Agha et al. 2012; Barros-Becker et al. 2012; Graham et al. 2010; Ji et al. 2010). Moreover, water quality standards are getting more stringent to reduce health risks from contaminated drinking water and meet increased demands for clean water for the public (Qu et al. 2013a). There is a critical global need for sustainable technologies to meet the demands for clean and safe drinking water.

Current technologies and relevant infrastructure for water treatment are costly and sometimes inefficient for the removal of low-concentration level highly toxic contaminants, which is a prerequisite so that the production of clean water and water reuse to meet the high quality standards and cover the increased needs of both humans and environment (Qu et al. 2013a). Nanomaterials, defined to have one dimension smaller than 100 nm, demonstrated unique optical, electrical, magnetic, and morphological properties compared to micro- or macromaterials (Shelley 2005), thus providing high efficiency, flexibility, and multiple functionality to specific treatment applications (Qu et al. 2013a,b). In fact, based on recent advances in nanoscale science and engineering, nanotechnology has been extensively applied to the development of highly effective and energy-saving operation units for water treatment, including adsorption, filtration, oxidation, and disinfection (Savage and Diallo 2005).

In this chapter, nanostructured membranes, nanoadsorbents, and nanocatalysts for water sustainability are presented. Different technologies for water treatment, nanofiltration, nanoadsorption, advanced oxidation, water disinfection, and soil and groundwater remediation are addressed herein. In addition, we put particular emphasis on the elucidation of the corresponding mechanisms, discussing the roles of reactive oxygen species (ROS) in photocatalysis-based advanced oxidation processes (AOPs).

5.2 CATALYSIS (ADVANCED OXIDATION FOR WATER TREATMENT)

Application of nanostructured catalysts in AOPs has been studied extensively for the remediation of different organic and inorganic water contaminants. The catalysts are attractive because they are environmentally benign and have high thermal stability, chemical inertness, and low toxicity (Comninellis et al. 2008; Han et al. 2011; Poyatos et al. 2010; Sharma et al. 2012). In this section, several AOPs using titanium dioxide (TiO_2)–based photocatalysts, ferrates, ferrites, and other novel catalysts will be discussed.

5.2.1 TiO_2 Photocatalyst

TiO_2, a well-known photocatalyst, has been studied extensively for environmental protection, air purification, soil and groundwater remediation, and drinking water and wastewater treatment (Higarashi and Jardim 2002; Pichat 2010; Vilar et al. 2009) since water splitting by TiO_2 photocatalysis was discovered (Fujishima and Honda 1972). TiO_2-based AOPs have been used to decompose water contaminants, including contaminants of emerging concern (e.g., cyanotoxins, pharmaceutical and personal care products, and pesticides) (Choi et al. 2006, 2007; Giraldo et al. 2010; Han et al. 2013; Khan et al. 2014). Conventional TiO_2 photocatalysts can produce ROS in the presence of UV, without the addition of expensive oxidants such as H_2O_2 required in current disinfection methods. AOPs typically lead to the production of superoxide anion radicals $\left(O_2^{-\bullet}\right)$ and hydroxyl radicals ($^\bullet OH$). Hydroxyl radicals readily oxidize most organic water contaminants and can ultimately lead to mineralization to carbon dioxide (CO_2), water, and mineral species (e.g., nitrates and sulfates) because of the high redox potential of $^\bullet OH$ (2.80 V) compared to that of ozone (2.08 V) (Pera-Titus et al. 2004). The photocatalytic activation of conventional TiO_2 photocatalysts requires UV light absorbance to generate a charge separated state; the electron can be captured by dissolved oxygen to produce a superoxide anion radical, which can act as a modest oxidant or disproportionate to yield hydrogen peroxide as shown in Figure 5.1. One of the major drawbacks of TiO_2 photocatalysis is charge recombination, which inhibits the production of $\bullet OH$ and superoxide anion radical. To obtain a high rate of ROS generation during TiO_2-based photocatalysis, it is critical to prevent recombining electrons and holes produced in the process, before the formation of ROS.

In addition to the decomposition of water contaminants, TiO_2 photocatalysis can be used for water disinfection because the ROS produced can damage the DNA and RNA of pathogenic microorganisms, which can cause waterborne diseases.

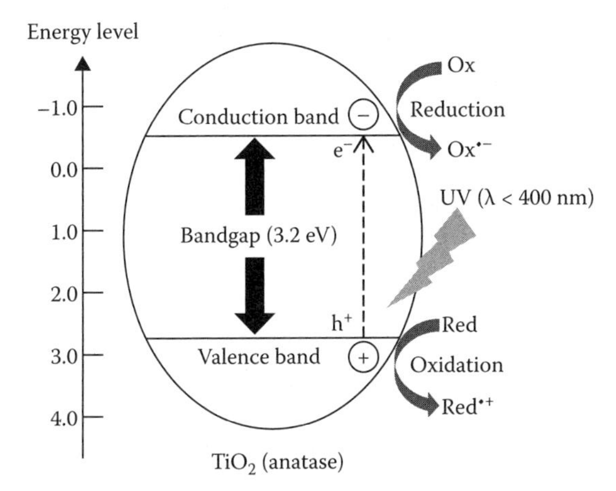

FIGURE 5.1 The photocatalytic activation mechanism of conventional TiO_2 photocatalysts.

Because of the fatal effects of ROS on DNA and RNA, pathogens can be effectively inactivated during TiO_2 photocatalysis. Effective inactivation of pathogenic microorganisms (e.g., *Escherichia coli*, *Staphylococcus aureus*, *Enterococcus faecalis*, *Candida albicans*, and *Aspergillus niger*) by TiO_2-based water disinfection has been reported. The efficacy of photocatalytic disinfection using TiO_2 in addition to UV is much higher compared to UV disinfection alone (Alrousan et al. 2009; Blake et al. 1999).

Additional studies have demonstrated improved activity of the titania photocatalysts for the degradation of water contaminants. An important approach is improving the morphological properties of TiO_2, including surface area, porosity, and pore size distribution because the efficiency of the photocatalytic process is strongly correlated with the materials' adsorption capabilities for target contaminants. In order to increase surface area and pore volume, and control pore size distribution, surfactants were employed in the sol-gel synthesis (i.e., a wet chemistry-based technique) of mesoporous materials (Choi et al. 2006; Han et al. 2011; Kresge et al. 1992; Pelaez et al. 2009). The surfactants form uniform micelles in the solution above the critical micelle concentration. In the solution, a TiO_2 framework forms around the micelles. After the removal of micelles by a calcination process, high surface area– and pore size–controlled TiO_2 photocatalysts are synthesized.

The use of conventional TiO_2 materials is limited only to UV irradiation (covering only 4%–5% of the whole solar spectrum) because of their large band gap (i.e., 3.0 eV for rutile phase and 3.2 eV for anatase phase). In order to improve the solar-activated performance of TiO_2 photocatalysts, materials have been developed to extend the photoresponse toward the visible light range (~45% of the solar spectrum). The development of nonmetal (e.g., carbon, nitrogen, sulfur, phosphorus, and fluorine)-doped TiO_2 has gained much attention because it circumvents the disadvantages of metal doping such as low thermal and chemical stability, potential hazards for metal leaching, and high rate of recombination of electron–hole pairs (Han et al. 2011, 2014a; Khan et al. 2014; Pelaez et al. 2009). With nonmetal doping, visible light–active TiO_2 photocatalysts were successfully synthesized with narrower band gaps or introduction of a localized energy states above the valance band of TiO_2. Because of the band gap modification, the synthesized non–metal-doped TiO_2 effectively produced ROS under visible light and solar irradiation leading to the degradation of water contaminants, including cyanotoxins, pharmaceuticals, and pesticides (Choi et al. 2007; Han et al. 2011, 2014a; Khan et al. 2014; Liu et al. 2012; Pelaez et al. 2009).

5.2.2 Ferrate

Ferrate(VI) ion, $Fe^{VI}O_4^{2-}$, has unique properties that have stimulated significant interest in its water and wastewater treatment applications (Jiang 2014; Lee et al. 2014; Yang and Ying 2013). The standard water treatment using chlorination can result in toxic or carcinogenic by-products in treated water, including chlorinated and brominated disinfection by-products (i.e., trihalomethanes, haloacetic acids, and bromate) (Heeb et al. 2014; Sharma et al. 2014). The by-products from the reduction of ferrate(VI), ferric oxide/hydroxides, are nontoxic and have the added benefit of acting as coagulants to remove additional water contaminants. Ferrate(VI) has no significant reactivity with bromide ion, and thus generation of carcinogenic bromate ion observed during ozone or chlorine treatments is avoided (Sharma 2011). The use of ferrate(VI) ion in a single-dosing treatment unit can simultaneously disinfect microorganisms, degrade inorganic and organic contaminants, including toxins, and remove heavy metals and colloidal/suspended particulate materials as summarized below (Casbeer et al. 2013; Jiang 2014; Prucek et al. 2013; Sharma 2013).

5.2.2.1 Disinfection and Detoxification

A number of studies have shown that ferrate(VI) ion is effective for the treatment of a wide range of viruses and bacteria such as MS2 coliphage, virus *f*2, virus *Q*β, *E. coli*, and total coliform (Hu et al. 2012; Jiang 2014). In the case of MS coliphage, Hu et al. (2012) demonstrated that ferrate(VI) damaged viral protein and genetic material through access to the interior of the virion. Ferrate(VI) was also effective in treating *Ichthyophthirius multifiliis*, an important freshwater teleost pathogen that can infect most species of freshwater fish worldwide (Ling et al. 2011). Recent results showed that ferrate(VI) can also effectively detoxify Microcystin-LR (MC-LR), a potent cyanotoxin in drinking water sources that poses a serious risk to public health (Jiang et al. 2014).

5.2.2.2 Oxidation

Several studies have shown the removal of inorganic and organic contaminants using ferrate(VI) (Sharma 2011, 2013). These include nitrogen- and sulfur-containing compounds such as cyanides, sulfide, thiols, amines, amino acids, anilines, and phenols (Casbeer et al. 2013; Lee and von Gunten 2010; Sharma 2013; Sharma et al. 2013). Most of the contaminants can be rapidly oxidized by ferrate(VI) in the second to minute time scale. The recent focus of the application of ferrate(VI) is on oxidizing micropollutants (Lee and von Gunten 2010). The kinetics of ferrate(VI) oxidation of endocrine disruptors (e.g., estrogenic phenols), bacteriostatic antibiotics (e.g., sulfamethoxazole and trimethoprim), β-lactam antibiotics (e.g., amoxicillin and ampicillin), and β-blockers (e.g., propranolol) have been carried out. Under neutral pH condition, these micropollutants are oxidized with half-lives in the seconds to minutes range. The analyses of oxidized products (OPs) of micropollutants are in progress. In the case of oxidation of propranolol, the OPs identified using liquid chromatography–tandem mass spectrometry were OP-292, OP-308, and OP-282 (Figure 5.2) (Anquandah et al. 2013).

The oxidation products shown in Figure 5.2 suggest that the Fe(VI) attacked both moieties, naphthalene and the secondary amine group of propranolol. Attacks on amine moieties of trimethoprim by ferrate(VI) were also observed (Anquandah et al. 2011). The results also showed that oxidative treatment of trimethoprim by ferrate(VI) resulted in the elimination of antibiotic activity.

5.2.2.3 Coagulation

Studies have shown the effectiveness of ferrate for removing metals and metals in cyanide complexes from water (Filip et al. 2011; Prucek et al. 2013). Examples

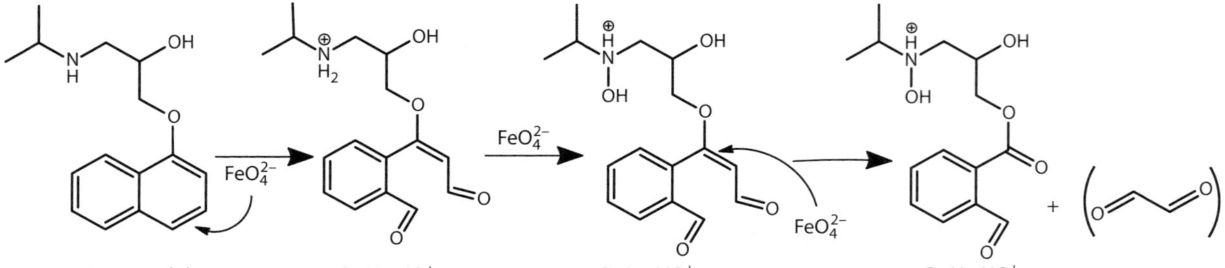

Propranolol

$C_{16}H_{22}NO_4^+$
m/z: 292

$C_{16}H_{22}NO_5^+$
m/z: 308

$C_{14}H_{20}NO_5^+$
m/z: 282

FIGURE 5.2 Oxidation of propranolol by ferrate(VI) ion. (Adapted from *Chemosphere*, 91, Anquandah, G. A. K., V. K. Sharma, V. R. Panditi, P. R. Gardinali, H. Kim, and M. A. Oturan, Ferrate(VI) oxidation of propranolol: Kinetics and products, 105–109, Copyright (2013), with permission from Elsevier, Inc.)

include removal of Cu in copper(I) cyanide $\left(Cu(CN)_4^{3-}\right)$ by first oxidizing Cu(I) to Cu(II) before coprecipitating it:

$$5HFeO_4^- + Cu(CN)_4^{3-} + 8H_2O \rightarrow 5Fe(OH)_3 + 4NCO^- + Cu^{2+} \downarrow + 6OH^- + 3/2O_2 \quad (5.1)$$

Similarly, As(III) can be oxidized by ferrate(VI) to result in As(V), which can subsequently be precipitated out by Fe(III) hydroxides (Prucek et al. 2013).

$$2HFeO_4^- + 3As(OH)_3 + 7OH^- \rightarrow 2Fe(OH)_3 + 3AsO_4^{3-} + 6H_2O \quad (5.2)$$

In this removal process, a significant portion of arsenic was embedded in tetrahedral sites of the Fe(III) hydroxide structure, thus minimizing the leaching of arsenic from the solid surfaces of the coagulant. The effect of Suwannee River natural organic matter (NOM) on the removal of As by the combination of Fe(VI) and Al(III) is demonstrated in Figure 5.3 (Jain et al. 2009). At low levels of Al(III), there was decrease in removal of As in 4 mg C L^{-1} NOM compared to the sample with <1 mg C L^{-1} NOM. However, at high levels of Al(III) (>100 μM), removals of As were similar. Overall, ferrate(VI) represents a sustainable strategy for removing multiple contaminants from water without producing disinfection by-products.

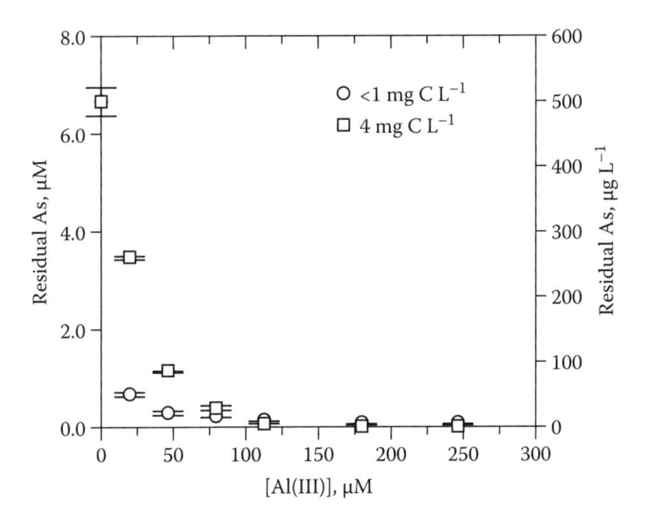

FIGURE 5.3 Effect of NOM on the removal of arsenite in drinking water by Fe(VI)/Al(III) salts at pH 6.5. [Fe(VI)] = 20 μM and initial arsenic concentration = 500 μg As(III) L^{-1}. (Adapted from *Journal of Hazardous Materials*, 169, Jain, A., V. K. Sharma, and M. S. Mbuya, Removal of arsenite by Fe(VI), Fe(VI)/Fe(III), and Fe(VI)/Al(III) salts: Effect of pH and anions, 339–344, Copyright (2009), with permission from Elsevier Inc.)

5.2.3 Ferrite for Water Sustainability

Ferrites (MFe_2O_4, M = Ni, Zn, Mg, Ca, Ti, Cu, etc.) have been widely used in the field of environmental remediation (Casbeer et al. 2012). They are chemically and thermally stable, and offer the photocatalytic and magnetic properties at the same time. Ferrites are capable of absorbing visible light because of their small band gap. Li et al. (2011) explored the photocatalytic activity of $ZnFe_2O_4$ nanotube arrays toward the degradation of 4-chlorophenol (4-CP) driven by visible light. The arrays were synthesized by a sol-gel–based method with anodized aluminum oxide template. After 6 h of irradiation, 100% of 4-CP was degraded by $ZnFe_2O_4$ nanotube arrays.

Composite photocatalysts with ferrite and other components have also attracted much attention for their improved photocatalytic reactivity. Several studies explored the photocatalytic activity of ferrites and carbonaceous nanomaterial composites. Hou et al. (2013) reported a hybrid of zinc ferrite multiporous microbricks and reduced graphene oxide (RGO) fabricated via a coprecipitation method. The introduction of RGO into $ZnFe_2O_4$ narrows the band gap of the semiconductor (photocatalyst). Under visible light irradiation ($\lambda > 420$ nm), complete removal of 4-CP was observed in the presence of $ZnFe_2O_4$/RGO composite after 1 h. The superior surface area and fast electron transfer ability of graphene can reduce the recombination of electron–hole pairs. Then, the active species, including radicals and holes generated in the photocatalytic process can be used to degrade target compounds. Another example of a light-activated photocatalysts is the $NiFe_2O_4$/multiwalled carbon nanotube composite (Xiong et al. 2012), prepared with a facile hydrothermal process in one step. Bi_2WO_6 was reported to be a visible light–activated photocatalyst with a band gap of 2.7–2.8 eV. Bi_2WO_6 was coated on the surface of $CoFe_2O_4$ to form a core–shell structure (Wang et al. 2013). The composite with a Bi_2WO_6-to-$CoFe_2O_4$ mass ratio of 10:1 was able to remove 90% of BPA in 2 h under simulated solar light and can be recovered by a magnet. Pelaez et al. (2013) investigated the performance of N-doped TiO_2 coated onto $NiFe_2O_4$ against MC-LR under visible light irradiation. With 1 g/L loading of composite material and 450 μg/L of initial concentration of MC-LR, complete degradation of MC-LR was observed after 5 h of irradiation. Comparatively, 75% of MC-LR removal was obtained with pure N-doped TiO_2. $NiFe_2O_4$ suppresses the recombination of photogenerated electrons and holes for the N-doped TiO_2 and exhibits a synergistic effect on the photocatalytic activity. The composite photocatalyst can be easily recovered with a magnet and achieved 70% of the initial degradation activity after three reuses.

Ferrite-based materials can also be involved in other AOPs. For example, with the addition of H_2O_2, ferrites can create a heterogeneous Fenton system, in which reactive species are formed. Wang et al. (2014) synthesized mesoporous $CuFe_2O_4$ and $CoFe_2O_4$ as heterogeneous Fenton catalysts using KIT-6 as hard template. The BET surface area of the meso-$CuFe_2O_4$ and meso-$CoFe_2O_4$ is 122 and 101 m^2/g, respectively, which are significantly higher compared to those of $CuFe_2O_4$ and $CoFe_2O_4$ fabricated by the conventional solid-state heat treatment. Ferrites with mesoporous structure have a large pore size, which is beneficial to the adsorption and mass transfer of large molecules. The meso-$CuFe_2O_4$ materials result in complete removal of imidacloprid with the addition of H_2O_2 in 5 h, which is superior to other reference catalysts. The high catalytic activity of meso-$CuFe_2O_4$ is credited to the Fe(III)/Fe(II) and Cu(I)/Cu(II) redox cycles. The active Fe(II) species in a Fenton reaction system can be generated by the reduction of Fe(III) by Cu(I). The meso-$CuFe_2O_4$ provided high stability with low iron leaching and superparamagnetic property, which make it possible to recover the catalyst with application of magnet field.

Ferrites can activate peroxymonosulfate (PMS). In a recent report by Zhang et al. (2013a), $CuFe_2O_4$ was prepared by a citrate combustion method and applied as a magnetic heterogeneous catalyst of PMS for the decomposition of iopromide, an iodinated contrast agent. The degradation tests were conducted at 20°C and pH 6.0 with 100 mg/L of oxides, 20 μM of PMS, and 1 μM of initial iopromide concentration. The iopromide degradation with PMS alone was negligible, while nearly 80% of iopromide was degraded with the PMS/$CuFe_2O_4$ in 10 min. The copper ferrite particles were recovered and reused seven times. The iopromide removal by PMS/$CuFe_2O_4$ remained almost the same during seven cycles and a low copper leaching was observed in the reactions. In contrast, the iopromide removal by PMS/CuO deteriorated with the increasing number of recycles. An apparent change of CuO crystal structure after the reaction was confirmed by x-ray diffraction. The PMS/$CuFe_2O_4$ exhibited highest degradation efficiency at neutral pH and decreased dramatically in acidic or alkaline solutions. The presence of NOM in water significantly reduced the iopromide degradation. Two radical scavengers, *tert*-butanol and ethanol, were used to identify the contribution of different radical species. It was shown that sulfate radicals in majority with a small fraction of hydroxyl radicals are responsible for the oxidation capacity. The reduction potential of Cu(III)/Cu(II) in solid state was reported to be 2.3 V, which is likely to account for the efficient sulfate radical generation from PMS by Cu(II)–Cu(III)–Cu(II) redox cycle on the surface of ferrites. Guan et al. (2013) prepared magnetic porous $CuFe_2O_4$ with the average particle diameter and pore size being 50–100 μm and 1–5 μm, by a sol-gel process involving egg white. The catalyst was capable of removing atrazine completely in 15 min by activating PMS. The degradation efficiency of atrazine by the PMS/$CuFe_2O_4$ increased as the pH increased from 4 to 9.5 and decreased from 9.5 to 10.6. The removal efficiency was inhibited in the presence of bicarbonate and NOM. The surface hydroxyl group of $CuFe_2O_4$ was shown to be able to activate PMS to generate hydroxyl radicals and sulfate radicals, which are both responsible for the removal of atrazine. The degradation test was also performed in the effluents of different stages at a water treatment plant, which showed possible application. A study on $CoFe_2O_4$ magnetic nanoparticles as a heterogeneous catalyst of PMS toward the degradation of diclofenac was also reported (Deng et al. 2013).

5.2.4 Other Novel Materials

With various methods employed within the scope of advanced oxidation technologies, numerous materials are utilized to provide oxidation reaction capabilities. Of these, some prominent materials are graphitic carbon nitride (g-C_3N_4), graphene-based compounds, and perovskites.

The *g*-C_3N_4 (graphitic carbon nitride) can usually be prepared by using a polymeric reaction of CH_2N_2 (cyanamide), $C_2H_2N_4$ (dicyanamide), and $C_3H_6N_6$ (melamine) (Thomas et al. 2008). Materials of different properties, reactivities, and degrees of condensation are produced from this reaction depending on the conditions used. Carbon nitrides are seen to be useful toward a variety of reactions including the benzene activation, carbon dioxide, and trimerization reactions attributed to unforeseen catalytic activity. This property is a result of the particular semiconductor properties of the *g*-C_3N_4 (Thomas et al. 2008). Of all the known allotropes of carbon nitride, *g*-C_3N_4 has the best stability under ambient conditions. This semiconductor material has a measured band gap value of 2.7 eV, consistent with an optical wavelength of 460 nm, thus making the material slightly yellow in color. With medium band gap as well as thermal and chemical stability in the ambient environment, it is becoming one of the most promising photocatalytic materials (Dong et al. 2014). When considering attacks, both chemically (e.g., acid, base, and organic solvents) and thermally (in air up to temperatures such as 600°C), this polymeric *g*-C_3N_4 possesses high stability because of its *tri-s*-triazine ring structure and high condensation (Bai et al. 2014). Several publications in recent years have highlighted the effectiveness of *g*-C_3N_4 as a photocatalyst in areas such as the selective oxidation of alcohols and hydrocarbons (Su et al. 2012; Zhang et al. 2013b), and as a good photocatalytic performer in relation to hydrogen or oxygen production via water splitting with the use of visible light irradiation (Wang et al. 2009). It has strong reduction reaction properties owing to the high potential of the conduction band but inferior oxidation capabilities because of its valence band located at approximately 1.4 eV, resulting in a small thermodynamic driving force for water or organic pollutant oxidation (Bai et al. 2014). To address the inferior oxidation-related issues of *g*-C_3N_4, various composites have been prepared in recent years to enhance its activities. One composite is the C_{60}/*g*-C_3N_4 (Bai et al. 2014), which exhibited a significant enhancement on the photocatalytic performance. This is attributed to a improved parting of the photo-induced electron–hole pairs and longer lifetime of the photo-generated charge carriers of the bulk *g*-C_3N_4. When tested against phenol and methylene blue (MB) dye, this effect promotes an improved photocatalytic activity of the *g*-C_3N_4. Another composite is one of tungsten (VI) oxide and *g*-C_3N_4 (WO_3/*g*-C_3N_4) (Jin et al. 2014) with the WO_3 used with the intention of providing a combination partner for the *g*-C_3N_4 as it is well known as an oxidation part photocatalyst for the Z-scheme photocatalytic water splitting (Abe 2011), where the Z-scheme involves a means for utilizing both high oxidation and reduction abilities under visible light irradiation (Chen et al. 2014; Kato et al. 2004). Ag@AgBr/*g*-C_3N_4 similarly exhibits higher levels of degradation with the presence of *g*-C_3N_4, with the formation of a Z-scheme reaction pathway when tested against organic dyes such as methyl orange (MO) along with a considerable mineralization capability with the active species being the hydroxyl radicals (Yang et al. 2014). CdS/*g*-C_3N_4 (Dai et al. 2014) is an effective photocatalyst used for the production of aldehydes from the oxidation of aromatic alcohols and the conversion of nitrobenzene into aniline through a reduction reaction. This is seen to be achieved through direct hole oxidation and direct electron reduction, respectively. Another promising photocatalytic material for pollutant purification is cobalt oxide and *g*-C_3N_4 (Co_3O_4-*g*-C_3N_4) (Han et al. 2014b), which was prepared on the basis that cobalt-based species are effective in enhancing the g-C_3N_4-based materials in regard to its photocatalytic ability. The group proposed that the Co_3O_4–g-C_3N_4 heterojunction structure enhances the separation of electron–hole pairs, thus reducing the recombination of charge carriers, which leads to the increase of the concentration of $O_2^{-\bullet}$ involved in the photodegradation process.

Apart from g-C_3N_4, there are many other graphene-based materials that have been studied. Graphene oxide (GO) is one such material. This material has numerous functional groups containing oxygen. These can be located in the basal plane with hydroxyls and epoxides present, along with carboxyl groups, which are situated at plane edges (Park and Ruoff 2009). Such surface functionalization allows GO to swell freely and easily disperse in water, enabling GO to be a perfect candidate for a catalyst supporting material (Si and Samulski 2008). One of these materials is a GO and cobalt oxide nanocomposite (Co_3O_4/GO) (Shi et al. 2012a,b; 2014). This material has excellent properties because of its high stability and a high surface area. In the presence of PMS, a widely known oxidizing agent, Co_3O_4/GO/PMS systems show the complete degradation of Orange II within a few minutes. This activity is attributed to preferential generation of sulfate radicals $\left(SO_4^{-\bullet}\right)$ from the PMS, which have a strong oxidizing potential (Shukla et al. 2010). Reduced graphene/MnO_2 (RGO/MnO_2) composites were also studied. These materials show an enhancement on their catalytic activity when compared to MnO_2. This enhancement of RGO/MnO_2 is mainly ascribed to the synergistic effects of RGO and MnO_2 with large surface area along with good electron transfer channels provided by the graphene resulting in efficient decomposition of MB dye (Qu et al. 2014). $MeFn_2O_4$–GO showed an increased rate of dye degradation when compared to $MeFn_2O_4$, with GO being attributed to activating PMS to produce sulfate radicals for degradation of organic pollutants. GO also plays a role in the enhanced degradation of dyes, which is attributed to the lowering of the required activation energy for the reactions to occur, with $MnFe_2O_4$–RGO (25.7 kJ/mol) being compared to $MnFe_2O_4$ (31.7 kJ/mol) (Shi et al. 2014).

Another group of material of interest is the perovskites, which are materials that have the same crystal structure as $CaTiO_3$. Perovskites have a cubic structure with a common formula of ABO_3, where rare earth cations are generally denoted by A and transition metal cations are denoted by B (Merino et al. 2005). This material type can undergo substitutions at both cation sites without structural modifications. Throughout the last few decades, there have been many propositions for the use of perovskites in the area of remediation, as such stable materials are catalytically active, with positive results seen in the complete oxidation of VOCs (volatile organic compounds) including chlorinated hydrocarbons using cordierite monolith–supported $LaMnO_3$, $LaCoO_3$, and $(La_{0.84}Sr_{0.16})(Mn_{0.67}Co_{0.33})O_3$ (Schneider et al. 2000). Less work has been published in the area of water remediation. The effectiveness of various lanthanum-based perovskite catalysts was investigated in regard to the ozonation of oxalic acid as a model organic pollutant and the dye, C.I. Reactive Blue 5. This study showed that $LaMnO_3$, which provided complete degradation of oxalic acid in less than 1 h, allowed oxidation via $HO^\bullet$ radicals in liquid bulk along with surface reactions. $LaCoO_3$ showed high performance because of its surface oxygen vacancies (Orge et al. 2013). Under catalytic experimental conditions, $La_{1-x}Ca_xCoO_3$

perovskites exhibit excellent structural stability and show neither surface area collapse nor structural change (Merino et al. 2005). Combination of ozone and a heterogeneous catalyst such as perovskite has shown to be a promising and effective route of ozonation of intractable molecules such as gallic acid (Carbajo et al. 2006) and pyruvic acid (Rivas et al. 2005); in such process, the perovskite demonstrated to be a stable catalyst. The literature search strongly indicates the critical need for further investigation on the development of highly efficient graphitic carbon nitride, graphene compounds, and perovskites.

5.3 PHOTOCATALYTICALLY DRIVEN WATER MEMBRANE FILTRATION

Until recently, hybrid advanced oxidation/membrane filtration processes incorporated two separate stages in series. A batch- or continuous-flow photocatalytic reactor utilized vigorously mixed slurries consisting of the photocatalyst nanoparticles and the polluted water. After sufficient period under irradiation, these slurries were driven to a membrane filtration unit to achieve separation of the photocatalyst from the purified water. Thereinafter, extensive engineering studies as well as membrane morphology characterization had been conducted and various types of membranes had been tested with respect to membrane fouling, frequently occurring via deposition and accumulation of the photocatalytic nanoparticles on the membrane surface and enhancing the pollutant removal and energy efficiency. Later, the concept of using a membrane coated with photocatalytic material and being irradiated during filtration was deployed as a way to address organic fouling or biofouling. The initial photocatalytic membranes were tested with naturally occurring macromolecules such as humic acid and under UV irradiation to demonstrate their antifouling properties but were not regarded as the key elements for the development of an efficient water treatment process.

A major hurdle in the development of a water purification process combining membrane filtration and an AOP in one single-stage continuous-flow module (Figure 5.4) was the design and implementation of an effectively irradiated filtration unit. Deploying and patenting such a technology in 2012 (Falaras et al. 2012) have resulted in efforts that focused on improving performance by controlling the flux properties and hydrophilicity of the membranes, developing double-sided visible light–activated photocatalytic membranes and incorporating fiber-stabilized photocatalysts by smart design of the membrane reactor's flow channels. These efforts and recent advances are described in detail below.

The development of hybrid materials exhibiting the simultaneous action of photocatalysis and membrane filtration can lead to improved water treatment processes. Photocatalysis has the potential to solve problems related to the fouling of membranes, the generation of toxic concentrates, and the existence of low concentrations and harmful organic pollutants in the permeate effluent. On the other hand, membranes, especially the ceramic ones, are appropriate supports for the deposition of thin photocatalytic layers because of their high affinity with the photocatalyst (e.g., TiO_2) and the possibility to further stabilize and activate the deposit with calcination. In addition, membranes exhibit two surfaces that come into contact with the polluted water and that can be exploited for the photocatalyst deposition. Thus,

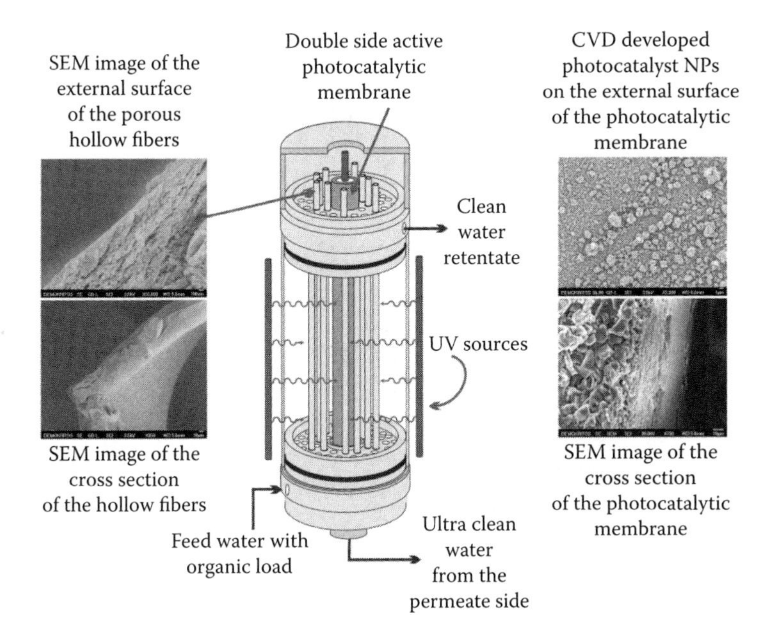

FIGURE 5.4 Side view of the photocatalytic membrane reactor cell showing the photocatalytic membrane and the fiber-stabilized TiO_2 nanoparticles.

with appropriate design, both membrane surfaces can be illuminated to develop very efficient photocatalytic ultrafiltration processes. Such processes must involve "double-sided active photocatalytic membranes," where the pollutant undergoes two sequential photodegradation steps, the first in contact with the feed surface and the second in contact with the permeate surface of the membrane. Moreover the asymmetric pore structure of ceramic membranes assures proper mixing of the fluid and better contact with the porous photocatalytic layers (Romanos et al. 2013).

An innovative photocatalytic device for water purification was recently fabricated (Falaras et al. 2012). The developed photocatalytic reactor permits efficient water treatment by removing organic contaminants and comprises a fluid delivery system, the membrane cell unit, irradiation sources, a pressure transducer, a stream selection three-way valve, and a backpressure regulator mounted at the retentate side of the membrane cell. The cell unit consists of three concentric tubes placed in the vertical direction. The outer tube is made of Plexiglas and has a length of 120 mm and an OD and ID of 60 and 50 mm, respectively (Figure 5.5). The inner tube is the membrane modified on both its sides. The reactor comprises a first flow channel for receiving fluid from an inlet means, a second flow channel for delivering fluid to an outlet means, a selectively permeable filtration membrane intermediate between the first and second flow channels (having a first surface that receives fluid from the first flow channel and an opposite second surface defining the second fluid flow channel), and one photocatalyst support disposed in the first flow channel. The first and second surfaces of the filtration membrane and the photocatalyst support each comprise an immobilized photocatalytic material capable of catalyzing photocatalytic breakdown of contaminants in the water, in use of the reactor.

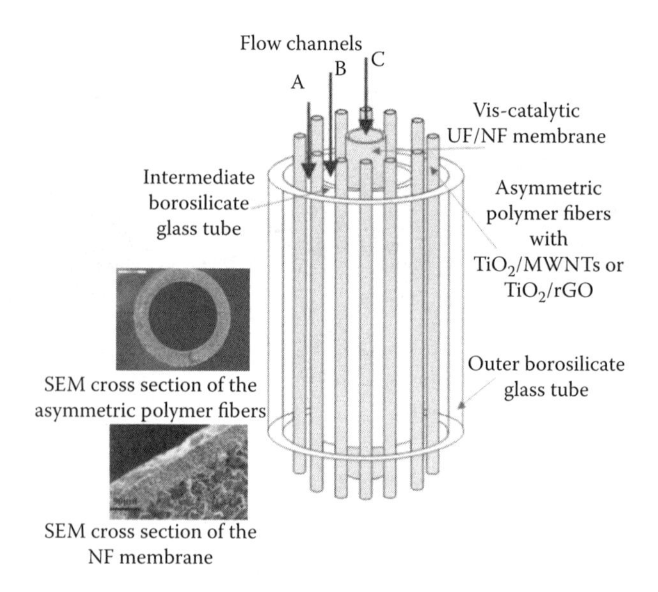

FIGURE 5.5 The three flow channels incorporated in the design of the novel photocatalytic membrane reactor for achieving maximum pollutant removal efficiency.

The contaminated water therefore undergoes three photocatalytic treatment stages and a filtration stage during its passage through the reactor.

This advanced oxidation technology device is scalable and was used for the implementation of sustainable and cost-effective continuous-flow water treatment. A number of crucial components were incorporated, with the most promising efforts focused on photocatalytic disinfection membranes modified by nanoengineered materials. In particular, the counterbalance between increasing the photocatalytic effectiveness by increasing the deposited catalyst and consequently decreasing the permeability was critical with respect to the development of an efficient and economically feasible photocatalytic membrane reactor process. For this reason, the effort focused on the development of ultrathin photocatalytically active TiO_2 layers, possessing high porosity and enhanced hydrophilic character. Thus, a chemical vapor deposition (CVD)–based innovative approach was applied to develop double-sided active TiO_2 photocatalytic nanofiltration (NF) membranes for continuous-flow photocatalytic reactor for effective water purification (Romanos et al. 2012). The method involved pyrolytic decomposition of titanium tetraisopropoxide (TTIP) vapor and formation of TiO_2 nanoparticles through homogeneous gas phase reactions and aggregation of the produced intermediate species (Figure 5.6a). The flow-through CVD techniques led to extended deposition of either granular or crystalline shapes depending on the sign of the temperature gradient, as the grown nanoparticles diffused and deposited on the surface of γ-alumina NF membrane tubes. Prompt control of the CVD reactor conditions allowed for online monitoring of the carrier gas permeability, providing the possibility to optimize the deposition rate and thickness, improving the homogeneity, and enhancing pore efficiency of the formed photocatalytic titania layers on both membrane sides, all these without sacrificing the high yield rates. The membrane efficiency to photocatalytically degrade typical azo-dye

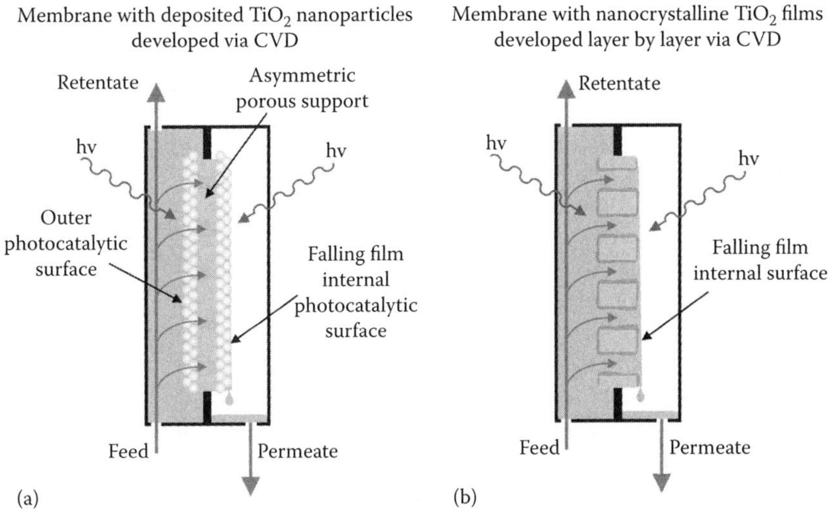

FIGURE 5.6 Cross section of the device focusing on the double-sided active photocatalytic membranes. (a) Membrane with deposited TiO_2 nanoparticles. (b) Membrane developed with layer-by-layer formation of nanocrystalline TiO_2 films.

water pollutants was evaluated in an innovative continuous-flow (rather a batch process) water purification device, operating in the common cross-flow membrane mode and applying UV irradiation on both membrane sides. The developed composite NF membranes were highly efficient in the photodecomposition of MO, exhibiting in parallel low adsorption-fouling tendency and high water permeability (permitting the treatment of an almost double amount of a common pollutant). In addition, the novel photocatalytic membrane purification device provided the possibility to effectively illuminate with UV light each membrane surface (external and internal) separately, and in this way, it was possible to discriminate between the fractions of pollutant that were removed as a result of adsorption on the alumina substrate and the TiO_2 layer, photolysis, and TiO_2 photocatalytic degradation. The newly developed purification device is very versatile and can also be used in hybrid ultrafiltration/photocatalysis water treatment processes (Athanasekou et al. 2012).

For this reason, very efficient composite TiO_2 photocatalytic ultrafiltration (UF) membranes were developed through chemical vapor layer-by-layer deposition (LBL/CVD) of TiO_2. The technique comprised chemisorption or physisorption of the TTIP vapor and a subsequent oxidative treatment in order to promote the precursor condensation and generate new adsorption sites for the accomplishment of the successive adsorption/surface reaction steps.

The membrane efficiency in photodegradation of MO was evaluated in the innovative continuous-flow reactor. Both membrane sides were covered with TiO_2 photocatalyst without affecting the high water recovery efficiency. Contrary to the highly temperature-dependent chemisorption process leading to growth of titania nanoparticles, the physisorption LBL led to the deposition of high TiO_2 amounts and the manufacture of very efficient photocatalytic NF membranes (Figure 5.6b). The

TiO_2-modified UF membranes developed through the physisorption path degraded almost double the amount of the azo-dye pollutant in the common cross-flow membrane operation mode, under UV irradiation of membrane (both annular and bore) surfaces.

Furthermore, the continuous-flow, hybrid photocatalytic/ultrafiltration water treatment process can be combined with photocatalytically active composite biopolymers consisting of AEROXIDE TiO_2 P25 effectively dispersed and stabilized on Ca alginate porous hollow fibers (Papageorgiou et al. 2012). It has been demonstrated that the presence of the TiO_2/Ca alginate fibers (Figure 5.7) as a pretreatment stage to the membrane process led to a threefold enhancement of the MO removal efficiency and to dilution rather than condensation in the membrane retentate as commonly observed in filtration processes, a fact that is very important especially in cases of highly toxic pollutants. Furthermore, the addition of fibers in the membrane ultrafiltration process increased the permeability (by almost 20%) of the photocatalytic membrane, leading to an increased recovery rate at steady state. This performance is achieved with 26 and 31 cm^2 of membrane and stabilized photocatalyst surfaces, respectively, and in this context, there is plenty of room for the upscaling of both membranes and fibers and the achievement of much higher water yields since the methods applied for the development of the involved materials (CVD and dry–wet phase inversion in a spinning setup) are easily upscalable and are not expected to add significant cost to the proposed water treatment process.

In an attempt to increase the efficiency and cost-effectiveness of the photocatalytic membrane filtration process, γ-alumina UF ceramic membranes modified with visible light active titania (nanostructured modified titania/m-TiO_2, synthesized using the gel combustion method based on the calcination of an acidified alkoxide solution mixed with urea; Moustakas et al. 2013) were for the first time developed using a sol-gel preparation technique combined with a dip-coating deposition

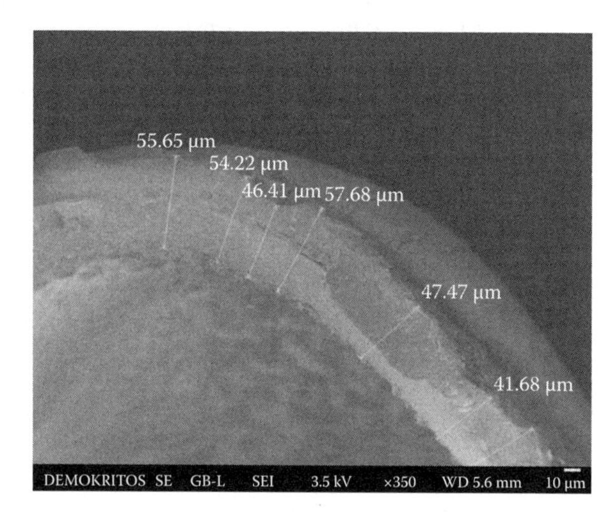

FIGURE 5.7 Scanning electron microscopy image showing the cross section of a TiO_2/Ca alginate fiber.

procedure (Moustakas et al. 2014). It has been confirmed that the structural, morphological, and physicochemical properties of the modified titania membranes strongly depend on the dip-coating and calcination rates. These highly hydrophilic modified membranes were incorporated in the water purification photocatalytic reactor under continuous-flow filtration conditions and tested for the photocatalytic degradation of azo-dye model compounds (namely, MO and MB) with very promising results (under ambient operating temperature and low pressure without any compromise on the efficiency of the membrane's permeate flux). It was observed that the photocatalytic efficiency depends on the effluent flow rate; however, under both UV and visible light, the permeability was continuously increasing because of the photoinduced hydrophilicity effect. Compared to MO, the MB pollutant was degraded at a much higher rate because of its better adsorption, independently of the type of the membrane. The permeability of the membranes increases with the volume treated because of the wettability of the m-TiO_2–treated membrane, rendering the need for regeneration or antifouling procedures unnecessary and making the process more energy efficient. Because of the low-temperature function and the photoinduced hydrophilic effect of the modified TiO_2 photocatalytic UF membranes, the photocatalytic reactor can efficiently work without any extra device, a fact that leads to low installation and operating costs and provides an energy-efficient procedure of cleaning polluted water using solar light.

Finally, visible light active composite membranes were also developed by the deposition of partially reduced graphene oxide/TiO_2 (GOT) composites on ceramic UF and NF monoliths via dip coating. The pore size of the monolith was crucial for the amount of the GOT composite stabilized on the substrate and for the visible light photocatalytic efficiency of the developed GOT membranes (Athanasekou et al. 2014). Cross-flow and dead-end filtration experiments were sequentially conducted in the dark, under UV and visible light, with the membrane surface irradiated for the elimination of two synthetic azo-dyes, MO and MB, from water solutions. The synergetic effects of GO on pollutant adsorption and the photocatalytic degradation capacity of TiO_2 were thoroughly studied, while the influence of the pore size of the monolithic substrate on deposition morphologies was also elucidated. It has been concluded that the membrane developed on the UF monolith with a 10-nm pore size exhibited the best photocatalytic performance under visible light. Additionally, the performance of the novel hybrid process was compared with that of standard nanofiltration with respect to pollutant removal efficiency and energy consumption, providing firm evidence for its economic feasibility and efficiency.

Thus, a highly efficient and easily upscalable (Figure 5.8) hybrid photocatalytic/filtration process is demonstrated for water purification using both UV and visible light. Composite UF and NF membranes can be incorporated into an innovative water purification device that combines membrane filtration with semiconductor photocatalysis offering significant advances for the treatment of polluted water. This includes operation in the flow-through mode without fouling, no generation of toxic concentrated retentate effluents, and the presence of two photocatalytically active surfaces on a one-membrane highly asymmetric element. Incorporation of visible light active functionality to the membranes enhances the combined photocatalytic

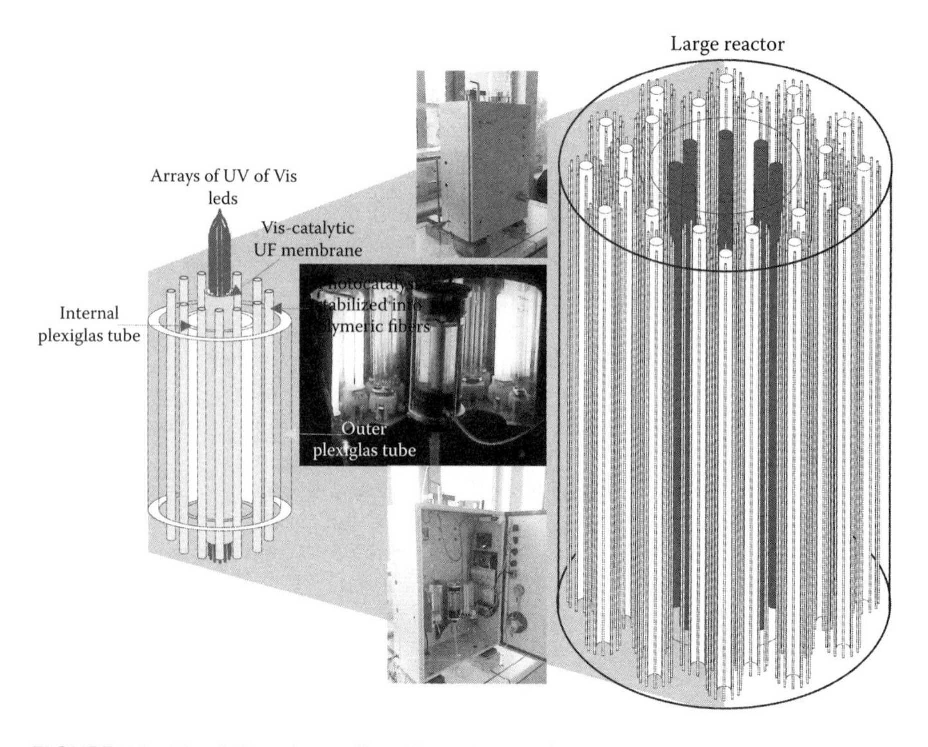

FIGURE 5.8 Feasibility of upscaling. Upscaling consists on incorporating a higher number of membrane and fiber elements having a higher length.

filtration process for remediation of organics using only solar light. Solar-driven processes are extremely attractive for energy-efficient alternatives to the typical NF process.

5.4 HYBRID ADSORPTION–CATALYTIC PROCESS

The synergistic coupling of catalysis and adsorption processes has attracted immense interest for an enhanced removal of water pollutants. Heterogeneous photocatalysis using TiO_2 is a viable option for degrading and mineralizing various recalcitrant organic pollutants without chemical addition. Adsorption using activated carbon (AC) is a proven technology for treatment of water and wastewater. AC, which typically possesses mesopores of a few nanometers, can serve as a good particulate support for TiO_2 whereon the TiO_2 nanoparticles are coated without causing pore blocking. AC has a strong adsorption affinity toward various organic pollutants to be degraded because of its high specific surface area, and it could retain the intermediates produced for further photocatalytic degradation, leading to complete mineralization. The lightweight AC would ensure adequate suspension of the TiO_2/AC composite particles in an aeration tank or upflow reactor. Thus, the coupled adsorption–solar photocatalysis process potentially presents an environmentally friendly and cost-effective treatment technology for pollutant removal. Various types of AC, such as powdered AC, granular AC, and AC fiber, have been used as TiO_2 supports. The disadvantage of using AC as TiO_2 support is that AC is opaque and that limits the application dose of a TiO_2/AC system in a photoreactor.

5.4.1 Techniques of TiO_2/AC Syntheses

Table 5.1 summarizes the techniques to synthesize TiO_2/AC composites. The sol-gel technique is the most commonly used technique for TiO_2/AC synthesis because of its versatility for TiO_2 coating (Lim et al. 2011). It also allows introducing various desirable functionalities into TiO_2/AC. Various types of Ti precursors, solvents, and acid/base catalysts can be used. In order to produce robust coating of TiO_2 nanoparticles on AC particles, several techniques can be used. These include (1) AC pretreatment with acids or bases, (2) viscosity-controlled immersion method, (3) UV-assisted irradiation, (4) formation of Ti sol during hydrolysis that serves as a binder for TiO_2 nanoparticle coating on AC, (5) pH control, and (6) repeated coating. A variety of TiO_2/AC composites have been synthesized. They include (1) visible light–responsive

TABLE 5.1
TiO_2/AC Synthesis Techniques

Synthesis Technique	Precursors Attempted	Synthesis Conditions	Remarks
Sol-gel	Ti precursor: titanium alkoxides and $TiCl_4$ Solvent: alcohols Catalysts: strong acids (HCl, H_2SO_4, HNO_3), strong bases (NH_4OH, TBAOH), weak bases	*T* (°C): 300 to 900 *D* (h): 1 to 10 *A*: Air, N_2, Air + N_2, CO_2	Anticalcination effect occurs: the TiO_2 coated onto TiO_2/AC was smaller as compared to that of the bare TiO_2.
Chemical vapor deposition	Ti precursor: titanium alkoxides and $TiCl_4$	*T* (°C): 200 to 600 *D* (h): 4 to 24 *A*: N_2	An interfacial film is formed between the deposition of TiO_2 on AC.
Hydrothermal	Ti precursor: P25, $TiCl_4$, $TiOSO_4$, peroxotitanate Mineralizer: NaOH, HNO_3	*T* (°C): 150 to 800 *D* (h): 2 to 15 *A*: N_2	Generally involves mild temperature (≤200°C), but subsequent calcination at higher temperature (300°C–800°C) have been reported.
Binder	Ti precursor: P25	*T* (°C): 400 to 580 *D* (h): <3 *A*: N_2	TiO_2 can form large clusters deposition on AC, and the surface area of TiO_2 may not be optimized.
Molecular adsorption-deposition	Ti precursor: $TiCl_4$	*D* (h): <2 h *A*: Ar	Ti precursor having low boiling point is preferable for ease of vaporization.

Note: *A*, atmosphere; *D*, duration of calcination; *T*, temperature of calcination.

composites with TiO_2 doped with metals and nonmetals (e.g., N-TiO_2/AC, F-TiO_2/AC, and V-TiO_2/AC), (2) bimodal TiO_2 coated on AC (Yap et al. 2012), and (3) TiO_2 coated on Fe_3O_4/AC composite, which allows magnetic separation (Ao et al. 2009). The formation of Ti–O–Ti chains in the sol is most apparent at low water content and low hydrolysis rate, and in excess of Ti-alkoxide (Chen and Mao 2007). Strong adherence of the TiO_2 nanoparticles on the surface of AC has been confirmed using various electron microscopic techniques such as scanning electron microscopy and transmission electron microscopy (Yap et al. 2010, 2011). The robust adherence facilitates charge transfer between the photoexcited TiO_2 and the AC that enhances photonic efficiency of the photocatalytic process.

Various mixing of TiO_2 and AC particles in the aqueous suspension has been reported (Cordero et al. 2007; Matos et al. 2007). It was anticipated that in the TiO_2–AC mixture, TiO_2–AC bonding could be formed via collisions of the particles. Since such synthesis technique consumes less energy and no chemicals, it appears as a more favorable option than wet synthesis of TiO_2/AC composites from a Ti precursor. Nevertheless, if the attachment of the TiO_2 and AC particles is weak and reversible, the short-lived hydroxyl radicals generated on the photoexcited TiO_2 nanoparticles could not be efficiently utilized for degradation of the pollutants adsorbed by the separated AC. In this context, it remains doubtful whether the synergism of adsorption–photocatalysis was truly established (Asenjo et al. 2013).

5.4.2 Performance and Potential Applications

TiO_2/AC composites provide better pollutant removal efficiency compared with bare TiO_2, AC, and mixtures of TiO_2–AC. The composites possess a dual functionality that promotes mass transfer of pollutants to the vicinity of TiO_2 and their degradation by the generated hydroxyl radicals. Compared to the bare TiO_2, the TiO_2/AC is less prone to deactivation. Liu et al. (2007) reported that at the eighth cycle of reuse, TiO_2/AC suffered only 10% reduction in activity while the bare TiO_2 suffered >70% decrease based on phenol removal compared to their respective initial activities. The high affinity of AC for organic pollutants (especially the hydrophobic) also reduces the phenomenon of catalyst deactivation because of the presence of dissolved ions in water such as sulfate (Yap and Lim 2011). Mineralization of various pollutants by TiO_2/AC composites has also been demonstrated, which resulted in a greater CO_2 evolution than that with the TiO_2–AC mixture (Torimoto et al. 1997). The TiO_2/AC composites can be easily recovered for reuse compared to the commercial nanosized TiO_2 particles. Inherently, the adsorption capacity of AC in the TiO_2/AC composite is reduced compared to that of the untreated AC because of blocking of the pores. There is an optimal TiO_2 loading in each TiO_2/AC composite for the optimal pollutant removal through combined adsorption–photocatalysis processes.

The TiO_2/AC can be used for aqueous-phase as well as gaseous-phase treatment in photoreactors. For water treatment and reclamation, the TiO_2/AC process can be coupled with biological treatment and membrane separation processes for optimal results. The TiO_2/AC can also be used as an adsorbent in the conventional carbon filter. Once its adsorption capacity is exhausted, it can be regenerated via solar photocatalysis. The solar photocatalytic regeneration (SPR) technique is a greener and cost-effective technique for regenerating exhausted AC. The SPR process combines solar heating and photocatalysis to desorb and degrade the pollutants in water, and thus it does not incur carbon footprint and chemical consumption and no secondary waste streams are produced, which presents advantages over the thermal and chemical regeneration techniques. Yap and Lim (2012) demonstrated that their N-TiO_2/AC composite, which had been preloaded with organic micropollutants, could be regenerated using solar light, and the SPR efficiency depends on the hydrophobicities of the pollutants, light intensity, N-TiO_2 loading in the composite, and temperature. They achieved the highest SPR efficiency of >80% within 8 h under one-sun irradiation. The rate-limiting process was the desorption–diffusion of pollutants from the interior adsorption sites of AC, which could be enhanced through solar heating. They proposed a kinetic model to predict the solar regeneration efficiency of a pollutant-loaded TiO_2/AC composite.

5.5 SOLAR DISINFECTION OF WATER

Solar disinfection of water (SODIS) is an easy-to-use and effectively zero-cost technique that can be used to decrease the pathogen loading in contaminated water. The contaminated water is put into UV and visible transparent bottles (PET is recommended). The bottles are then placed in direct sunlight for 6 h (effectively 1 day) or 12 h if conditions are cloudy (2 days). After exposure to solar radiation, the effective pathogen loading is reduced and the water is "safer" to drink. Although the SODIS process is not guaranteed to be 100% effective, it is recognized and promoted by WHO. It has been estimated that around 4.5 million people throughout the world use SODIS regularly, predominately in Africa, Latin America, and Asia (McGuigan et al. 2012). SODIS is effective against a range of pathogens, is effectively a zero-cost process, and is simple to use. However, there are a number of disadvantages or issues. The efficacy of the SODIS process is dependent on different environmental parameters, which include solar irradiance (depending on latitude, daylight hours, and atmospheric conditions); water quality parameters, including organic and inorganic contamination, suspended solids, and turbidity; and the level and type of microbiological contamination. Furthermore, different microorganisms have varying levels of resistance to solar disinfection, and this leads to variation in the required treatment times.

Malato et al. reviewed published data concerning the time taken for the inactivation of different microorganisms using solar disinfection, at approximately 1 kW/m^2 global irradiance (Malato et al. 2009). The time taken for inactivation varied substantially, for example, 20 min for *Campylobacter jejuni* and up to 8 h for *Cryptosporidium parvum* oocysts (a resistant form of the protozoan parasite). No inactivation was observed for *Bacillus subtilis* endospores after 8 h of solar disinfection. Also, the SODIS process is user-dependent and requires the end user to measure the time exposed to solar radiation. There is no simple quality assurance for the SODIS process and, subsequently, the lack of compliance with the recommended protocol is a major issue that decreases the efficacy (Du Preez et al. 2010).

There are several approaches to improve or enhance the SODIS process, for example, the use of specifically designed solar reactors that increase the solar dose (CPC reactors) and the use of heterogeneous photocatalysis (Byrne et al. 2011).

With heterogeneous photocatalysis, a semiconductor material (e.g., TiO_2) is irradiated with electromagnetic radiation of wavelength equal to or greater than the semiconductor band gap energy. The photon energy is absorbed, promoting an electron from the valence band to the conduction band. These charge carriers, referred to as electron–hole pairs $\left(e^-_{cb} \text{ and } h^+_{vb}\right)$, can recombine with the loss of energy as light or heat, or they can migrate to the catalyst surface. If the electrons and holes survive and reach the particle surface, they can participate in redox reactions at the surface (Cassano and Alfano 2000). The hole can oxidize water or a hydroxyl ion to give a hydroxyl radical and the electron can reduce molecular oxygen to form a superoxide radical anion, with subsequent reduction reactions giving rise to hydrogen peroxide and hydroxyl radicals. ROS are powerful and indiscriminate oxidants that can degrade chemical contaminants and inactivate pathogenic microorganisms (Lee et al. 2006). Matsunaga et al. were the first to report the application of photocatalysis for the inactivation of bacteria using UV-excited TiO_2 (Matsunaga et al. 1985). Over the last three decades, there has been a substantial increase in the number of research papers investigating the inactivation of microorganisms. Photocatalysis has been shown (at least at the lab scale) to be effective in the inactivation of bacteria (including spores), viruses, protozoa (including oocysts), fungi (including spores), and algae. In 1999, Blake et al. carried out an extensive review of the microorganisms reported to be inactivated by photocatalysis (Blake et al. 1999). McCullagh et al. (2007) subsequently reviewed the published work focused on the photocatalytic disinfection of water. Malato et al. (2009) undertook an extensive review of works concerning the decontamination and disinfection of water using solar-driven photocatalysis, and Dalrymple et al. (2010) have reviewed the mechanisms of photocatalytic disinfection.

In most studies concerning the mechanism of photocatalytic disinfection, hydroxyl radical is reported to be the main ROS responsible for the inactivation of microorganisms. Of course, other ROS species generated by photocatalysis, including hydrogen peroxide and superoxide radical anion, play a part in the disinfection process (Huang et al. 2000). ROS cause fatal damage to microorganisms through disruption of the cell membrane and by damage to DNA and RNA (Blake et al. 1999). Other effects of oxidative damage include damage to the respiratory system within the cells (Maness et al. 1999) and damage to the cell membranes leading to loss of fluidity and increased ion permeability in the cell membrane (Wainwright 2000). There is evidence that cell death is attributed to lipid peroxidation within bacterial cell membranes (Huang et al. 2000). The peroxidation of the unsaturated phospholipids in the bacterial cell membrane will cause a loss in the respiratory activity of the cells (Maness et al. 1999) and may result in a loss of fluidity and increased ion permeability (Wainwright 2000). Damage to the cell membrane will increase the likelihood for additional oxidative attack on internal cellular components (Rincon and Pulgarin 2004).

Alrousan et al. (2009) investigated the photocatalytic inactivation of *E. coli* as a model organism in distilled water and real surface water (obtained from a catchment in Northern Ireland) using immobilized films of TiO_2 nanoparticles. The rate of inactivation of *E. coli* was higher with UVA-excited TiO_2 as compared to direct photolytic inactivation with UVA alone in both distilled water and real surface water. The optimal catalyst loading was determined to be 0.5 mg cm^{-2}, which is approximately half of the optimal loading determined for the photocatalytic degradation of formic acid and atrazine under the same conditions. Dunlop et al. (2008) reported on the photocatalytic and electrochemically assisted photocatalytic inactivation of *E. coli* and *Clostridium perfringens* spores on TiO_2 electrodes. The photocatalytic inactivation of *E. coli* and *C. perfringens* spores was observed on all immobilized TiO_2 films (under open circuit) under UVA irradiation. The rate of photocatalytic inactivation of *E. coli* was determined to be an order of magnitude greater than that observed for the *C. perfringens* spores. With the application of an external electrical bias (electrochemically assisted photocatalysis), the rate of inactivation was markedly increased with the *C. perfringens* spores. Sunnotel et al. (2010) investigated the photocatalytic inactivation of *Cryptosporidium parvum* oocysts on immobilized nanoparticle TiO_2 films. Scanning electron microscopy analysis of the oocysts before and after treatment showed physical damage of the oocyst cell walls and revealed large numbers of ghost cells (empty cells) after photocatalytic treatment. No significant inactivation of the oocysts was observed with exposure to UVA radiation alone over the same time frame. Other research has reported the photocatalytic inactivation of *C. parvum* with TiO_2 immobilized on flexible plastic inserts in 1.5-ml bottles under natural sunlight (Méndez-Hermida et al. 2007).

Novel photocatalytic materials are being investigated to enhance the solar efficiency of photocatalytic disinfection. With wide-band gap semiconductors that absorb in the UV domain (such as TiO_2), the solar efficiency is limited as only 4% of solar photons are in the UV region (<400 nm). Therefore, there is currently significant research activity focusing on making visible light active photocatalytic materials so that more solar photons can be utilized. Different approaches have been attempted to increase activity of wide–band gap semiconductors in the visible region. For example, many researchers have attempted to create visible light active materials by doping TiO_2 with metal ions (Hamilton et al. 2008) or nonmetal elements including N and S. In 2001 Asahi et al. (2001) reported that doping TiO_2 with nitrogen gave a visible light active photocatalyst. Since then, many researchers have investigated the nonmetal doping of TiO_2 and showed that the optical absorbance is extended into the visible light region. However, the number of publications concerned with the solar photocatalytic disinfection of water with visible light active photocatalytic materials is limited.

For solar applications, novel photocatalyst materials should be tested under simulated solar irradiation or preferably under real sun conditions. Also, the doped materials should be tested against a known research standard with good UV activity (e.g., *Evonik Aeroxide P25*) as there is a small but significant proportion of UV in the solar spectrum. Rengifo-Herrera and Pulgarin investigated the UV/Vis photocatalytic activity of N, S co-doped, and N-doped commercial anatase (Tayca TKP 102) TiO_2 for the oxidation of phenol and the inactivation of *E. coli* using solar simulated conditions (Rengifo-Herrera and Pulgarin 2010). The non–metal-doped TiO_2 did not show any significant enhancement for the degradation of phenol or the inactivation of *E. coli*, as compared to the commercial undoped *Evonik Aeroxide P25*. While the N or the N and S co-doped TiO_2 may absorb some of the visible light photons, the visible light activity of the materials is small compared to the UV activity of the undoped TiO_2. Recent investigations showed that Cu-doped TiO_2 films accelerated bacterial inactivation (of *E. coli* and *E. faecalis*) in the presence of natural sunlight (Fisher et al. 2013). Turki et al. (2013) evaluated the effect of four different TiO_2 morphologies, namely, nanotubes, nanoplates, nanorods, and nanospheres, for solar photocatalytic inactivation of fungi spores of *Fusarium solani* in water. The authors demonstrated that at very low concentrations of photocatalyst, the inactivation of *F. solani* over TiO_2 nanospheres had the best disinfection efficiency with respect to the other morphologies. The results on simultaneous degradation of formic acid and fungi inactivation showed that TiO_2 nanotubes permitted both decontamination and disinfection of water (Turki et al. 2013).

Recently, Fernández-Ibáñez et al. (2014) reported on the photocatalytic disinfection of water using titania–graphene composites. The composites were created by the photocatalytic reduction of graphene oxide (GO) in the presence of titania and methanol as a hole acceptor. The materials were tested under real sun for the disinfection of water contaminated with *E. coli* and *F. solani* spores (fungal plant pathogen). Both the unmodified titania (*Evonik Aeroxide P25*) and titania–graphene composites showed very rapid inactivation of *E. coli* and *F. solani* spores. A small increase in efficiency was observed for the titania–graphene composites for the inactivation of *E. coli* as compared to the unmodified titania. Surprisingly, when the main UV component of the solar spectrum was screened out, the disinfection efficiency for titania–graphene composites with *E. coli* was maintained, whereas there was a large decrease in the efficiency observed for the unmodified titania. This effect was proven with either both water pathogens, bacteria cells and fungal spores. They found some evidence of singlet oxygen ($^1O_2$) production by the composites under visible light irradiation.

While photocatalysis is effective for the inactivation of microorganisms in water, there are many challenges to be addressed before the technology can be deployed for reducing the incidence of waterborne disease in the developing world.

5.6 REMEDIATION OF CONTAMINATED GROUNDWATER

Groundwater contamination is caused by corrosion of underground chemical storage tanks, use and release of chemical compounds, direct discharge of wastes to the environment, and, more recently, shale gas development. Common chemicals found in groundwater include organic solvents (e.g., trichloroethylene [TCE] and perchloroethylene [PCE] also known as degreasing chemicals), perfluoroalkyl compounds (e.g., perfluorooctanesulfonic acid and perfluorooctanoic acid used in fire-retardant and nonstick cookware), synthetic organic chemicals and agricultural compounds (e.g., pesticides, herbicides, and fertilizers), and heavy metals (Leeson et al. 2013).

5.6.1 Remediation Challenges

Along with the high stability of chemicals found in groundwater, the physiographical features of contaminated sites make them difficult to treat. For example, because of its sparing solubility and higher density than water, TCE forms a dense non–aqueous-phase liquid site in groundwater and soil pore space, of which treatment has been challenging. Identifying contaminated sites and follow-up cleaning processes consumes extensive time and cost and requires skills in various areas including hydrology, geology, biology, and chemistry other than environmental engineering. Many biological, chemical, and physical remediation strategies have been developed including bioaugmentation, phytoremediation, chemical oxidation, chemical precipitation, ozonation, electrochemical treatment, thermal remediation, membrane separation, air sparging, and physical capping (Ma and Jiao 2012). Recently, in situ remediation technologies have been proposed to be more sustainable and less eco-disruptive.

5.6.2 Advanced Approaches Using Nanomaterials

Remediation of groundwater particularly contaminated with recalcitrant halogenated chemicals remains a scientific and technical challenge, which in fact triggered the development of highly effective nanomaterials and their applications (Matlochova et al. 2013). Metal oxide nanoparticles have been reported to be effective to remove many toxic metals in groundwater. Basically, oxide particles in nanosize possess ample sorption and reaction sites. Many nanoparticles have been proposed to remove arsenic (Jegadeesan et al. 2010; Kanel et al. 2006). Along with its high sorption capacity, interestingly, TiO_2 can change the oxidation state of arsenic from As(III) to As(V) under ultraviolet raddiation (Ferguson et al. 2005). Arsenic (V) is less toxic and easier to remove by conventional treatment technology. The TiO_2 technology can combine physical sorption with chemical oxidation. Nanostructured iron-cerium mixed oxide was also applied for the treatment of groundwater containing arsenic at high concentration in the presence of many coexisting ions (Basu and Ghosh 2013):

$$Fe^0 \rightarrow Fe^{2+} + 2e^- \tag{5.3}$$

$$\text{C-X} + H^+ + 2e^- \rightarrow \text{C-H} + X^- \tag{5.4}$$

Recently, research efforts have also been given to reactive zerovalent nanoparticles such as Fe, Mg, and Al (Fu et al. 2014). Among them, zerovalent iron (ZVI, Fe^0) nanoparticles are popular thanks to the abundance of iron. ZVI provides electrons to halogenated chemicals, while initiating electrochemical reduction, as shown in Equations 5.3 and 5.4. The dehalogenation reaction is important because chemicals, once dehalogenated, become less toxic and thus more biodegradable, and are susceptible to subsequent chemical oxidation. Zerovalent metals have been proven to be effective to electrochemically reduce nitro-organic compounds and to dechlorinate many chlorinated chemicals such as TCE and PCE (Raychoudhury and Scheytt 2013). Synthesizing well-defined ZVI nanoparticles with high specific surface area has been extensively studied to increase ZVI overall reactivity (Li et al. 2006). Adding a small amount of surface modifiers such as electrolytes and surfactants keeps nanoparticles segregated. Instead of the borohydride reduction of dissolved Fe^{3+} ions to Fe^0, solid-phase reduction of abundant Fe oxides to ZVI under hydrogen gas at high temperature was proposed to reduce ZVI production

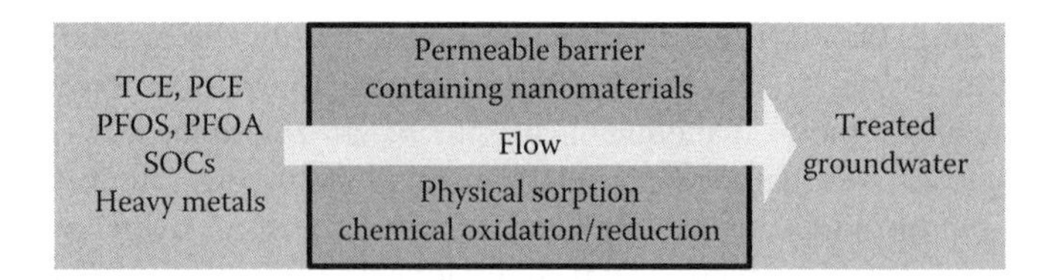

FIGURE 5.9 Applications of nanomaterials for groundwater remediation. A permeable barrier composed of reactive nanomaterials is deployed for flow-through removal of many groundwater contaminants such as halogenated compounds, synthetic organic chemicals, and heavy metals via physical adsorption and chemical oxidation and reduction.

cost. Coupling of two metals to form a galvanic cell was documented to enhance the electrochemical reduction via hydrodehalogenation (Agarwal et al. 2007). A catalytic amount of Pd embedded onto a ZVI surface can significantly improve the kinetics (Choi et al. 2009). Overall reactivity and longevity of ZVI can be greatly controlled by coupling it with other less active metals such as Cu, Co, and Ni, which is a huge advantage for responding to the characteristics of groundwater contamination (e.g., low level and long duration). Along with organic chemicals, inorganic contaminants such as nitrate, arsenic, and cadmium can also be electrochemically reduced.

5.6.3 Application and Prospects

For real field applications, researchers have attempted to immobilize such nanomaterials into porous support materials (Mackenzie et al. 2012). ZVI nanoparticles can be integrated into the porous structure of AC to couple the physical adsorption of contaminants with their reductive chemical decomposition (Choi et al. 2008). In the presence of an oxidant injected intentionally, Fe ions released from ZVI nanoparticles can activate the oxidant to produce strong transient oxidizing species such as hydroxyl radicals and sulfate radicals, which decompose contaminants oxidatively (the so-called Fenton reaction) (Liang and Lai 2007). Nanomaterials can be integrated into a permeable barrier for flow-through treatment of groundwater, as shown in Figure 5.9 (Henderson and Demond 2007). However, release of nanoparticles and dissolved metal ions to the environment is of great concern. Understanding of these technologies is currently mainly based on lab-scale tests and thus more field applications should be demonstrated.

5.7 ROS IN TiO_2 PHOTOCATALYSIS

TiO_2 has received extensive attention in the area of photocatalysis during the past few decades because of its superior photoactivity, low toxicity, high physical and chemical stability, low cost, and good corrosion resistance (Hoffmann et al. 1995; Pugazhenthiran et al. 2013). TiO_2 naturally exists in three different polymorphic forms: anatase, rutile, and brookite with band gaps of 3.2, 3.0, and ~3.2 eV, respectively. Rutile is the most common and thermodynamically stable. Anatase has been studied and found to exhibit higher photocatalytic activity compared with rutile and brookite (Banerjee et al. 2014; Pelaez et al. 2012). In 1972, Fujishima and Honda first discovered the photoelectrochemical splitting of water using TiO_2 as a photoanode and platinum as a counter electrode by UV light (Fujishima and Honda 1972). This breakthrough prompted other promising applications on various problems of environmental interests in air and wastewater purification (de la Cruz et al. 2013; Hoffmann et al. 1995).

TiO_2 photocatalysis is initiated when the TiO_2 semiconductor absorbs light energy equal to or greater than its band gap between the conduction and valence bands, promoting an electron to the conduction band $\left(e^-_{cb}\right)$ and leaving a "hole" in valence band $\left(h^+_{vb}\right)$ shown in Equation 5.5. In the case of anatase TiO_2, the energy band gap is 3.2 eV; therefore, it can be activated by UV illumination (less than 387 nm). The excited-state conduction band electrons and valence band holes can undergo recombination and dissipate the energy by Equation 5.6. Molecular oxygen is generally used to scavenge the e^-_{cb} at the TiO_2 surface, yielding a superoxide radical anion (Equation 5.7). The pK_a of the superoxide radical anion occurs at pH = 4.6; thus, a hydroperoxide radical $\left(HO^{\bullet}_2\right)$ is formed under more acidic conditions (Equation 5.8), and subsequent disproportionation results in the formation of hydrogen peroxide (H_2O_2) (Equation 5.9). The h^+_{vb} produced during photoexcitation of the TiO_2 semiconductor has the potential to oxidize surface-absorbed H_2O or hydroxyl groups to generate hydroxyl radical (Equation 5.10). The hydroxyl radical is a powerful oxidant and can react with most organic pollutants (Thakur et al. 2010), and after a series of reactions, it can lead to the formation of nontoxic final products H_2O and CO_2.

$$TiO_2 + h\nu \rightarrow e^-_{cb} + h^+_{vb} \tag{5.5}$$

$$e^-_{cb} + h^+_{vb} \rightarrow \text{heat} \tag{5.6}$$

$$e^-_{cb} + O_2 \rightarrow O_2^{-\bullet} \tag{5.7}$$

$$O_2^{-\bullet} + H^+ \leftrightarrow HO^{\bullet}_2 \tag{5.8}$$

$$2O_2^{-\bullet} + 2H_2O \rightarrow H_2O_2 + 2OH^- + O_2 \tag{5.9}$$

$$h^+_{vb} + H_2O \rightarrow H^+ + {}^{\bullet}OH \tag{5.10}$$

$${}^{\bullet}OH + \text{Pollutant} \rightarrow\rightarrow\rightarrow\rightarrow CO_2 + H_2O \tag{5.11}$$

A number of recent studies have been reported on the $^{\bullet}OH$ formation and determination in UV/visible light–activated (VLA) TiO_2 photocatalysis. The spin trap and electron spin resonance (ESR) technique was first applied to detect the formation of $^{\bullet}OH$ during the illumination of TiO_2 by Jaeger and Bard in 1979 (Jaeger and Bard 1979). They used α-phenyl *N-tert*-butyl nitrone and α-(4-pyridyl *N*-oxide) *N-tert*-butyl nitrone as ESR spin traps that efficiently scavenged a reactive free radical ($^{\bullet}OH$) to form an ESR signal of the nitroxide adduct. Alternatively, ESR spin traps such as dimethyl pyrroline-*N*-oxide (DMPO) and 3-carboxyproxyl were also reported for determination of $^{\bullet}OH$ in UV TiO_2 photocatalysis (Schwarz et al. 1997; Uchino et al. 2002). Besides the ESR technique for measurement of $^{\bullet}OH$, the applications using fluorescence probe methods with coumarin and terephthalic acid have also been widely studied (Louit et al. 2005; Samuni et al. 2002; Yang et al. 2009; Zhang and Nosaka 2013). Zhang and Nosaka (2013) have reported the quantitative detection of $^{\bullet}OH$ using coumarin for various modified VLA TiO_2 including nitrogen-doped, Pt complex–deposited, Fe(III)–grafted, and Fe(III)–grafted Ru-doped TiO_2. They found that Fe(III)–grafted Ru-doped TiO_2 produced the highest quantum yield under 470-nm LED light irradiation. Measurement of $^{\bullet}OH$ using terephthalic acid was also studied to test the photoactivity of solvothermal synthesized anatase TiO_2 nanosheet by Yang et al. (2009). The high-quality anatase TiO_2 nanosheet performs superior photoactivity (more than five times) compared with P25 TiO_2. Quencher methods using *t*-butanol/*iso*-propanol have also been studied to elucidate the role of $^{\bullet}OH$ in UV/VLA TiO_2 photocatalysis (Zhao et al. 2014; Zheng et al. 2014).

Superoxide radical anions can lead to the degradation of the target compound through direct or indirect chemical reactions. The formation of $O_2^{-\bullet}$ during TiO_2 irradiation has been detected by ESR technique using the spin trapping agent DMPO (Konaka et al. 1999). UV/Vis probe methods using 2,3-bis(2-methoxy-4-nitro-5-sulfophenyl)-2*H*-tetrazolium-5-carboxanilide (XTT) and nitroblue tetrazolium (NBT) have been studied to quantify the formation of $O_2^{-\bullet}$ during TiO_2 photocatalysis. Auffan et al. (2010) measured the amount of $O_2^{-\bullet}$ generated by a newly synthesized TiO_2 nanosheet (TiO_2 core coated with two successive protective layers of $Al(OH)_3$ and polydimethylsiloxane). They monitored an absorption peak at 470 nm, corresponding to XTT formazan produced through XTT reduction by $O_2^{-\bullet}$. There was no measurable $O_2^{-\bullet}$ generated by this new TiO_2 nanosheet. Quantitation of $O_2^{-\bullet}$ using NBT as a probe has also been determined by measuring the decreased intensity of the absorbed NBT in the solution (Goto et al. 2004). Goto et al. demonstrated $O_2^{-\bullet}$ as the main product from molecular oxygen in rutile TiO_2 photocatalyzed aqueous solution containing 2-propanol. Superoxide dismutase as the quencher of $O_2^{-\bullet}$ was used to investigate the role of $O_2^{-\bullet}$ in UV and VLA TiO_2 photocatalysis. Recent studies have reported that $O_2^{-\bullet}$ plays a significant role in VLA TiO_2 photocatalysis (Zhao et al. 2014) and a minimal role in UV TiO_2 photocatalysis (Zheng et al. 2010).

Singlet oxygen exhibits high reactivity because of its energy (22.5 kcal/mol), which is greater than that of a triplet ground-state oxygen molecule (Kearns 1971). The generation of $^1O_2$ during TiO_2 irradiation in ethanol has been demonstrated by ESR using 2,2,6,6-tetramethyl-4-piperidone as a $^1O_2$-sensitive trapping agent. Konaka et al. (1999) revealed that singlet oxygen is produced directly at the irradiated TiO_2 surface but not by superoxide radical anion–mediated reactions. Zhao et al. (2014) have studied the role of $^1O_2$ in VLA NF-TiO_2 photocatalysis using furfuryl alcohol (FFA) as a quencher. They found that $^1O_2$ does not play an important role based on the quencher experiments. To further test the role of $^1O_2$ in VLA NF-TiO_2 photocatalysis, they carried out the degradation process in D_2O solution. Since the lifetime of $^1O_2$ in D_2O is longer than that in H_2O, the singlet oxygen–mediated processes are enhanced in D_2O. The results show that the degradations in solution of H_2O and D_2O were similar, which was consistent with the studies using FFA as the quencher.

ROS ($^{\bullet}OH$, $O_2^{-\bullet}$, and $^1O_2$) produced in TiO_2 photocatalysis are important for the remediation of problematic pollutants in drinking water and also critical for assessing TiO_2 photocatalysis as a potential water treatment for pollutants.

5.8 CONCLUSIONS

This chapter summarizes different nanotechnologies for water sustainability such as catalysis (i.e., AOPs), nanofiltration, adsorption, water disinfection, and groundwater remediation. In order to decompose water contaminants, several catalysts are used or being tested including TiO_2 photocatalysts, ferrate(VI), ferrites, graphitic carbon nitride, and perovskite. Results show that such materials effectively decompose various water contaminants such as alcohols, hydrocarbons, dyes, endocrine disruptors, cyanotoxins, pharmaceuticals, and pesticides. Moreover, TiO_2 and ferrate(VI) demonstrate efficient pathogen inactivation for water disinfection. In addition to the contribution of catalytic process for sustainable water treatment, the photocatalytic membrane filtration and hybrid adsorption–catalytic process were discussed in this chapter. The photocatalytic membrane process significantly improves membrane fouling, which frequently occurs via deposition and accumulation of the photocatalytic nanoparticles on the surface of membranes, and enhances pollutant removal and energy efficiency. TiO_2/AC composites demonstrated enhanced removal of water pollutants. AC has not only a high specific surface area but also a strong adsorption affinity toward various organic contaminants and TiO_2 is able to achieve complete mineralization of the contaminants. The enhanced removal of water contaminants was attributed to the combination of AC adsorption and TiO_2 photocatalysis. Additionally, SODIS was discussed for water sustainability for developing countries because it is a simple and cost-effective method for drinking water disinfection. SODIS could inactivate different microorganisms efficiently because of the presence of UV in the solar spectrum. With the use of TiO_2 photocatalysts, the efficacy of solar disinfection can be significantly improved because of the formation of ROS, which can damage the cell membrane, DNA, and RNA of microorganisms. Moreover, the development of highly effective nanomaterials and their deployment strategies was discussed for groundwater remediation. In order to comply with recent in situ groundwater remediation needs, reactive metal nanoparticle-based remediation technologies, including ZVI nanoparticles, iron/palladium bimetallic nanoparticles, TiO_2 nanoparticles, and AC-supported metal nanoparticles, are applied by various groups and they have been briefly discussed in this chapter. Finally, ROS were discussed in terms of their formation and roles in AOPs. Their formation and roles depend on types of photocatalysts, reaction conditions, and targeted water contaminants. For sustainable water treatment management, it is important to determine ROS formation and understand the interaction between ROS and water contaminants in such engineered processes.

ACKNOWLEDGMENTS

D. D. Dionysiou would like to acknowledge support from (a) the Cyprus Research Promotion Foundation through Desmi 2009–2010, which was co-funded by the Republic of Cyprus and the European Regional Development Fund of the EU under contract number NEA IPODOMI/STRATH/0308/09; (b) the United States National Science Foundation (CBET-1236209); and (c) the University of Cincinnati through a UNESCO co-chair professor position on "Water Access and Sustainability." C. Han acknowledges support from the Postgraduate Research Program at the National Risk Management Research Laboratory administered by the Oak Ridge Institute for Science and Education through an interagency agreement between the US Department of Energy and the US Environmental Protection Agency. G. E. Romanos and P. Falaras acknowledge financial support from the European Social Fund and Greek national funds through the Operational Program "Education and Lifelong Learning" of the National Strategic Reference Framework—Research Funding Programs: ARISTEIA (SolMeD-3635) and THALES (AOP-NanoMat/MIS 379409). P. Falaras also acknowledges funding of this work by Prince Sultan Bin Abdulaziz International Prize for Water—Alternative Water Resources Prize 2014. V. K. Sharma and N. Johnson acknowledge the support of the United States National Science Foundation (CBET-1439314). H. Choi is grateful to the Texas Higher Education Coordinating Board for the Norman Hackerman Advanced Research Program fund (THECB13311).

DISCLAIMER

The US Environmental Protection Agency, through its Office of Research and Development, funded and managed, or partially funded and collaborated in, the research described herein. It has been subjected to the agency's administrative review and has been approved for external publication. Any opinions expressed in this paper are those of the author(s) and do not necessarily reflect the views of the agency; therefore, no official endorsement should be inferred. Any mention of trade names or commercial products does not constitute endorsement or recommendation for use.

REFERENCES

Abe, R. 2011. Development of a new system for photocatalytic water splitting into H_2 and O_2 under visible light irradiation. *Bulletin of the Chemical Society of Japan* 11:1000–1030.

Agarwal, S., S. R. Al-Abed, and D. D. Dionysiou. 2007. Enhanced corrosion-based Pd/Mg bimetallic systems for dechlorination of PCBs. *Environmental Science & Technology* 41:3722–3727.

Agha, R., S. Cirés, L. Wörmer, J. A. Domínguez, and A. Quesada. 2012. Multi-scale strategies for the monitoring of freshwater cyanobacteria: Reducing the sources of uncertainty. *Water Research* 46:3043–3053.

Alrousan, D. M. A., P. S. M. Dunlop, T. A. McMurray, and J. A. Byrne. 2009. Photocatalytic inactivation of *E. coli* in surface water using immobilised nanoparticle TiO_2 films. *Water Research* 43:47–54.

Anquandah, G. A. K., V. K. Sharma, D. A. Knight, S. R. Batchu, and P. R. Gardinali. 2011. Oxidation of trimethoprim by ferrate (VI): Kinetics, products, and antibacterial activity. *Environmental Science & Technology* 45:10575–10581.

Anquandah, G. A. K., V. K. Sharma, V. R. Panditi, P. R. Gardinali, H. Kim, and M. A. Oturan. 2013. Ferrate (VI) oxidation of propranolol: Kinetics and products. *Chemosphere* 91:105–109.

Ao, Y., J. Xu, D. Fu, and C. Yuan. 2009. Photocatalytic degradation of X-3B by titania-coated magnetic activated carbon under UV and visible irradiation. *Journal of Alloys and Compounds* 471:33–38.

Asahi, R., T. Morikawa, T. Ohwaki, T. Aoki, and K. Taga. 2001. Visible-light photocatalysis in nitrogen-doped titanium oxides. *Science* 293:269–271.

Asenjo, N. G., R. Santamaría, C. Blanco, M. Granda, P. Álvarez, and R. Menéndez. 2013. Correct use of the Langmuir-Hinshelwood equation for proving the absence of a synergy effect in the photocatalytic degradation of phenol on a suspended mixture of titania and activated carbon. *Carbon* 55:62–69.

Athanasekou, C. P., G. E. Romanos, F. K. Katsaros, K. Kordatos, V. Likodimos, and P. Falaras. 2012. Very efficient composite titania membranes in hybrid ultrafiltration/photocatalysis water treatment processes. *Journal of Membrane Science* 392–393:192–203.

Athanasekou, C., S. Morales-Torres, V. Likodimos et al. 2014. Prototype composite membranes of partially reduced grapheneoxide/TiO_2 for photocatalytic ultrafiltration water treatment under visible light. *Applied Catalysis B: Environmental* 58–159:361–372.

Auffan, M., M. Pedeutour, J. Rose et al. 2010. Structural degradation at the surface of a TiO_2-based nanomaterial used in cosmetics. *Environmental Science & Technology* 44:2689–2694.

Bai, X., L. Wang, Y. Wang, W. Yao, and Y. Zhu. 2014. Enhanced oxidation ability of g-C_3N_4 photocatalyst via C_{60} modification. *Applied Catalysis B: Environmental* 152–153: 262–270.

Banerjee, S., S. C. Pillai, P. Falaras, K. E. O'Shea, J. A. Byrne, and D. D. Dionysiou. 2014. New insights into the mechanism of visible light photocatalysis. *The Journal of Physical Chemistry Letters* 5:2543–2554.

Barros-Becker, F., J. Romero, A. Pulgar, and C. G. Feijóo. 2012. Persistent oxytetracycline exposure induces an inflammatory process that improves regenerative capacity in zebrafish larvae. *PLoS ONE* 7:e36827.

Basu, T., and U. C. Ghosh. 2013. Nano-structured iron(III)-cerium(IV) mixed oxide: Synthesis, characterization and arsenic sorption kinetics in the presence of co-existing ions aiming to apply for high arsenic groundwater treatment. *Applied Surface Science* 283:471–481.

Blake, D. M., P. C. Maness, Z. Huang, E. J. Wolfrum, J. Huang, and W. A. Jacoby. 1999. Application of the photocatalytic chemistry of titanium dioxide to disinfection and the killing of cancer cells. *Separation and Purification Methods* 28:1–50.

Byrne, J.A., P. Fernández-Ibáñez, P. S. M. Dunlop, D. M. A. Alrousan, and J. W. J. Hamilton. 2011. Photocatalytic enhancement for solar disinfection of water: A review. *International Journal of Photoenergy* 2011:798051.

Carbajo, M., F. J. Beltran, F. Medina, O. Gimeno, and F. J. Rivas. 2006. Catalytic ozonation of phenolic compounds: The case of gallic acid. *Applied Catalysis B: Environmental* 67:177–186.

Casbeer, E., V. K. Sharma, and X. Z. Li. 2012. Synthesis and photocatalytic activity of ferrites under visible light: A review. *Separation and Purification Technology* 87:1–14.

Casbeer, E. M., V. K. Sharma, Z. Zajickova, and D. D. Dionysiou. 2013. Kinetics and mechanism of oxidation of tryptophan by ferrate (VI). *Environmental Science & Technology* 47:4572–4580.

Cassano, A. E., and O. M. Alfano. 2000. Reaction engineering of suspended solid heterogeneous photocatalytic reactors. *Catalysis Today* 58:167–197.

Chen, X., and S. S. Mao. 2007. Titanium dioxide nanomaterials: Synthesis, properties, modifications, and applications. *Chemical Reviews* 107:2891–2959.

Chen, S., Y. Hu, S. Meng, and X. Fu. 2014. Study on the separation mechanisms of photogenerated electrons and holes for composite photocatalysts g-C_3N_4-WO_3. *Applied Catalysis B: Environmental* 150–151:564–573.

Choi, H., E. Stathatos, and D. D. Dionysiou. 2006. Synthesis of nanocrystalline photocatalytic TiO_2 thin films and particles using sol–gel method modified with nonionic surfactants. *Thin Solid Films* 510:107–114.

Choi, H., S. R. Al-Abed, and S. Agarwal. 2009. Catalytic role of palladium and relative reactivity of substituted chlorines during adsorption and treatment of PCBs on reactive activated carbon. *Environmental Science & Technology* 43:7510–7515.

Choi, H., S. R. Al-Abed, S. Agarwal, and D. D. Dionysiou. 2008. Synthesis of reactive nano Fe/Pd bimetallic system-impregnated activated carbon for the simultaneous adsorption and dechlorination of PCBs. *Chemistry of Materials* 20:3649–3655.

Choi, H., M. G. Antoniou, M. Pelaez, A. A. de la Cruz, J. A. Shoemaker, and D. D. Dionysiou. 2007. Mesoporous nitrogen-doped TiO_2 for the photocatalytic destruction of the cyanobacterial toxin microcystin-LR under visible light, *Environmental Science & Technology* 41:7530–7535.

Comninellis, C., A. Kapalka, S. Malato, S. A. Parsons, I. Poulios, and D. Mantzavinos. 2008. Advanced oxidation processes for water treatment: Advances and trends for R&D. *Journal of Chemical Technology and Biotechnology* 83:769–776.

Cordero, T., J.-M. Chovelon, C. Duchamp, C. Ferronato, and J. Matos. 2007. Surface nano-aggregation and photocatalytic activity of TiO_2 on H-type activated carbons. *Applied Catalysis B: Environmental* 73:227–235.

Dai, S., M. Xie, S. Meng, X. Fu, and S. Chen. 2014. Coupled systems for selective oxidation of aromatic alcohols to aldehydes and reduction of nitrobenzene into aniline using CdS/g-C_3N_4 photocatalyst under visible light irradiation. *Applied Catalysis B: Environmental* 158–159:382–390.

Dalrymple, O.K., E. Stefanakos, M. A. Trotz, and D. Y. Goswami. 2010. A review of the mechanisms and modeling of photocatalytic disinfection. *Applied Catalysis B: Environmental* 98:27–38.

de la Cruz, A. A., A. Hiskia, T. Kaloudis et al. 2013. A review on cylindrospermopsin: The global occurrence, detection, toxicity and degradation of a potent cyanotoxin. *Environmental Science: Processes & Impacts* 15:1979–2003.

Deng, J., Y. Shao, N. Gao, C. Tan, S. Zhou, and X. Hu. 2013. $CoFe_2O_4$ magnetic nanoparticles as a highly active heterogeneous catalyst of oxone for the degradation of diclofenac in water. *Journal of Hazardous Materials* 262:836–844.

Dong, G., Y. Zhang, Q. Pan, and J. Qiu. 2014. A fantastic graphitic carbon nitride (g-C_3N_4) material: Electronic structure, photocatalytic and photoelectronic properties. *Journal of Photochemistry and Photobiology C: Photochemistry Reviews* 20:33–50.

Du Preez, M., K. G. McGuigan, and R. M. Conroy. 2010. Solar disinfection of drinking water in the prevention of dysentery in South African children aged under 5 years: The role of participant motivation. *Environmental Science & Technology* 44:8744–8749.

Dunlop, P. S. M., T. A. McMurray, J. W. J. Hamilton, and J. A. Byrne. 2008. Photocatalytic inactivation of *Clostridium perfringens* spores on TiO_2 electrodes. *Journal of Photochemistry and Photobiology A: Chemistry* 196, 113–119.

Falaras, P., G. Romanos, and P. Aloupogiannis. 2012. Photocatalytic Purification Device. European Patent, EP2409954 (A1).

Ferguson, M. A., M. R. Hoffmann, and J. G. Hering. 2005. TiO_2-photocatalyzed As (III) oxidation in aqueous suspensions: Reaction kinetics and effects of adsorption. *Environmental Science & Technology* 39:1880–1886.

Fernández-Ibáñez, P., M. I. Polo-López, S. Malato et al. 2014. Solar photocatalytic disinfection of water using titanium dioxide graphene composites. *Chemical Engineering Journal* doi: 10.1016/j.cej.2014.06.089.

Filip, J., R. A. Yngard, K. Siskova et al. 2011. Mechanisms and efficiency of the simultaneous removal of metals and cyanides by using ferrate (VI): Crucial roles of nanocrystalline iron (III) oxyhydroxides and metal carbonates. *Chemistry—A European Journal* 17:10097–10105.

Fisher, M. B., D. A. Keane, P. Fernández-Ibáñez et al. 2013. Nitrogen and copper doped solar light active TiO_2 photocatalysts for water decontamination. *Applied Catalysis B: Environmental* 130–131:8–13.

Fu, F., D. D. Dionysiou, and H. Liu. 2014. The use of zero-valent iron for groundwater remediation and wastewater treatment: A review. *Journal of Hazardous Materials* 267:194–205.

Fujishima, A., and K. Honda. 1972. Electrochemical photolysis of water at a semiconductor electrode. *Nature* 238:37–38.

Giraldo A.L., G. A. Peñuela, R. A. Torres-Palma, N. J. Pino, R.A. Palominos, and H. D. Mansilla. 2010. Degradation of the antibiotic oxolinic acid by photocatalysis with TiO_2 in suspension. *Water Research* 44:5158–5167.

Goto, H., Y. Hanada, T. Ohno, and M. Matsumura. 2004. Quantitative analysis of superoxide ion and hydrogen peroxide produced from molecular oxygen on photoirradiated TiO_2 particles. *Journal of Catalysis* 225:223–229.

Graham, J. L., K. A. Loftin, M. T. Meyer, and A. C. Ziegler. 2010. Cyanotoxin mixtures and taste-and-odor compounds in cyanobacterial blooms from the Midwestern United States. *Environmental Science & Technology* 44:7361–7368.

Guan, Y.-H., J. Ma, Y.-M. Ren et al. 2013. Efficient degradation of atrazine by magnetic porous copper ferrite catalyzed peroxymonosulfate oxidation via the formation of hydroxyl and sulfate radicals. *Water Research* 47:5431–5438.

Hamilton, J. W. J., J. A. Byrne, C. McCullagh, and P. S. M. Dunlop. 2008. Electrochemical investigation of doped titanium dioxide. *International Journal of Photoenergy* 2008:631597.

Han, C., J. Andersen, V. Likodimos, P. Falaras, J. Linkugel, and D. D. Dionysiou. 2014a. The effect of solvent in the sol-gel synthesis of visible light-activated, sulfur-doped TiO_2 nanostructured porous films for water treatment. *Catalysis Today* 224:132–139.

Han, C., L. Ge, C. Chen et al. 2014b. Novel visible light induced Co_3O_4-g-C_3N_4 heterojunction photocatalysts for efficient degradation of methyl orange. *Applied Catalysis B: Environmental* 147: 546–553.

Han, C., V. Likodimos, J. A. Khan et al. 2013. UV-visible light-activated Ag-decorated, monodisperse TiO_2 aggregates for treatment of a pharmaceutical oxytetracycline. *Environmental Science and Pollution Research* doi: 10.1007/s11356-013-2233-5.

Han, C., M. Pelaez, V. Likodimos et al. 2011. Innovative visible light-activated sulfur doped TiO_2 films for water treatment. *Applied Catalysis B: Environmental* 107:77–87.

Heeb, M. B., J. Criquet, S. G. Zimmermann-Steffens, and U. Von Gunten. 2014. Oxidative treatment of bromide-containing waters: Formation of bromine and its reactions with inorganic and organic compounds—A critical review. *Water Research* 48:15–42.

Henderson, A. D., and A. H. Demond. 2007. Long-term performance of zero-valent iron permeable reactive barriers: A critical review. *Environmental Engineering Science* 24:401–423.

Higarashi, M. M., and W. F. Jardim. 2002. Remediation of pesticide contaminated soil using TiO_2 mediated by solar light. *Catalysis Today* 76:201–207.

Hoffmann, M. R., S. T. Martin, W. Choi, and D. W. Bahnemann. 1995. Environmental applications of semiconductor photocatalysis. *Chemical Reviews* 95:69–96.

Hou, Y., X. Li, Q. Zhao, and G. Chen. 2013. $ZnFe_2O_4$ multi-porous microbricks/graphene hybrid photocatalyst: Facile synthesis, improved activity and photocatalytic mechanism. *Applied Catalysis B: Environmental* 142:80–88.

Hu, L., M. A. Page, T. Sigstam, T. Kohn, B. J. Mariñas, T. J. Strathmann. 2012. Inactivation of bacteriophage MS2 with potassium ferrate (VI). *Environmental Science & Technology* 46:12079–12087.

Huang, Z., P. Maness, D. M. Blake, E. J. Wolfrum, S. L. Smolinski, and W. A. Jacoby. 2000. Bactericidal mode of titanium dioxide photocatalysis. *Journal of Photochemistry and Photobiology A: Chemistry*, 130:163–170.

Jaeger, C. D., and A. J. Bard. 1979. Spin trapping and electron spin resonance detection of radical intermediates in the photodecomposition of water at titanium dioxide particulate systems. *The Journal of Physical Chemistry* 83:3146–3152.

Jain, A., V. K. Sharma, and M. S. Mbuya. 2009. Removal of arsenite by Fe (VI), Fe (VI)/Fe (III), and Fe (VI)/Al (III) salts: Effect of pH and anions. *Journal of Hazardous Materials* 169:339–344.

Jegadeesan, G., S. R. Al-Abed, S. Vijayakumar et al. 2010. Arsenic sorption on TiO_2 nanoparticles: Size and crystallinity effects. *Water Research* 44:965–973.

Ji, K., K. Choi, S. Lee et al. 2010. Effects of sulfathiazole, oxytetracycline and chlortetracycline on steroidogenesis in the human adrenocarcinoma (H295R) cell line and freshwater fish *Oryzias latipes*. *Journal of Hazardous Materials* 182:494–502.

Jiang, J. Q. 2014. Advances in the development and application of ferrate (VI) for water and wastewater treatment. *Journal of Chemical Technology and Biotechnology* 89:165–177.

Jiang, W., L. Zhu, V. K. Sharma et al. 2014. Oxidation of microcystin-LR by ferrate (VI): Intermediates, degradation pathways, and toxicity assessments. *247th Meeting Abstract*, Dallas.

Jin, Z., N. Murakami, T. Tsubota, and T. Ohno. 2014. Complete oxidation of acetaldehyde over a composite photocatalyst of graphitic carbon nitride and tungsten (VI) oxide under visible-light irradiation. *Applied Catalysis B: Environmental* 150–151:479–485.

Kanel, S. R., J. M. Grenche, and H. Choi. 2006. Arsenic (V) removal from groundwater using nano scale zero-valent iron as a colloidal reactive barrier material. *Environmental Science & Technology* 40:2045–2050.

Kato, H., M. Hori, R. Konto, Y. Shimodaira, and A. Kudo. 2004. Construction of Z-scheme type heterogeneous photocatalysis systems for water splitting into H_2 and O_2 under visible light irradiation. *Chemistry Letters* 33:1348–1349.

Kearns, D. R. 1971. Physical and chemical properties of singlet molecular oxygen. *Chemical Reviews* 71:395–427.

Khan, J. A., C. Han, H. M. Khan et al. 2014. Ultraviolet–visible light–sensitive high surface area phosphorous-fluorine–co-doped TiO_2 nanoparticles for the degradation of atrazine in water. *Environmental Engineering Science* 31:435–446.

Konaka, R., E. Kasahara, W. C. Dunlap, Y. Yamamoto, K. C. Chien, and M. Inoue. 1999. Irradiation of titanium dioxide generates both singlet oxygen and superoxide anion. *Free Radical Biology & Medicine* 27:294–300.

Kresge, C. T., M. E. Leonowicz, W. J. Roth, J. C. Vartuli, and J. S. Beck. 1992. Ordered mesoporous molecular sieves synthesized by a liquid-crystal template mechanism. *Nature* 359:710–712.

Lee, Y., Y. Kissner, and U. von Gunten. 2014. Reaction of ferrate (VI) with ABTS and self-decay of ferrate(VI): Kinetics and mechanisms. *Environmental Science & Technology* 48:5154–5162.

Lee, C., Y. Lee, and J. Yoon. 2006. Oxidative degradation of dimethylsulfoxide by locally concentrated hydroxyl radicals in streamer corona discharge process. *Chemosphere* 65:1163–1170.

Lee, Y., and U. von Gunten. 2010. Oxidative transformation of micropollutants during municipal wastewater treatment: Comparison of kinetic aspects of selective (chlorine, chlorine dioxide, ferrate^{VI}, and ozone) and non-selective oxidants (hydroxyl radical). *Water Research* 44:555–566.

Leeson, A., H. F. Stroo, and P. C. Johnson. 2013. Groundwater remediation: Today and challenges and opportunities for the future. *Groundwater* 51:175–179.

Li, L., M. H. Fan, R. C. Brown et al. 2006. Synthesis, properties, and environmental applications of nanoscale iron-based materials: A review. *Critical Review in Environmental Science & Technology* 36:405–431.
Li, X., Y. Hou, Q. Zhao, W. Teng, X. Hu, and G. Chen. 2011. Capability of novel $ZnFe_2O_4$ nanotube arrays for visible-light induced degradation of 4-chlorophenol. *Chemosphere* 82:581–586.
Liang, C. and M. Lai. 2007. Trichloroethylene degradation by zerovalent iron activated persulfate oxidation. *Environmental Engineering Science* 25:1071–1077.
Lim, T.-T., P.-S. Yap, M. Srinivasan, and A. G. Fane. 2011. TiO_2/AC Composites for synergistic adsorption-photocatalysis processes: Present challenges and further developments for water treatment and reclamation. *Critical Reviews in Environmental Science and Technology* 41:1173–1230.
Ling F., J. G. Wang, G. X. Wang, and X, N. Gong. 2011. Effect of potassium ferrate (VI) on survival and reproduction of *Ichthyophthirius multifiliis* tomonts. *Parasitology Research* 109:1423–1428.
Liu, S.X., X. Y. Chen, and X. Chen. 2007. A TiO_2/AC composite photocatalyst with high activity and easy separation prepared by a hydrothermal method. *Journal of Hazardous Materials* 143:257–263.
Liu, G., C. Han, M. Pelaez et al. 2012. Characterization and photocatalytic evaluation of visible light activated C-doped TiO_2 nanoparticles. *Nanotechnology* 23:294003 (10 pp).
Louit, G., S. Foley, J. Cabillic et al. 2005. The reaction of coumarin with the OH radical revisited: Hydroxylation product analysis determined by fluorescence and chromatography. *Radiation Physics and Chemistry* 72:119–124.
Ma, W. J., and B. Q. Jiao. 2012. Review of contaminated sites remediation technology. *Research Journal of Chemistry and Environment* 16:137–139.
Mackenzie, K., S. Bleyl, A. Georgi et al. 2012. Carbo-iron—An Fe/AC composite—As alternative to nano-iron for groundwater treatment. *Water Research* 46:3817–3826.
Malato, S., P. Fernández-Ibáñez, M. I. Maldonado, J. Blanco, and W. Gernjak. 2009. Decontamination and disinfection of water by solar photocatalysis: Recent overview and trends. *Catalysis Today* 147:1–59.
Maness, P. C., S. Smolinski, D. M. Blake, Z. Huang, E. J. Wolfrung, and W. A. Jacoby. 1999. Bactericidal activity of photocatalytic TiO_2 reaction: Toward an understanding of its killing mechanism. *Applied and Environmental Microbiology* 65:4094–4098.
Matlochova, A., D. Placha, and N. Rapantova. 2013. The application of nanoscale materials in groundwater remediation. *Polish Journal of Environmental Studies* 22:1401–1410.
Matos, J., J. Laine, J. M. Herrmann, D. Uzcategui, and J. L. Brito. 2007. Influence of activated carbon upon titania on aqueous photocatalytic consecutive runs of phenol photodegradation. *Applied Catalysis B: Environmental* 70:461–469.
Matsunaga, T., R. Tomoda, T. Nakajima, and H. Wake. 1985. Photoelectrochemical sterilization of microbial cells by semiconductor powders. *FEMS Microbiology Letters* 29:211–214.
McCullagh, C., J. M. Robertson, D. W. Bahnemann, and P. J. K. Robertson. 2007. The application of TiO_2 photocatalysis for disinfection of water contaminated with pathogenic micro-organisms: A review. *Research on Chemical Intermediates* 33:359–375.
McGuigan, K. G., R. M. Conroy, H. J. Mosler, M. du Preez, E. Ubomba-Jaswa, and P. Fernández-Ibáñez. 2012. Solar water disinfection (SODIS): A review from bench-top to roof-top. *Journal of Hazardous Materials* 235–236:29–46.
Méndez-Hermida, F., E. Ares-Mazás, K. G. McGuigan, M. Boyle, C. Sichel, and P. Fernández-Ibáñez. 2007. Disinfection of drinking water contaminated with *Cryptosporidium parvum* oocysts under natural sunlight and using the photocatalyst TiO_2. *Journal of Photochemistry and Photobiology B: Biology* 88:105–111.

Merino, N. A., B. P. Barbero, P. Grange, and L. E. Cadus. 2005. $La_{1-x}Ca_xCoO_3$ perovskite-type oxides: Preparation, characterisation, stability, and catalytic potentiality for the total oxidation of propane. *Journal of Catalysis* 231:232–244.

Moustakas, N. G., F. Katsaros, A. G. Kontos, G. E. Romanos, D. D. Dionysiou, and P. Falaras. 2014. Visible light active TiO_2 photocatalytic filtration membranes with improved permeability and low energy consumption. *Catalysis Today* 224:56–69.

Moustakas, N. G., A. G. Kontos, V. Likodimos et al. 2013. Inorganic–organic core–shell titania nanoparticles for efficient visible light activated photocatalysis. *Applied Catalysis B: Environmental* 130–131:14–24.

Orge, C. A., J. J. M. Orfao, M. F. R. Pereira, B. P. Barbero, and L. E. Cadus. 2013. Lanthanum-based perovskites as catalysts for the ozonation of selected organic compounds. *Applied Catalysis B: Environmental* 140–141:426–432.

Papageorgiou, S. K., F. K. Katsaros, E. P. Favvas et al. 2012. Alginate fibers as photocatalyst immobilizing agents applied in hybrid photocatalytic/ultrafiltration water treatment processes. *Water Research* 46:1858–1872.

Park, S. and R. S. Ruoff. 2009. Chemical methods for the production of graphenes. *Nature Nanotechnology* 29:217–224.

Pelaez, M., B. Baruwati, R. S. Varma, R. Luque, and D. D. Dionysiou. 2013. Microcystin-LR removal from aqueous solutions using a magnetically separable N-doped TiO_2 nanocomposite under visible light irradiation. *Chemical Communications* 49:10118–10120.

Pelaez, M., A. A. de la Cruz, E. Stathatos, P. Falaras, and D. D. Dionysiou. 2009. Visible light-activated N-F-codoped TiO_2 nanoparticles for the photocatalytic degradation of microcystin-LR in water. *Catalysis Today* 144:19–25.

Pelaez, M., N. T. Nolan, S. C. Pillai et al. 2012. A review on the visible light active titanium dioxide photocatalysts for environmental applications. *Applied Catalysis B: Environmental* 125:331–349.

Pera-Titus, M., V. García-Molina, M. A. Baños, J. Giméneza, and S. Esplugas. 2004. Degradation of chlorophenols by means of advanced oxidation processes: A general review. *Applied Catalysis B: Environmental* 47:219–256.

Pichat, P. 2010. Some views about indoor air photocatalytic treatment using TiO_2: Conceptualization of humidity effects, active oxygen species, problem of C_1–C_3 carbonyl pollutants. *Applied Catalysis B: Environmental* 99:428–434.

Poyatos, J. M., M. M. Muñio, M. C. Almecija, J. C. Torres, E. Hontoria, and F. Osorio. 2010. Advanced oxidation processes for wastewater treatment: State of the art. *Water, Air, & Soil Pollution* 205:187–204.

Prucek R., J. Tuček, J. Kolařík et al. 2013. Ferrate (VI)-induced arsenite and arsenate removal by in situ structural incorporation into magnetic iron (III) oxide nanoparticles. *Environmental Science & Technology* 47:3283–3292.

Pugazhenthiran, N., S. Murugesan, and S. Anandan. 2013. High surface area Ag-TiO_2 nanotubes for solar/visible-light photocatalytic degradation of ceftiofur sodium. *Journal of Hazardous Materials* 263:541–549.

Qu, X., P. J. J. Alvarez, and Q. Li. 2013a. Applications of nanotechnology in water and wastewater treatment. *Water Research* 47:3931–3946.

Qu, X., J. Brame, Q. Li, and P. J. J. Alvarez. 2013b. Nanotechnology for a safe and sustainable water supply: Enabling integrated water treatment and reuse. *Accounts of Chemical Research* 46:834–843.

Qu, J., L. Shi, C. He et al. 2014. Highly efficient synthesis of graphene/MnO_2 hybrids and their application for ultrafast oxidative decomposition of methylene blue. *Carbon* 66:485–492.

Raychoudhury, T., and T. Scheytt. 2013. Potential of zerovalent iron nanoparticles for remediation of environmental contaminants in water: A review. *Water Science Technology* 68:1425–1439.

Rengifo-Herrera, J. A. and C. Pulgarin. 2010. Photocatalytic activity of N, S co-doped and N-doped commercial anatase TiO_2 powders towards phenol oxidation and *E. coli* inactivation under simulated solar light irradiation. *Solar Energy* 84:37–43.

Rincon, A. and C. Pulgarin. 2004. Field solar *E. coli* inactivation in the absence and presence of TiO_2: Is UV solar dose an appropriate parameter for standardization of water solar disinfection? *Solar Energy* 77:635–648.

Rivas, F. J., M. Carbarjo, F. J. Beltran, B. Acedo, and O. Gimeno. 2005. Perovskite catalytic ozonation of pyruvic acid in water: Operating conditions influence and kinetics. *Applied Catalysis B: Environmental* 62:93–103.

Romanos, G., C. Athanasekou, F. K. Katsaros et al. 2012. Double-side active TiO_2-modified nanofiltration membranes in continuous flow photocatalytic reactors for effective water purification. *Journal of Hazardous Materials* 211–212:304–316.

Romanos, G., C. Athanasekou, V. Likodimos, P. Aloupogiannis, and P. Falaras. 2013. Hybrid ultrafiltration/photocatalytic membranes for efficient water treatment. *Industrial & Engineering Chemistry Research* 52:13938–13947.

Samuni, A., S. Goldstein, A. Russo, J. B. Mitchell, M. C. Krishna, and P. Neta. 2002. Kinetics and mechanism of hydroxyl radical and OH-adduct radical reactions with nitroxides and with their hydroxylamines. *Journal of the American Chemical Society* 124:8719–8724.

Savage, N. and M. S. Diallo. 2005. Nanomaterials and water purification: Opportunities and challenges. *Journal of Nanoparticle Research* 7:331–342.

Schneider, R., D. Kiebling, and G. Wendt. 2000. Cordierite monolith supported perovskite-type oxides-catalysts for the total oxidation of chlorinated hydrocarbons. *Applied Catalysis B: Environmental* 28:187–195.

Schwarz, P. F., N. J. Turro, S. H. Bossmann, A. M. Braun, A.-M. A. A. Wahab, and H. Dürr. 1997. A new method to determine the generation of hydroxyl radicals in illuminated TiO_2 suspensions. *The Journal of Physical Chemistry B* 101:7127–7134.

Sharma, V. K. 2011. Oxidation of inorganic contaminants by ferrates (Fe (VI), Fe (V), and Fe (IV)—Kinetics and mechanisms—A review. *Journal of Environmental Management* 92:1051–1073.

Sharma, V. K. 2013. Ferrate (VI) and ferrate (V) oxidation of organic compounds: Kinetics and mechanism. *Coord. Chemical Reviews* 257:495–510.

Sharma, V. K., F. Liu, S. Tolan, M. Sohn, H. Kim, and M. A. Oturan. 2013. Oxidation of β-lactam antibiotics by ferrate (VI). *Chemical Engineering Journal* 221:446–451.

Sharma, V. K., T. M. Triantis, M. G. Antoniou et al. 2012. Destruction of microcystins by conventional and advanced oxidation processes: A review. *Separation and Purification Technology* 91:3–17.

Sharma, V. K., R. Zboril, and T. J. McDonald. 2014. Formation and toxicity of brominated disinfection byproducts during chlorination and chloramination of water: A review. *Journal of Environmental Science and Health—Part B Pesticides, Food Contaminants, and Agricultural Wastes* 49:212–228.

Shelley S. A. 2005. Nanotechnology: Turning basic science into reality. In *Nanotechnology: Environmental Implications and Solutions*, eds. L. Theodore and R. G. Kunz, 61–107. John Wiley & Sons, Inc.

Shi, P., R. Su, F. Wana, M. Zhu, D. Li, and S. Xu. 2012a. Co_3O_4 nanocrystals on graphene oxide as a synergistic catalyst for degradation of Orange II in water by advanced oxidation technology based on sulfate radicals. *Applied Catalysis B: Environmental* 123–124: 265–272.

Shi, P., R. Su, S. Zhu, M. Zhu, D. Li, and S. Xu. 2012b. Supported cobalt oxide on graphene oxide: Highly efficient catalysts for the removal of Orange II from water. *Journal of Hazardous Materials* 229–230:331–339.

Shi, P., X. Dai, H. Zheng, D. Li, W. Yao, and C. Hu. 2014. Synergistic catalysis of Co_3O_4 and graphene oxide on Co_3O_4/GO catalysts for degradation of Orange II in water by advanced oxidation technology based on sulfate radicals. *Chemical Engineering Journal* 240:264–270.

Shukla, P. R., S. Wang, H. Sun, H. M. Ang, and M. Tadé. 2010. Activated carbon supported cobalt catalysts for advanced oxidation of organic contaminants in aqueous solution. *Applied Catalysis B: Environmental* 100:529–534.

Si, Y, and E. T. Samulski. 2008. Synthesis of water soluble graphene. *Nano Letters* 8: 1679–1682.

Su, F. Z., S. C. Mathew, G. Lipner et al. 2012. mpg-C_3N_4-catalyzed selective oxidation of alcohols using O_2 and visible light. *Journal of the American Chemical Society* 132: 16299–16301.

Sunnotel, O., R. Verdoold, P. S. M. Dunlop et al. 2010. Photocatalytic inactivation of *Cryptosporidium parvum* on nanostructured titanium dioxide films. *Journal of Water and Health* 8:83–91.

Thakur, R. S., R. Chaudhary, and C. Singh. 2010. Fundamentals and applications of the photocatalytic treatment for the removal of industrial organic pollutants and effects of operational parameters: A review. *Journal of Renewable Sustainable Energy* 2:042701.

Thomas, A., A. Fischer, F. Goettmann et al. 2008. Graphitic carbon nitride materials: Variation of structure and morphology and their use as metal-free catalysts. *Journal of Materials Chemistry* 18:4893–4908.

Torimoto, T., Y. Okawa, N. Takeda, and H. Yoneyama. 1997. Effect of activated carbon content in TiO_2-loaded activated carbon on photodegradation behaviors of dichloromethane. *Journal of Photochemistry and Photobiology A: Chemistry* 103:153–157.

Turki, A., H. Kochkara, I. García-Fernández et al. 2013. Solar photocatalytic inactivation of *Fusarium solani* over TiO_2 nanomaterials with controlled morphology-Formic acid effect. *Catalysis Today* 209:147–152.

Uchino, T., H. Tokunaga, M. Ando, and H. Utsumi. 2002. Quantitative determination of OH radical generation and its cytotoxicity induced by TiO_2–UVA treatment. *Toxicology in Vitro* 16:629–635.

Vilar, V. J. P. A. I. E. Gomes, V. M. Ramos, M. I. Maldonado, and R. A. R. Boaventur. 2009. Solar photocatalysis of a recalcitrant coloured effluent from a wastewater treatment plant. *Photochemical & Photobiological Sciences* 8:691–698.

Wainwright, M. 2000. Methylene blue derivatives—Suitable photoantimicrobials for blood product disinfection? *International Journal of Antimicrobial Agents* 16:381–394.

Wang, X. C., K. Maeda, A. Thomas et al. 2009. A metal-free polymeric photocatalyst for hydrogen production from water under visible light. *Nature Materials* 8:76–80.

Wang, Y., H. Zhao, M. Li, J. Fan, and G. Zhao. 2014. Magnetic ordered mesoporous copper ferrite as a heterogeneous Fenton catalyst for the degradation of imidacloprid. *Applied Catalysis B: Environmental* 147:534–545.

Wang, C., L. Zhu, C. Chang, Y. Fu, and X. Chu. 2013. Preparation of magnetic composite photocatalyst $Bi_2WO_6/CoFe_2O_4$ by two-step hydrothermal method and its photocatalytic degradation of bisphenol A. *Catalysis Communications* 37:92–95.

WHO. 2014. Progress on drinking water and sanitation. 2014 update.

Xiong, P., Y. Fu, L. Wang, and X. Wang. 2012. Multi-walled carbon nanotubes supported nickel ferrite: A magnetically recyclable photocatalyst with high photocatalytic activity on degradation of phenols. *Chemical Engineering Journal* 195:149–157.

Yang, Y., W. Guo, Y. Guo, Y. Zhao, X. Yuan, and Y. Guo. 2014. Fabrication of Z-scheme plasmonic photocatalyst Ag@AgBr/g-C_3N_4 with enhanced visible-light photocatalytic activity. *Journal of Hazardous Materials* 271:150–159.

Yang, H. G., G. Liu, S. Z. Qiao et al. 2009. Solvothermal synthesis and photoreactivity of anatase TiO_2 nanosheets with dominant {001} facets. *Journal of the American Chemical Society* 131:4078–4083.

Yang, B., and G. Ying. 2013. Oxidation of benzophenone-3 during water treatment with ferrate(VI). *Water Research* 47:2458–2466.

Yap, P.-S., Y.-J. Cheah, M. Srinivasan, and T.-T. Lim. 2012. Bimodal N-doped P25-TiO_2/AC composite: Preparation, characterization, physical stability, and synergistic adsorptive-solar photocatalytic removal of sulfamethazine. *Applied Catalysis A: General* 427–428:125–136.

Yap, P.-S. and T.-T. Lim. 2011. Effect of aqueous matrix species on synergistic removal of bisphenol-A under solar irradiation using nitrogen-doped TiO_2/AC composite. *Applied Catalysis B: Environmental* 101:709–717.

Yap, P.-S. and T.-T. Lim. 2012. Solar regeneration of powdered activated carbon impregnated with visible-light responsive photocatalyst: Factors affecting performances and predictive model. *Water Research* 46:3054–3064.

Yap, P.-S., T.-T. Lim, M. Lim, and M. Srinivasan. 2010. Synthesis and characterization of nitrogen-doped TiO_2/AC composite for the adsorption–photocatalytic degradation of aqueous bisphenol-A using solar light. *Catalysis Today* 151:8–13.

Yap, P.-S., T.-T. Lim, and M. Srinivasan. 2011. Nitrogen-doped TiO_2/AC bi-functional composite prepared by two-stage calcination for enhanced synergistic removal of hydrophobic pollutant using solar irradiation. *Catalysis Today* 161:46–52.

Zhang, J. and Y. Nosaka. 2013. Quantitative detection of OH radicals for investigating the reaction mechanism of various visible-light TiO_2 photocatalysts in aqueous suspension. *The Journal of Physical Chemistry C* 117:1383–1391.

Zhang, P. F., Y. T. Gong, H. R. Li, Z. R. Chen, and Y. Wang. 2013b. Selective oxidation of benzene to phenol by $FeCl_3$/mpg-C_3N_4 hybrids. *RSC Advances* 3:5121–5126.

Zhang, T., H. Zhu, and J.-P. Croué. 2013a. Production of sulfate radical from peroxymonosulfate induced by a magnetically separable $CuFe_2O_4$ spinel in water: Efficiency, stability, and mechanism. *Environmental Science & Technology* 47:2784–2791.

Zhao, C., M. Pelaez, D. D. Dionysiou, S. C. Pillai, J. A. Byrne, and K. E. O'Shea. 2014. UV and visible light activated TiO_2 photocatalysis of 6-hydroxymethyluracil, a model compound for the potent cyanotoxin cylindrospermopsin. *Catalysis Today* 224:70–76.

Zheng, S., Y. Cai, and K. E. O'Shea. 2010. TiO_2 photocatalytic degradation of phenylarsonic acid. *Journal of Photochemistry and Photobiology A* 210:61–68.

Zheng, S., W. Jiang, Y. Cai, D. D. Dionysiou, and K. E. O'Shea. 2014. Adsorption and photocatalytic degradation of aromatic organoarsenic compounds in TiO_2 suspension. *Catalysis Today* 224:83–88.

6 Industrial Water Usage and Wastewater Treatment/Reuse

Tapas K. Das

CONTENTS

6.1 INTRODUCTION

Industry and communities around the world are increasingly turning to water recycling and reuse for a wide range of purposes. The drivers are varied and include the need to augment limited water supplies, reduce nutrients in treated effluent, maintain ecological balance, and reduce the costs of purchased and treated water, among others. In the industry, another incentive for water reuse is energy efficiency, as part of the water–energy nexus. Water reuse can be energy intensive depending on the level of treatment required, and a life cycle cost (LCC) analysis is required to compare overall resource costs with the cost of alternative water supplies.

6.1.1 Water Purification Treatment Technology Options

Treatment technologies are available to achieve virtually any desired level of water quality, and the level of treatment required depends on the reuse application. For industrial uses of reclaimed water, conventional processes involving secondary treatment, filtration, and disinfection steps are sufficient to achieve the necessary water quality. In applications with the potential for human contact or with sensitive equipment, advanced treatment may be required.

Water treatment or *purification* is the process of removing undesirable chemicals, biological contaminants, suspended solids, and gases from contaminated water. The goal of this process is to produce water fit for a specific purpose. Most water is disinfected for human consumption (drinking water), but water purification may also be designed for a variety of other purposes, including meeting the requirements of medical, pharmacological, chemical, and industrial applications. In general, the methods used include physical processes such as filtration, sedimentation, and distillation; biological processes such as slow sand filters or biologically active carbon; and chemical processes such as flocculation and chlorination and the use of electromagnetic radiation such as ultraviolet (UV) light.

The purification process of water may reduce the concentration of particulate matter including suspended particles, parasites, bacteria, algae, viruses, fungi, and a range of dissolved and particulate material derived from the surfaces that water may have made contact with after falling as rain.

Wastewater treatment for industrial reuse often employs processes include screening, equalization, and primary clarification serve as pretreatment steps that allow subsequent treatment processes to operate more efficiently.

Dissolved air floatation (DAF). When particles to be removed do not settle out of solution easily, DAF is often used. Water supplies that are particularly vulnerable to unicellular algae blooms and supplies with low turbidity and high color often employ DAF. After coagulation and flocculation processes, water flows to DAF tanks where air diffusers on the tank bottom create fine bubbles that attach to floc resulting in a floating mass of concentrated floc. The floating floc blanket is removed from the surface and clarified water is withdrawn from the bottom of the DAF tank.

Membrane filtration. Membrane filtration technologies are becoming much more acceptable for use as solid separation processes upstream of biological treatment systems. Clarifiers depend on biomass settling; if the biomass does not settle well or if hydraulic flows vary, clarifier operation is upset and becomes inefficient. Also, membranes are widely used for filtering both drinking water and municipal wastewater. For drinking water, membrane filters can remove virtually all particles larger than 0.2 μm—including *giardia* and *cryptosporidium*. Membrane filters are an effective form of tertiary treatment when it is desired to reuse the water for industry, for limited domestic purposes, or before discharging the water into a river that is used by towns further downstream. They are widely used in industry, particularly for beverage preparation (including bottled water). However, no filtration can remove substances that are actually dissolved in the water such as phosphorus, nitrates, and heavy metal ions.

The membranes used in industrial water reuse include microfilters (MF), ultrafilters (UF), nanofilters (NF), and reverse osmosis (RO) membrane. Ultrafiltration membranes use polymer membranes with chemically formed microscopic pores that can be used to filter out dissolved substances avoiding the use of coagulants. The type of membrane media determines how much pressure is needed to drive the water through and what sizes of microorganisms can be filtered out.

Ion exchange. Ion exchange systems use ion exchange resin- or zeolite-packed columns to replace unwanted ions. The most common case is water softening consisting of removal of Ca^{2+} and Mg^{2+} ions and replacing them with benign (soap friendly) Na^+ or K^+ ions. Ion exchange resins are also used to remove toxic ions such as nitrite, lead, mercury, arsenic, and many others.

Precipitative softening. Water rich in hardness (calcium and magnesium ions) is treated with lime (calcium oxide) or soda ash (sodium carbonate) to precipitate calcium carbonate out of solution, utilizing the common-ion effect.

Electrodeionization. Water is passed between a positive electrode and a negative electrode. Ion exchange membranes allow only positive ions to migrate from the treated water toward the negative electrode and permit only negative ions to migrate toward the positive electrode. High-purity deionized water is produced with a little worse degree of purification in comparison with ion exchange treatment. Complete removal of ions from water is regarded as electrodialysis. The water is often pretreated with an RO unit to remove nonionic organic contaminants.

Reverse osmosis. RO is a water purification technology that uses a semipermeable membrane. This membrane technology is not a proper filtration method. In RO, an applied pressure is used to overcome osmotic pressure, a colligative property, that is driven by chemical potential, a thermodynamic parameter. RO can remove many types of molecules and ions from solutions and is used in both industrial processes and the production of potable water. The result is that the solute is retained on the pressurized side of the membrane and the pure solvent is allowed to pass to the other side. To be "selective," this membrane should not allow large molecules or ions through the pores, but should allow smaller components of the solution (such as the solvent) to pass freely.

In the normal osmosis process, the solvent naturally moves from an area of low solute concentration (high water potential), through a membrane, to an area of high solute concentration (low water potential). The movement of a pure solvent is driven to reduce the free energy of the system by equalizing solute concentrations on each side of a membrane, generating osmotic pressure. Applying an external pressure to reverse the natural flow of pure solvent, thus, is RO. The process is similar to other membrane technology applications. However, key differences are found between RO and filtration. The predominant removal mechanism in membrane filtration is straining, or size exclusion; hence, the process can theoretically achieve perfect exclusion of particles regardless of operational parameters such as influent pressure and concentration. Moreover, RO involves a diffusive mechanism, so that separation efficiency is dependent on solute concentration, pressure, and water flux rate. RO is most commonly known for its use in drinking water purification from seawater, removing the salt and other effluent materials from the water molecules.

Advanced oxidation processes, in a broad sense, refer to a set of chemical treatment procedures designed to remove organic (and sometimes inorganic) materials in water and wastewater by oxidation through reactions with hydroxyl radicals (·OH). In real-world applications of wastewater treatment, however, this term usually refers more specifically to a subset of such chemical processes that employ ozone (O_3), hydrogen peroxide (H_2O_2) or UV light with titanium dioxide (UV/Ti), and a variety of Fenton reactions using Fe/H_2O_2, Fe/Ozone, and Fe/H_2O_2/UV.

6.1.2 Use of Treated Municipal Wastewater as Power Plant Cooling System Makeup Water: Tertiary Treatment versus Expanded Chemical Regimen for Recirculating Water Quality Management

6.1.2.1 Introduction

Every day, water-cooled thermoelectric power plants in the United States withdraw from 60 billion to 170 billion gallons of freshwater from rivers, lakes, streams, and aquifers, and consume from 2.8 billion to 5.9 billion gallons of that water. Freshwater withdrawals for cooling in thermoelectric power production account for approximately 40% of all withdrawals, essentially the same amount of withdrawals for agricultural irrigation, as documented by the US Geological Survey. Sustained droughts nationwide underscore the critical need to think about using treated municipal wastewater for use in cooling in electric power generation. It needs a great deal of water for electric power production, to condense stream in the power plant stream cycle. Air cooling is possible, but it is more costly and less efficient. Water will continue to be the preferred coolant for new thermoelectric power plants (Dzombak 2013).

6.1.2.2 Motivation for the Project

Increase in demand for electricity brings with it an increase in water needed for cooling. The cooling of thermoelectric power plants accounts for 41% of all freshwater withdrawal in the United States, that is, approximately the same amount of water as is withdrawn for agricultural irrigation. Some areas of the United States have little or no freshwater available for use. Alternative sources of water are needed for new electric power production. The US Department of Energy (DOE) has been conducting and sponsoring research to investigate the feasibility and costs of using alternative sources of water for power plant cooling, especially in recirculating cooling systems, which are required for most new power production in the United States.

6.1.2.3 Goals and Highlights of the Project

Treated municipal wastewater is a common, widely available alternative source of cooling water for thermoelectric power plants across the United States, as determined in a predecessor DOE project (2006–2009) by the project team. However, the biodegradable organic matter, ammonia-nitrogen, carbonate, and phosphates in the treated wastewater pose challenges with respect to enhanced biofouling, corrosion, and scaling, respectively. The overall objective of this study was to evaluate the benefits and LCCs of implementing tertiary treatment of secondary treated municipal wastewater before use in recirculating cooling systems.

The study comprised bench- and pilot-scale experimental studies with three different tertiary treated municipal wastewaters and life cycle costing and environmental analyses of various tertiary treatment schemes. Sustainability factors and metrics for reuse of treated wastewater in power plant cooling systems were also evaluated. The three tertiary treated wastewaters studied were secondary treated municipal wastewater subjected to acid addition for pH control (MWW_pH), secondary treated municipal wastewater subjected to nitrification and sand filtration (MWW_NF), and secondary treated municipal wastewater subjected to nitrification, sand filtration, and GAC adsorption (MWW_NFG).

Tertiary treatment was determined to be essential to achieve appropriate corrosion, scaling, and biofouling control for use of secondary treated municipal wastewater in power plant cooling systems. The ability to control scaling, in particular, was found to be significantly enhanced with tertiary treated wastewater compared to secondary treated wastewater. MWW_pH treated water (adjustment to pH 7.8) was effective in reducing scale formation, but increased corrosion and the amount of biocide required to achieve appropriate biofouling control. Corrosion could be adequately controlled with tolytriazole addition (4–5 ppm TTA), however, which was the case for all of the tertiary treated waters. For MWW_NF treated water, the removal of ammonia by nitrification helped reduce the corrosivity and biocide demand. Additional GAC adsorption treatment, MWW_NFG, yielded no net benefit. For all of the tertiary treatments, biofouling control was achievable, and most effectively with preformed monochloramine (2–3 ppm) in comparison with NaOCl and ClO_2.

LCC analyses were performed for the tertiary treatment systems studied experimentally and for several other treatment options. A public domain conceptual costing tool (LC^3 model) was developed for this purpose. MWW_SF (lime softening and sand filtration) and MWW_NF were the most cost-effective treatment options among the tertiary treatment alternatives considered because of the higher effluent quality with moderate infrastructure costs and the relatively low doses of conditioning chemicals required (Dzombak 2013).

Life cycle inventory (LCI) analysis along with integration of external costs of emissions with direct costs was performed to evaluate relative emissions to the environment and external costs associated with construction and operation of tertiary treatment alternatives. Integrated LCI and LCC analysis indicated that MWW_NF and MWW_SF alternatives exhibited moderate external impact costs with moderate infrastructure and chemical conditioner dosing, which makes them (especially MWW_NF) better treatment alternatives from the environmental sustainability perspective since they exhibited minimal contribution to environmental damage from emissions (Dzombak 2013).

6.1.2.4 Key Points

1. This study undertook a comprehensive, integrated approach by looking at all aspects of the water quality control problem in a recirculating cooling system employing treated municipal wastewater as makeup water, and we determined optimal approaches for tertiary treatment considering both direct economic costs and environmental impacts of alternative water treatment/management approaches in an integrated manner.
2. The work included pilot-scale demonstrations in the field, in addition to laboratory and modeling work, in partnership with a municipal wastewater treatment facility.
3. In regard to originality and innovation, this study was the first comprehensive research study in the public domain of the use of treated municipal wastewater as makeup water in recirculating cooling systems of electric power plants.
4. The challenges of using treated municipal wastewater in power plant cooling systems are many, and include the technical challenges of controlling corrosion, scaling, and biofouling in a recirculating cooling system employing a low-quality makeup water; the challenge of determining capital and operating costs in a complex operational environment in which the water is being concentrated four times or more in the recirculating cooling system; and the challenge of assessing environmental and social risks and benefits for use of treated wastewater as cooling system makeup water. It was a complex problem and further complicated by the economic, social acceptance, and sustainability issues involved.
5. Alternative sources of water are needed for new power production in regions without new sources of freshwater available. Our research will help advance the use of treated municipal wastewater in electric power production and thus help contribute to economic development. Further, in this research, they evaluated explicitly the relative sustainability of various water treatment/management alternatives by considering environmental impact and social acceptance factors in addition to direct economic costs of the alternatives.

6.2 WHY AIM FOR ZERO WATER DISCHARGE?

Zero discharge is applied industrial ecology at the manufacturing level: a practical approach with a concrete methodology to redesign industrial processes so that they have no discharges. This section describes some specific zero discharge processes and technologies that have been successfully operating.

The increasing scarcity of water coupled with the escalating cost of freshwater and its treatment has prompted the industry to think of water conservation and recycle in most industries; the case for zero discharge is neither compelling nor far-fetched, though many a time regulatory authorities dictate the implementation of a zero-discharge system. It is prudent for existing facilities to develop a systematic approach to effective and efficient plant-wide water management rather than to implement full-fledged zero-discharge systems. Later in the chapter (Section 6.3), we present a case study of a cement plant utilizing captive power plant (CPP) effluents.

Industrial operations use water for processing, conveying, heating, cooling, steam production, and housekeeping. The bulk of the water consumed (85%–90%) by the industry is discharged as process wastewater. The rising price of freshwater and stringent environmental regulations with respect to effluent discharge are now compelling industries to consider reduction in water consumption, as well as recovery and recycling of water.

While there are several practical definitions of zero discharge, a zero-discharge system is most commonly defined as one from which no water effluent stream is discharged by the processing site. All the wastewater after secondary or tertiary treatment is converted to solid waste by evaporation processes, such as brine concentration followed by crystallization or drying. The solid waste may then be landfilled. It is, however, important to be clear in one's definition of zero discharge, as and when mentioned by any regulatory body (Byers 1995; Dalan 2000; Goldblatt et al. 1993; Kiranmayee and Manian 2000; Rosain 1993).

6.2.1 Advantages and Disadvantages

Zero-discharge systems offer many advantages. The principal ones are as follows:

- Minimize consumption of freshwater
- Allow the recovery of valuable resources
- Reduce volumes for sludge handling
- Improve product quality by yielding water of better quality than raw water
- Facilitate site selection (since site location is less limited if a receiving waterway is not needed for wastewater effluent)

The principal disadvantages can be described briefly as maintenance problems, reduced plant reliability, and trace chemicals.

Scaling, especially on heat transfer surfaces, is quite common as the salt concentration of the water increases. It is also quite common for the resulting water quality to be incompatible with metallurgy selected for different conditions. Efforts made to combat scale and corrosion through local pH adjustments or changed flow configurations can be handled only temporarily.

With respect to reduced plant reliability, it must be borne in mind that a failure or shutdown in the plant could curtail water availability or change water quality in a way that affects operation in another part of the plant.

Finally, water reuse may cause a buildup of trace metals and organic solvents in the water system, again necessitating increased maintenance.

6.2.2 Design Principles

To implement a successful zero-discharge system, the design must accomplish the following:

Minimization of raw water consumption—This can be accomplished by reusing plant effluents or using secondary treated effluent from a nearby municipal wastewater treatment plant (WWTP), and so on, resulting in reduced freshwater makeup for the plant.

Source reduction—Vary the process design parameters and raw materials (wherever applicable) to minimize the wastewater generation. A careful planning in the selection of process, equipment, raw materials, and operating conditions can reduce the wastewater flows.

Segregation and reuse of wastewater streams—Segregated treatment may be particularly effective if the removal of only one or two contaminants will allow the wastewater stream to be reused directly or will reduce the size or complexity of the end-of-the-pipe treatment system. Usually, an integrated water reuse system will likely employ a combination of segregated and end-of-pipe treatment system to achieve cost-effective water reuse.

Advanced treatment and processing—Advanced processing that removes suspended solids as well as dissolved solids may produce boiler quality water from the wastewater. These treatments may include precipitation softening, multimedia filtration, carbon adsorption, deionization, RO, or distillation.

Disposal—Once all steps to minimize and reuse wastewater streams are taken, the remaining wastewater is normally treated and then disposed of. However, if an ideal zero liquid discharge (ZLD) system is to be implemented, the treated water is completely reused.

With these general principles in mind, we are ready to look at three case studies involving wastewater recovery in three very different industrial areas: the manufacture of cement, automobiles, and chemicals.

Also essential to good design is consideration of the economics of water reuse. This includes the availability and cost of supply water, restrictions on and costs of discharge water, recycle stream characteristics and effects on production or product quality, and purchase and operating costs of a water purification system. In the case of an existing plant, complete implementation of true ZLD entails extensive repiping or costly unit operations.

6.3 CASE STUDY: A CEMENT PLANT IN INDIA

Industrial wastewater recovery and reuse systems are cropping up all over the world. This chapter's first case study explores reuse of effluents from a CPP used in the making of cement in India.

The unit studied is a 3 × 23 MW coal-based CPP. The environmental clearance document for the project stated that all effluents generated in the plant activities were to be collected in the central effluent treatment plant and treated to ensure adherence to specified standards. It was expected that the concept of zero discharge would be adopted.

The detailed engineering specifications revealed that the quantity of effluent, mainly the cooling tower blowdown, was too large to be used entirely in the cement plant and the CPP together; therefore, adopting a zero-discharge system would be extremely difficult. Other alternatives, such as using the effluent in neighborhood industries, were evaluated, but were not practical. The installation of a complete zero-discharge system would have involved adopting technologies such as RO and brine concentrators. While the RO process recovers approximately 70%–85% water based on feed water characteristics, it leaves behind a very high total dissolved solids (TDS) effluent, a briny substance that can be disposed of only by means of brine concentrators, which are expensive devices that would have exceeded the project's budget (Erkman and Ramaswamy 2001).

6.3.1 Review of Existing Water Balance at the Plant

An extensive review was conducted on the baseline information on existing cement plant water use, the wastewater generation, and the existing wastewater treatment facilities and the water quality required for various plant processes and equipment. A large water tank receives makeup water from a nearby irrigation canal for the cement plant. The CPP receives water from a nearby river. There are mainly two areas in the cement plant where large quantities of water are required. These are the gas conditioning towers (GCTs) and the cement mills. The plant owner provided the following rates of water consumption and loss:

- Gas conditioning tower in phase 1—1000 m^3/day
- Gas conditioning tower in phase 2—1100 m^3/day
- Flow rate for four cement mills—576 m^3/day

All the other water used by the facility (e.g., for kiln bearing cooling, packing plant compressors, etc.) is under circulation. Also to be considered are domestic, horticultural, and nondocumented water uses (service waters, washing, etc.) and evaporative losses from the large open water tank and cooling towers. Since the wastewaters were intended to be reused after appropriate treatment in critical process operations such as in GCTs, the effluents were restricted to cooling tower blowdown, the boiler blowdown, and the neutralized effluent from demineralizer regeneration; strict monitoring and control over the quality of combined effluents was possible. The use of service wastewater and any other effluent for horticulture or dust suppression was possible because these discharges were small and well within the norms laid out for discharge onto land. The first approach was to segregate all the streams so that recycling could be accomplished with as little treatment as possible and the effluents could be reused with a minimum of operational complexities. Combining all the effluents in a single effluent treatment plant would have the following disadvantages:

- The effluent treatment plant would have to be large. This is because a large amount of effluents (2350 m^3/day) was to be handled with only three cycles of concentration (COC).
- Operational complexities would be severe.
- Maintaining and monitoring water quality for a particular process unit would be difficult.

Although the cooling tower blowdown quality was sufficient for use directly in the cement mills, it would have to be treated in accordance with the process licensor's specifications before it could be used in the GCTs. Analysis of the licensor's data indicated that while the TDS of cooling tower blowdown (the chlorides as well as the sulfates) were well within specifications, the hardness and alkalinity in the stream were of concern. To validate the consequences of this variation, water samples from GCTs of various cement plants were analyzed. Although GCT water analysis results from some cement plants were off spec, no cement plant had serious operating problems.

6.3.2 Effluent Reuse Options

Two approaches to the reuse of effluents were considered: the use of scaling/corrosion inhibitors and the use of a softener. Both measures are intended to reduce the hardness and alkalinity of the cooling tower blowdown. Blowdown was an area that could not be overlooked because the values developed in the process licensor's analysis were too high to meet the specifications.

The idea of using scaling and corrosion inhibitors was discarded because the vendors lacked experience with this kind of service. Thus, bids were obtained for softening equipment, which would allow the blowdown to be treated by both ion exchange and lime soda softening. However, the ion exchange system would produce a regeneration effluent of approximately 180 m^3/day with a very high TDS (11,000 mg/L), introducing, in turn, a very difficult disposal problem. The lime soda system, on the other hand, results in huge quantities of sludge, which would have to be removed to a landfill.

In addition to hardness and alkalinity, frequent problems with excessive chlorides (>13 mg/L as Cl^-) had led cement plant process personnel to suggest that chloride reduction be incorporated into the design. An RO system could have served this purpose, but it would have been prohibitively expensive. Moreover, the creation of another waste stream was undesirable because the environmental clearance mandated a common effluent collection point from which all the power plant effluents could be recycled or reused.

Ultimately, the softener option was selected, but without RO, to accommodate the requirement to use a single treatment plant for all the effluents. In addition, the COC for the cooling tower, originally envisaged as 3, was set at 6. As a result, the amount of effluent was greatly reduced, but the concentration of pollutants increased accordingly.

6.3.3 Effluent Treatment Plant Design

In view of the intermittent nature of blowdowns, it was necessary to ensure that all the effluents had sufficient retention time to attain uniform concentration, and therefore, two RCC effluent treatment pits of 500 m^3 each were provided. The combined effluent quality is well within specified norms for discharge into inland surface water or for irrigation. Only a pH adjustment was found to be necessary. The final expected quantity and quality of effluents is given in Table 6.1. The final effluent quality is maintained such that 2100 mg/L TDS is never exceeded for discharge onto land for irrigation, especially for purposes of gardening. The effluents from the CPP could be utilized as summarized in Table 6.2. Seasonal variations were also considered.

TABLE 6.1
Water Balance for 3 × 23 MW Coal-Based Captive Power Plant

Effluent	Flow (m^3/Day)	pH	Temperature (°C)	TSS (mg/L)	TDS (mg/L)	Hardness (mg/L)	Alkalinity (mg/L)	Cl (mg/L)
Typical raw water quality		8	Ambient (30–35)	5	286	146	194	19
Cooling tower blowdown	1080	9	40	3	2100	1686	1164	115
Boiler blowdown	206	10.8–11.8	100 max	–	200	0	20	0
DM regeneration	21	7	30	7	4700	1800	1950	2272
Total effluents (mixed)	1307	9.0–11.0	47.00	2.59	1842.31	1422.10	996.32	131.53

Source: Kiranmayee, V. and Manian, C. V. (2000). Zero discharge systems—A practical approach. Indian Chemical Engineering Congress, Science City, Kolkata, India, Paper # WTR21 (Dec. 18–21).

TABLE 6.2
Water Requirements in Various Places

Effluent	Flow (m^3/Day)		pH	Temperature (°C)	TSS (mg/L)	TDS (mg/L)	Hardness (mg/L)	Alkalinity (mg/L)	Cl (mg/L)
Gas conditioning tower	2100		Neutral	Amb.	NIL	1000	130	200	55
Cement mills	560		Neutral	Amb.	No limits		NA	NA	NA
Dust suppr. in mines	120								
Coal quench	Summer	Winter							
In cement plant	90	45	Neutral	Amb.	No limits			NA	
In power plant	10	5	Neutral	Amb.	No limits			NA	
Total—coal quench	100	50							
Horticulture—total	592	295	5.5–9	Amb.	200	2100	No limits		600
Total, max	3472	3125							
Water req. without GCT	1372	1025							
Power plant effluents	1307	1307							

Source: Kiranmayee, V. and Manian, C. V. (2000). Zero discharge systems—A practical approach. Indian Chemical Engineering Congress, Science City, Kolkata, India, Paper # WTR21 (Dec. 18–21).

Thus, for a 3 × 23 MW CPP, the effluent quantity would be approximately 1307 m^3/day, with the cement plant water requirement being the same at 1372 m^3/day in summer. In summer in this part of India, all power plant effluents from cement mills can be used; in winter, there is an excess of effluent, which may be diverted for use in GCTs with additional dilution with freshwater. Thus, the cement plant at Mithapur is an ideal example of an economically viable zero-discharge system.

6.3.4 Case Study: DaimlerChrysler's Zero-Discharge Wastewater Treatment Plant in Mexico

DaimlerChrysler's production complex in Toluca, Mexico, home of the Chrysler PT Cruiser, has received much attention not only because of its in-demand product but also because of its state-of-the-art zero-discharge WWTP. Located 37 miles north of Mexico City, Toluca suffered for years from a worsening water shortage owing to urban sprawl, regional drought, and increased industrial activity. The city is one of the leading producers of beverages, textiles, and automobiles in Mexico, as well as a center for food processing.

DaimlerChrysler, one of Mexico's largest manufacturers, mindful of the mounting strain on the world's natural resources, has consistently sought ways to decrease operational waste, reduce costs, and increase process efficiencies. Upon locating in Mexico, the automaker began to study the region's rapidly dropping aquifer, hoping to be able to minimize the stress on this valuable resource, yet keep its operations in compliance with the federal government's water quality standards.

In 1999, the company hit upon a solution. It would build its own $17 million WWTP that would treat sanitary and manufacturing-process water generated by the facility's four separate plants—engine, transmission, stamping, and assembly. To make this WWTP truly state of the art, a comprehensive ZLD system would be installed. By using a ZLD system, the Toluca complex would avoid further depleting the local aquifer and the environmentally friendly and cost-efficient system would discharge no process water, but rather would recycle it to use throughout the facility. It was projected that implementing a ZLD solution and thus reusing water could extend the facility's life without disrupting production and causing costly overhauls.

6.3.5 System Design

DaimlerChrysler put in operation two kinds of ZLD systems. The first uses RO to produce a concentrate of TDS, which is sent to a large evaporator and, eventually, onto a lagoon or solar evaporator pond. Used in dry, arid areas of low elevation, this system is frequently found in the WWTPs of Northern Mexico's automotive facilities. The other system, used at the Toluca facility, softens and removes silica from the RO concentrate through microfiltration before sending the water onto another RO unit where it is further concentrated. Water is then returned and blended with the water from the first-stage water, where the concentrate is sent onto either an evaporator or a crystallizer to dry TDS to powder and eliminate the need to dispose of liquid.

In essence, industrial plants with ZLD installations can expect to recover nearly 100% of water that would otherwise be discharged to the environment as wastewater. At the Toluca facility, the WWTP recovers 95% or more of the water used for processing, with a recovery rate of up to 237,500 gallons per day (gpd). In actuality, the ZLD installation at the Toluca facility is two separate systems: a sanitary water system that biologically treats wastewater from the complex's restrooms, showers, cafeterias, and other domestic areas, and a manufacturing-process water system that chemically treats wastewater mixed with heavy metals and paint from the assembly plant. The latter also treats wastewater containing emulsified and soluble oils from the facility's stamping, transmission, and engine plants.

In the sanitary water system, domestic water is collected and sent through a screening mechanism before moving onto the biological treatment system's equalization tank, ensuring a constant, even flow of water through the system. This water is then passed through jet aeration sequential batch reactors that treat the water with microorganisms and air to reduce the biological and chemical oxygen demands (BOD and COD), as well as suspended solids. The complex uses 150,000 to 200,000 gpd of disinfected water to irrigate its landscape. The microorganisms and solids recovered from the batch reactors are then sent through a sludge digester and eventually a filter press that eliminates the water. While the dewatered sludge is used as fertilizer, the filtered water reenters the system.

Wastewater from the Toluca facility's three machining plants is directed through the manufacturing-process system where it is first chemically treated, passing through a filtering screen. In a separate tank, chemicals are used to de-emulsify the free-floating oils that comprise most of the waste. Afterward, the oils are removed and stored in another tank before disposal. The process water from the machining plants is then mixed with water from the assembly plant that contains residue from the spray painting, phosphating, E-coating, and body-wash operations. Upon being mixed with a combination of ferric chloride, lime, and magnesium oxide, metal pollutants and silica are rendered insoluble and turned into sludge that is removed and sent to a landfill. Then, to further lower the proportion of unwanted organic compounds, the water is pumped to a biological system that reduces the BOD to 20 to 30 ppm.

6.3.6 Results

Since installing the wastewater recovery system, the Toluca facility has noted several benefits, including decreased production and operation costs, reduced aquifer use, better environmental friendliness, and greater employee safety (Zacerkowny 2002). Moreover, the integrated system helps preserve the environment, is safe for employees to work with, and provides almost 7000 jobs to local residents. The Toluca industrial complex uses approximately 250,000 gpd of water, recovering more than 95% of its processing water. The ZLD system allows the facility to treat more than 550,000 gpd, significantly reducing the amount of water that must be drawn from the local aquifer. Using treated water might also extend the life of the facility's equipment, as the salt content of the processed industrial water is much lower than that of the aquifer.

6.4 CASE STUDY: ZERO EFFLUENT SYSTEMS AT FORMOSA PLASTICS MANUFACTURING, TEXAS

The Formosa Plastics Complex is a wholly owned subsidiary of the Formosa Plastics Corporation, USA, with operations at Point Comfort, Texas. The company is a vertically integrated plastics manufacturer whose core business is the production and processing of common chemicals and plastic resins.

6.4.1 Complying with ISO 14001

Formosa Plastics' complex in Point Comfort was the first major chemical plant in the United States to be certified as complying with ISO 14001, the series of international standards developed for managing environmental impacts. The standards address six distinct, but related, components that together form the basis of a comprehensive environmental management system. To be in compliance, the company's environmental management program must include a specific plan that describes actions proposed to meet each objective and target, the person(s) responsible for meeting each objective, and the time schedule for attaining each target (Delaney and Schiffman 1997).

In addition to meeting the ISO requirements, the company entered into a pact that has become a model for good industry–community relations, the "Wilson–Formosa Zero Discharge Agreement" (Ford et al. 1994a,b). The parties to this agreement are Formosa Plastics-Texas, community activist Diane Wilson, the US Environmental Protection Agency (EPA), the Texas Natural Resource Conservation Commission, and the Formosa Technical Review Commission (TRC). By this agreement, the company made a commitment to studying and implementing alternative methods to reduce, recycle, or remove the wastewater generated at its Point Comfort facility. The goal was to create a process to resolve disagreements between the parties regarding the feasibility of wastewater recycling, reduction, or removal programs to be studied and implemented at its Point Comfort facility with a goal of zero discharge to Lavaca Bay.

6.4.2 The Quest for Zero Discharge

The Wilson–Formosa Agreement launched the company on a comprehensive analysis of the possibility of a "zero-discharge" system for the facility. Although complete recycling in a complex chemical facility is virtually impossible, serious attempts to reduce pollution from hazardous wastes must be made. Thus, a list of candidate solutions was developed, all based on water quality realities and cost-effectiveness. According to the agreement, the system selected was to be "economically beneficial, environmentally superior, and technically proven to be effective in a similar industrial applications" (Ford et al. 1994a,b). It is noted that a successful "zero-discharge" scenario eliminates much of the costly monitoring and offers other potential cost savings.

Meetings were held between the parties to the Agreement to identify potential alternatives. A summary of these candidate systems and the associated estimated costs (1999) is presented in Table 6.3 (Blackburn and Ford 1998). These alternatives concentrated on comprehensive concepts for removal of the wastewater from Lavaca Bay.

TABLE 6.3
Summary of Costs and Environmental Issues for Viable Zero Discharge Alternatives

System No.	Description	Capital ($MM)	O&M $/1000 Gal Removed from Discharge	Economic and Environmental Issues: Advantages	Economic and Environmental Issues: Disadvantages
1	No additional action	0	4.15	Meets permit requirements No additional capital costs	Not zero discharge Continuous monitoring costs Continued permitting costs Local public opposition Potential damage to Lavaca Bay
2	Reverse osmosis with deep well injection	37.1	7.12	Low energy cost No additional solids generated or atmospheric emissions Proven technology	Potential negative public opinion Holding capacity or spare wells Water lost to well Significant capital investment
3	Reverse osmosis with deep well injection	164.9 (72.9)	18.92	Meets zero discharge Low energy consumption Potential for $CaSO_4$ by-product recovery	Significant land sacrifices to dead salt lake Technology unproven in full scale Large quantities of chemicals input to system Technology unproven in full scale Large quantities of chemicals input to system Variable brine quality may impact production

(Continued)

TABLE 6.3 (CONTINUED)
Summary of Costs and Environmental Issues for Viable Zero Discharge Alternatives

System No.	Description	Capital ($MM)	O&M $/1000 Gal Removed from Discharge	Economic and Environmental Issues: Advantages	Economic and Environmental Issues: Disadvantages
4	Reverse osmosis with wastewater cooling towers with sulfate removal	29.6	12.27	Meets zero discharge Potential for $CaSO_4$ by-product recovery	Large quantities of chemicals input to system Technology unproven at 26% salt concentration
4A	Reverse osmosis with wastewater cooling towers and vapor compression concentrator with sulfate removal	35.5	11.05	Meets zero discharge Wastewater recycled not lost to atmosphere or in well Proven technology Potential for $CaSO_4$ by-product recovery	High energy requirement Large quantities of chemicals input to system Large quantities of sludge generated requiring disposal
5	Reverse osmosis with vapor compression concentrate and sulfate removal	33.2	13.27	Meets zero discharge Wastewater recycled not lost to atmosphere or in wall Proven technology Potential for $CaSO_4$ by-product recovery	High energy requirement Large quantities of chemicals input to system Large quantities of sludge generated requiring disposal Significant capital investment

While the work on the zero-discharge alternatives was proceeding, studies were made of the various contributing waste streams and the ability to recycle or reuse them.

At the initiation of the Wilson–Formosa Agreement, the Point Comfort facility was recycling its treated sanitary wastewater to the cooling towers. Additional analysis of the water use patterns of the plant identified three waste streams that could be recycled and reused to further reduce water use and wastewater generation: (1) low-strength organic wastewater from the polyvinyl chloride process, (2) cooling tower discharges, and (3) evaporator process condensate recycle.

The low-strength organic waste stream can be segregated from the other organic wastes and biologically treated to be suitable for reuse, a step that will reduce water use and effluent production by approximately 1 million gpd. By increasing the cycles on the cooling tower, both water consumption and the wastewater it generates can be reduced by another million gallons per day. The third stream, the recycled evaporator process condensate, can itself be recycled, thus eliminating a waste stream of approximately 600,000 gpd. Together, these three alternatives represent a reduction in the volume of effluent from the plant of 2.6 million gpd, approximately 32%, with a comparable reduction in water use. These alternatives were recommended by the TRC and adopted by both Wilson and Formosa.

Until the late 1990s, however, a total zero-discharge option has not been adopted, largely because complications associated with the concentrated brine system could not be quickly resolved.

Nevertheless, water conservation by recycle and reuse has been successful at Formosa Plastics. The plant only uses 17 million gallons per day (mgd) of the 26.6 contract allowance and discharges less than 8 mgd to Lavaca Bay, despite the permitted allowance of 9.7 mgd average and 15.1 mgd maximum. The path to this significant reduction in effluent flow has been extensively documented in the literature (Formosa Plastics Corporation 1991; Ford 1996, 1997; Ford and Blackburn 1997; Ford et al. 1994a,b; Morris 1993; Parsons Engineering-Science 1993; The University of Texas at Austin 1993). Indeed, the goal of "zero discharge" to Lavaca Bay could well be realized in the future.

The TRC has considered two other options for the disposal of the brine stream. One of these options is to return the brine to the salt dome. This option is technically feasible but expensive. The second option is to chemically treat the brine stream and reuse the brine stream as a feedstock. This alternative is technically the best solution. However, it presents difficult chemical engineering challenges. Designers of the chlorine plant have concurred that this recycle concept is technically feasible and its feasibility is receiving more detailed scrutiny.

6.4.3 Summary

The attempt to approach zero discharge at Formosa Plastics, a chemical manufacturing facility, has been a complex, multidisciplinary, regulation-sensitive, and technically challenging decades-long project. By virtually all yardsticks, progress is being made, demonstrating how industrial expansion and economic growth can coexist with environmental control and enhancement, how constructive and creative agreements allow a process to move forward concomitant with oversight and controls, how allocation of resources can be better applied for scientific evaluation and evolution as compared with litigation, and how an overall higher probability of achieving better human health and the environment can be attained.

6.5 UV DISINFECTION

In many parts of the world, waterborne diseases such as typhoid, cholera, hepatitis, and gastroenteritis infect and kill many infants and children each day (WHO 1996). Wastewaters or water can contain an incredibly large variety of microorganisms. Some are harmless; many, however, are disease causing. These microorganisms must be destroyed before wastewaters can be safely discharged into a receiving body of water or reclaimed or reused. With increasing emphasis on promoting a sustainable ecological future and concern over the presence of toxic chemicals in the water, the modern disinfection processes are increasingly leaning toward technologies that destroy pathogens while balancing the effects of introducing the disinfected wastewater into populations of aquatic biota or drinking water supplies.

6.5.1 The Move toward UV Disinfection in North America

In most areas of the United States and Canada, wastewater is disinfected by means of irradiation with UV light instead of by such older methods as chlorination and chlorination/dechlorination.

A traditional method of purifying drinking water, the addition of chlorine, long ago became impractical for large municipalities. Not only did fire codes begin to limit the amount of liquid chlorine that could be stored, but US regulations in place since 1985 restrict the amount of chlorine that may be discharged into receiving waters and established limits on the total residual chlorine permissible in wastewater effluents. Compliance with these regulations could be achieved by incorporating a dechlorination step into the disinfection process, and indeed, this was the route usually chosen (USEPA 1985).

Chlorination followed by dechlorination has ceased to be the treatment of choice however. Not only does the process destroy the aquatic biota in receiving waters and produce compounds that may be carcinogens, but the sulfur dioxide that removes chlorine from the effluent stream is itself an environmental pollutant.

Thus, across the United States, environmental protection agencies and corporations began to look for alternative ways to disinfect wastewater. These efforts produced literature indicating that UV disinfection systems were both effective and economical (Das 2004; Das and Ekstrom 1999; Loge et al. 1996a,b; LOTT 1994; Scheible et al. 1986; USEPA 1992; Washington Department of Ecology 1998; White 1999). We shall discuss some of these alternative technologies, including a development from the early 1980s: parallel-flow, open-channel modular UV systems, usable both in the retrofit market and for new WWTPs.

Mini-Case Study: The LOTT System

After considering the available alternatives, the city of Olympia, Washington, decided to install and operate the first comprehensive UV disinfection system in the WWTP on the west coast. Details of the system for the cities and counties of Lacey, Olympia, Tumwater, and Thurston (LOTT) will be presented later (see Figure 6.2). During the pilot study and through completion of the project, the state Department of Ecology worked with the counties, providing assistance and support (LOTT 1994).

The LOTT system treats wastewater coming primarily from more than 100,000 residences but also from a brewery and some light industries. The secondary treatment process is a biological nutrients removal system able to remove from the water more than 90% of the biodegradable organic material (BOD) and total suspended solid (TSS), as well as nutrients including phosphorus, before discharging an average of 22.0 mgd into Puget Sound's Budd Inlet.

LOTT also treats storm water, and during the winter months, the WWTP receives high storm water flow, sometimes totaling 55 mgd. Because of higher flow through the WWTP between November and February, total retention time in the clarifiers goes down and consequently TSS and turbidity level go up slightly. The modular components of the system can be adjusted to provide adequate year-round disinfection. Effluent grab samples are taken daily and analyzed for fecal coliform counts to determine the compliance with the permit from the EPA's National Pollutant Discharge Elimination System (NPDES). Twenty-four–hour composite samples are taken daily and analyzed for TSS to determine the compliance with the permit. Some results of these studies are presented in Section 6.5.

6.5.2 UV Light and Its Mechanism of Germicidal Action

The power of sunlight to destroy microbial life has long been known and appreciated. Effective disinfection in air, on surfaces, and in water has been accomplished by exposure to the direct rays of the sun. Sunlight is an important factor in the self-purification of water in streams and in impounding reservoirs. The effect of sunlight at destroying bacteria, particularly intestinal bacteria, has been reported upon many times. The ordinary rays of sunlight play little part in this bactericidal action. The results are caused by UV rays. Sources of high-intensity UV light have been developed, which can be used to disinfect water, wastewater, air, and so on.

The term *ultraviolet light* is applied to electromagnetic radiation emitted from the region of the spectrum lying beyond the visible light and before x-rays. The upper wavelength limit is 400 nm (1 nm = 10^{-9} meter) and the lower wavelength limit is 100 nm, below which radiation ionizes virtually all molecules. The narrow band of UV light lying between wavelengths of 200 and 300 nm has often been called the germicidal region because UV light in this region is lethal to microorganisms including bacteria, protozoa, viruses, molds, yeasts, fungi, nematode eggs, and algae. Figure 6.1 shows that the most destructive wavelength is 260 nm, which is very close to the wavelength of 254 nm produced by germicidal low-pressure UV lamps. Figure 6.1 also shows the similarity between UV light's ability to kill the fecal coliform bacterium *Escherichia coli* and the ability of its genetic material (i.e., nucleic acid) to absorb UV light. UV light causes molecular rearrangements in the genetic material of microorganisms, and this prevents them from reproducing. If a microorganism cannot reproduce, then it is considered to be dead (Das 2001, 2004, 2005; USEPA 1999).

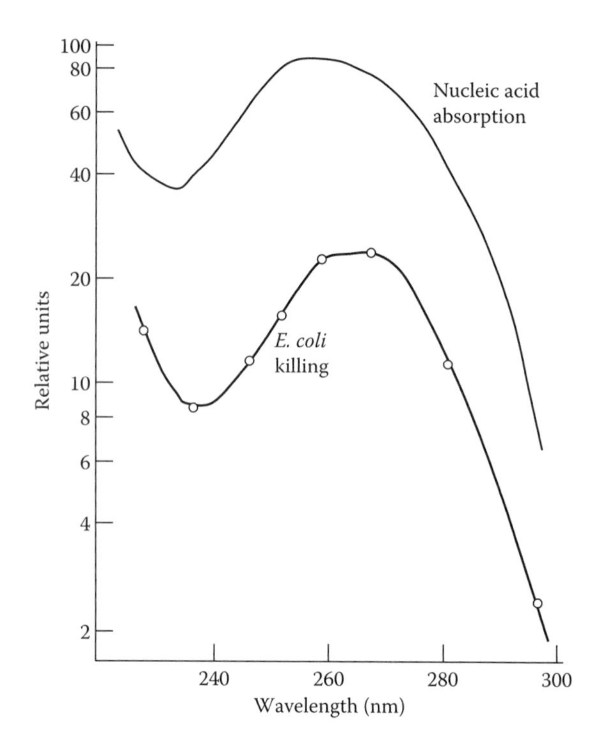

FIGURE 6.1 Comparison of the action spectrum for inactivation of *E. coli* to the absorption spectrum of nucleic acids. (From Harm, W. (1980). *Biological Effects of Ultraviolet Radiation*. IUPAB Biophysics Series, Cambridge University Press, 29 pp.)

Thus far, we have spoken only of killing cells, using unqualified words like "germicidal," "lethal," and "dead." As intuition suggests, however, not all pathogens in a UV-treated effluent stream are killed, even in the most efficient WWTP.

We shall discuss germicidal efficiency later (see Section 6.5.6). For the moment, we merely point out that photochemical damage caused by UV may be repaired by some organisms. Studies show that the amount of cell damage and subsequent repair is directly related to the UV dose. The amount of repair will also depend on the dose (intensity) of photoreactivating light. For low UV doses, the resulting minimal damage can be more readily repaired than for high doses where the number of damaged sites is greater (Lindenauer and Darby 1994).

6.5.3 UV Lamps

Germicidal lamps operate electrically on the same principle as fluorescent lamps. UV light is emitted as a result of an electron flow through the ionized vapor between the electrodes of the lamps. The glass of the germicidal lamp is made of quartz, which transmits UV light, and the glass of a fluorescent lamp is made of soft glass, which absorbs all of the UV light at a wavelength of 254 nm. The bulb of the fluorescent lamp is coated with a phosphor compound that converts UV to visible light.

A germicidal lamp produces approximately 86% of its total radiant intensity at a wavelength of 254 nm and approximately 1% at other germicidal wavelengths. Germicidal lamps with high-quality quartz also produce UV light at a wavelength of 185 nm. This wavelength produces ozone, which is corrosive to the UV equipment and the ends of the lamps. The UV lamps in UV equipment should not produce ozone. The medium-pressure mercury lamp spectrum produces most of its light in the visible range. The medium-pressure mercury lamp operates at very high temperatures (600°C–800°C), and the lifetime is approximately one-third that of a low-pressure mercury lamp (Trojan Technologies Inc. 2000).

6.5.3.1 Ballasts and Power Supplies to UV Lamps

The principal function of a ballast is to limit current to a lamp. A ballast also supplies sufficient voltage to start and operate the lamp. In the case of rapid start circuits, a ballast supplies voltage to heat the lamp cathodes continuously. A UV lamp is an arc discharge device. The more current in the arc, the lower the resistance becomes. Without a ballast to limit current, the lamp would draw so much current that it would destroy itself.

The most practical solution to limiting current is an inductive ballast. The simplest inductive ballast is a coil inserted into the circuit to limit current. This works satisfactorily for low-wattage lamps. For most lamps, the line voltage must be increased to develop sufficient starting voltage. Rapid start circuits require low voltage to heat the electrodes continuously to reduce the starting voltage. The pilot study report provides descriptions of different types of ballast systems, including details of construction, operation, and efficiency (LOTT 1994).

6.5.4 Open-Channel Modular UV Systems

Figure 6.2 is a schematic diagram of the modular disinfection system at the LOTT treatment plant introduced in the mini-case.

The design features racks of UV lights placed in an open channel so that the water flows parallel to the radiation source. Each rack is independent of every other rack and has its own group of ballasts. Every group of ballasts has an individual ground fault interrupter circuit. The level of the effluent over the lamps is controlled by a flow-sensitive device.

The benefits of open-channel modular UV systems are as follows:

- Because each major component is modular, there is no need to shut down the entire UV system to replace or clean any part. This eliminates the need for a backup system.
- The flow of water by the UV lamps is by gravity, thereby eliminating pumps.
- The effluent flows parallel to the UV lamps so that debris can only catch on the lamp holders and not on the UV lamps.
- The system can be sited outdoors or indoors.
- The UV system can be installed in an existing channel or contact chamber.
- Increases in system size can be accommodated by simply making the original channel long enough to contain more than one bank of UV racks.

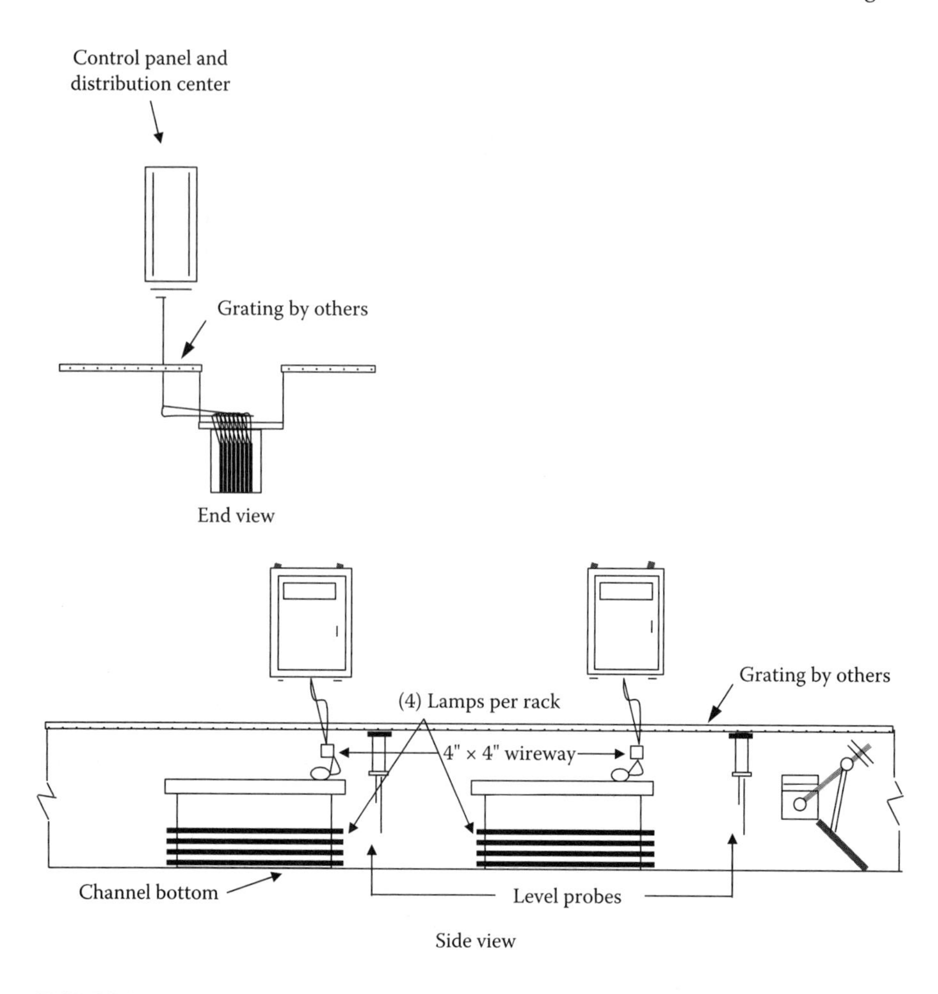

FIGURE 6.2 Simple diagram of an open-channel modular UV system.

6.5.4.1 UV System for Wastewater at LOTT

The subsections that follow describe major components of the most popular UV system for treating wastewater.

6.5.4.1.1 UV Channel

The channel is built to accommodate one or more banks of UV racks in series along with a water level control device. The optimum distance between UV banks is 4 ft.

If large variations in flow are anticipated, it is better to have more than one channel. Having multiple channels saves on electrical costs and increases lamp life because the channels can be turned on and off above or below predetermined flow rates. This is important because above a certain range, the depth of water over the lamps will be too great, and below the proper flow range, UV lamps will be exposed to the air.

6.5.4.1.2 UV Lamp Racks

The lamp racks that hold the UV lamps parallel to the flow of wastewater also protect people from the UV light. The lamp racks at the LOTT facility are made of stainless steel. Anodized aluminum is sometimes used to hold electronic ballasts, but this metal is not resistant to acids used in cleaning.

The racks are sturdy enough to permit WWTP personnel to walk on them for maintenance purposes. Each UV rack has its own power/communication cable, with a connector either on the rack itself or at the control panel/power distribution center.

6.5.4.1.3 Level Controller

The level controller serves to maintain a constant water depth of 1.9 to 2.54 cm (0.75 to 1.0 inch) over the top of the highest protective quartz sleeve at all the anticipated flow rates. If the wastewater were to exceed this depth, the UV intensity would be too low to destroy all the pathogens.

The two primary types of level control devices are the sharp-crested weir and flap gate. A flow proportional valve or sluice gate and a combination of a weir and flap gate could also be used. Normally weirs are used for small UV systems of less than 20 UV lamps and flap gates are used for all the larger UV systems.

6.5.4.1.4 Flap Gate

A flap gate operates through the use of gravity: as the water flowing through the channel hits the face of the gate, the gate opens, allowing water to pass. Weights are placed on the gate to limit the opening of the gate for a given flow. As flow increases, the force on the gate is greater and the gate opens further. A properly designed gate will maintain water levels within the specified limits over a wide range of flow rates. The disadvantage of a flap gate is that it leaks at or near zero flow, thereby allowing UV lamps to come into contact with the air. When this happens, contaminants bake onto the protective quartz sleeve, forming a coating that prevents UV disinfection when the normal flow returns.

6.5.4.1.5 Weir

A weir will guarantee a maximum water level at peak flows owing to the predictability of water crest elevation over a weir at a given flow. A weir can also be designed to keep the lamps submerged at zero flow. The main disadvantage of a weir is the considerable space required and the tendency for solids to accumulate at the bottom upstream side of the weir. A valve can be installed to flush the solids from the weir. Configuring the weir in a serpentine fashion will save on space, but even so, the space between the edges of the weir must be large enough to prevent flooding.

6.5.4.1.6 Power Distribution and Control Center

For every bank of UV lamp racks, there must be a power distribution and control (PDC) center to house the components that interface with any remote process control equipment or other banks of UV lamp racks.

If a chlorination building already exists beside a channel or chlorine contact chamber, that building can be used as the PDC center.

6.5.5 Parameters Affecting the UV Disinfection of Wastewater

The efficiency of a UV disinfection system strongly depends on effluent quality. The higher the level of contaminants, the more drastic is the intensity of the irradiation in wastewater. The following are the major parameters that must be considered in designing a UV disinfection system for wastewater:

UV transmittance (T) or absorbance
Total suspended solids
Particle size distribution
Flow rate
Iron
Hardness
Wastewater source
Equipment maintenance and worker safety

The customer or the consultant must provide this information to the UV manufacturer because each UV system is designed on an individual basis.

6.5.5.1 UV Transmission or Absorbance

The ability of UV lights to penetrate wastewater is measured in a spectrophotometer at the same wavelength (254 nm) that is produced by germicidal lamps. This measurement is called the *percent transmission* or *absorbance* and it is a function of all the factors that absorb or reflect UV light. As the percent transmission gets lower (higher absorbance), the ability of the UV light to penetrate the wastewater and reach the target organisms decreases.

The UV transmission of wastewater must be measured because it cannot be estimated simply by looking at a sample of wastewater with the naked eye. The system designer must either obtain samples of wastewater during the worst conditions or carefully attempt to calculate the expected UV transmission by testing wastewaters from plants that have a similar influent and treatment process. The designer must also strictly define the disinfection limits because this determines the magnitude of the UV dose.

The range of effective transmittances (T) will vary depending on the secondary treatment systems. In general, suspended growth-treatment processes produce effluent with T varying from 60% to 65%. Fixed film processes range from 50% to 55% T and lagoons range from 35% to 40% T. Industries that influence UV transmittance include textile, printing, pulp and paper, food processing, meat and poultry processing, photo developing, and chemical manufacturing.

Figure 6.3 illustrates the effect of a UV absorbing soluble compound on the disinfection ability of a parallel flow UV system with two banks of UV lamps in series. As the UV transmission decreases, the number of fecal coliform counts increases. Therefore, the applied dose of UV light required is dependent on the disinfection standard and the UV transmission. Figure 6.3 also shows the results of doubling the UV dose as the wastewater passes from one bank of UV lights through a second identical bank of UV lamps. By doubling the UV dose, a UV transmission of 7.5%

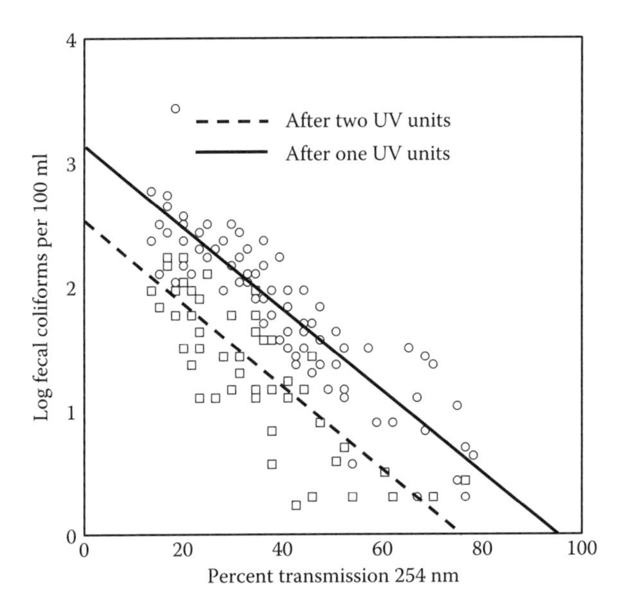

FIGURE 6.3 The effect of UV transmission on the fecal coliforms after one bank (○) and two banks (□) of UV lights.

as compared to 24% can be treated to reach a fecal coliform limit of 200 per 100 ml. Therefore, the UV system must be designed for the minimum UV transmission.

6.5.5.2 Suspended Solids

Suspended solids in biologically treated effluents are typically composed of bacteria-laden particles of varying number and size. Some of the suspended solids in wastewater will absorb or reflect the UV light before it can penetrate the solids to kill any occluded microorganisms. With longer contact times and higher intensities, UV light can penetrate suspended solids, but its germicidal ability is limited.

Obtaining the proper information about the level of suspended solids is very important for the sizing of the UV system. If a WWTP producing high levels of suspended solids is already in operation, a pilot study will show how often the quartz sleeves must be cleaned to eliminate fouling by the suspended solids. Pilot testing will also determine whether the fecal coliform limit can be attained.

Filtration can reduce TSS levels, thus lowering the level of UV irradiation necessary to achieve a given disinfection target. If wastewaters were devoid of suspended solids, UV disinfection could be used almost universally.

6.5.5.3 Particle Size Distribution

Particle size distribution (PSD) measurements of wastewater effluent are used as an indicator of filter and clarifier performance. Typically, particle sizes are related to the type of wastewater process and level of treatment, which, in turn, results in a decrease in both the number and mean size of particles. Table 6.4 illustrates the effect of large particle size on UV demand.

TABLE 6.4
An Increase in Particle Size Directly Affects the UV Demand

Particle Size (µm)	UV Demand
<10	Easily penetrated, low UV demand
	Can be penetrated, UV demand increased
>40	Will not be completely penetrated, high UV demand

6.5.5.4 Flow Rate

The US EPA provided an in-depth analysis of the effect of hydraulics on the UV disinfection of wastewater (USEPA 1992).

The degree of inactivation by UV radiation is directly related to the UV dose applied to the water or wastewater. Dose is described as the product of the rate at which the energy is emitted (intensity) and the time the organism is exposed to the energy.

$$D = It, \tag{6.1}$$

where D = dose, in microwatt-seconds per square centimeter; I = intensity or irradiation, in microwatts per square centimeter; and t = time, in seconds.

As the flow rate increases, the number or size of the UV lamps must be proportionately increased to maintain the required level of disinfection. Therefore, the UV system must be designed for the maximum flow rate at the end of lamp life.

To ensure that every microorganism is exposed to the specified average dose of UV light, the UV unit must be designed to provide as much sideways motion as possible with very little forward mixing. This is especially important when the water has a low UV transmission or high suspended solids. The open-channel UV system where the wastewater flows parallel to the submerged lamps has a very good hydraulic profile (LOTT 1994).

Since the height of the wastewater above the top row of UV lamps is rigidly controlled by a flap gate or weir at all flow rates, the system must be designed for the maximum flow rate. This is especially important if the WWTP receives runoff water after storms.

The UV system design must also accommodate the minimum flow rate. Many smaller WWTPs approach zero flow at night. During this period, the wastewater has a greater chance to warm up around the quartz sleeves and produce deposits on the sleeves. If the quartz sleeves are exposed to the air, not only will any compounds left on the sleeves bake onto the warm lamps, but water splashing onto the sleeves also will result in UV absorbing deposits. When the flow returns to normal, a layer of water will pass through the UV unit without being properly disinfected. The designer must select the flow device very carefully to compensate for this situation. A flap gate has a normal flow range of 1:5, and as mentioned earlier, all these gates leak at low flow rates. It is possible to reach 1:10, but it is better to use two or more channels. A weir, which keeps the lamps fully submerged at zero flow, may be a much better solution (LOTT 1994).

6.5.5.5 Iron

Iron affects UV disinfection by absorbing UV light. It does this in three ways. If the concentration of dissolved iron is high enough in the wastewater, the UV light will be adsorbed before it can kill any microorganisms. Regardless of concentration, however, some iron will precipitate out on the quartz sleeves and absorb the UV light before it enters the wastewater. In addition, iron that is absorbed onto suspended solids, clumps of bacteria, and other organic compounds prevent UV light from piercing the suspended solids and killing the entrapped microbes. The UV industry has adopted a level of 0.3 ppm as the maximum allowable level of iron, but there are no data to substantiate this limit. The level of iron should be measured in the wastewater, and if it approaches 0.3 ppm, a pilot study should be instituted to determine whether the disinfection level can be attained and what the cleaning frequency should be. An in-place cleaning system can be incorporated in the UV design. If possible, a WWTP should be designed with a non-iron method of precipitating phosphate.

6.5.5.6 Hardness

Calcium and magnesium salts, which are generally present in water as bicarbonates or sulfates, cause water hardness, which, in turn, produces the formation of mineral deposits. For example, when water containing calcium and bicarbonate ions is heated, insoluble calcium carbonate is formed:

$$Ca^{2+} + 2HCO_3 \rightarrow CaCO_3\ (\text{precipitate}) + CO_2 + H_2O.$$

This product precipitates and coats on any warm or cold surfaces. The optimum temperature of the low-pressure mercury lamp is 40°C or 104°F. At the surface of the protective quartz sleeve, there will be a molecular layer of warm water where calcium and magnesium salts will be precipitated, preventing UV light from entering the wastewater.

Unfortunately, no rule exists for determining when hardness will become a problem. Table 6.5 shows the classification of water hardness. Waters containing around 300 mg/L of $CaCO_3$ deposits may require pilot testing of a UV system. This is especially important if very low flow or no-flow situations are anticipated because the water will warm up around the quartz sleeves, and excessive coating will result.

TABLE 6.5
Classification of Water Hardness

Hardness Range (mg/L as $CaCO_3$)	Hardness Description
0–75	Soft
75–150	Moderately hard
150–300	Hard
>300	Very hard

6.5.5.7 Wastewater Source

It should be determined whether the WWTP receives periodic influxes of industrial wastewater, which may contain UV absorbing organic compounds, iron, or hardness, which may affect UV performance. These industries may be required to pretreat their wastewater.

For example, a textile mill may be periodically discharging low concentrations of dye into the municipal wastewater system. By the time this dye reaches the treatment plant, it may be too diluted to detect without using a spectrophotometer. Yet, even low levels of dye can readily absorb UV light, thereby preventing UV disinfection.

6.5.5.8 Equipment Maintenance, Lamp Life, and Workers' Safety

Equipment maintenance factors affecting UV intensity include lamp age and sleeve fouling. The intensity of the radiation supplied by a lamp gradually decreases with use of the device, and this is factored into the design. The recommended low-pressure lamp replacement time is approximately 5000 h, but some plants have disinfected successfully using lamps up to 8000 h. Medium-pressure lamp replacement time is approximately 5000 h. The lamp life depends on the number of ON and OFF cycles used for flow pacing during disinfection. Uniform intensity in a system can be managed with a staged lamp replacement schedule.

Because accumulations of inorganic and organic solids on the quartz sleeve decrease the intensity of UV light that enters the surrounding water, conventional low-pressure technology systems include a fouling factor in the design. These systems require cleaning and maintenance by plant operators on a regular basis. The Trojan System UV4000 has an automatic wiping system in place that combines chemical and mechanical cleaning and does not require operator maintenance time (Trojan Technologies Inc. 2000).

UV is generated on-site and does not raise significant safety concerns in surrounding communities. Worker safety requirements are directed to protection from exposure (primarily of the eyes and skin) from UV light, electrical hazards, and safe handling and disposal of the expended lamps, quarts, ballasts, and cleaning chemicals.

6.5.6 Germicidal Efficiency

Investigations have shown that microorganisms vary widely in their sensitivity to UV energy. Kawabata and Harada (1959) reported the following contact times required to achieve a 99.9% kill (3-log reduction) at a fixed UV intensity for the following organisms:

E. coli	60 s
Shigella	47 s
Salmonella typhosa	49 s
Streptococcus faecalis	165 s
Bacillus subtilis	240 s
Bacillus subtilis spores	369 s

UV radiation has also been shown to be effective in the inactivation of viruses. Huff (1965) reported satisfactory results, which included studies of several strains of polio virus, Echo 7, and Coxsackie 9 viruses. The intensities varied from 7000 to 11,000 μW · s/cm².

Current designs are for high intensity and lower exposure times—6–10 s. There is no doubt that the germicidal efficiency of UV is predictable for a given species of organism on the basis of the UV intensity–exposure time product (Equation 6.1). In practice, it will be necessary to prove and evaluate a given installation to confirm the design parameters (White 1999).

The relationship between UV dose and bactericidal kill is characterized by a mathematical model that assumes second-order kinetics when the coliform concentrations are in the range where disinfection usually takes place, as follows:

$$\frac{dN}{dt} = kN^2I. \tag{6.2}$$

Integrated, this becomes

$$\frac{1}{N} - \frac{1}{N_o} = kIt, \tag{6.3}$$

where N = coliform counts, most probable number (MPN)/100 ml, at time t; N_o = influent coliform concentration (MPN/100 ml); k = rate constant (counts/s); I = the average UV intensity (or irradiation) in the exposure chamber; and t = exposure time (s).

The influent coliform concentration is usually so much greater than the final concentration that the term $1/N_o$ becomes negligible and Equation 6.3 can be simplified to

$$\frac{1}{N} = kIt. \tag{6.4}$$

Equation 6.4 can be used to calculate coliform count at the end of UV exposure at a varying UV intensity and time of exposure for a known rate constant.

On the basis of 350 samplings conducted throughout a 1-year pilot program, Scheible and Bassell (1981) and Scheible (1987) have been able to show quite favorable correlation between UV dose and coliform kill by the following empirical equation:

$$\text{Effluent fecal coliform} = (1.26 \times 10^{13})\,(\text{UV dose})^{-2.27}. \tag{6.5}$$

Loge et al. (1996a) used a probabilistic design approach in their pilot studies and developed an empirical formula with four coefficients that can be used to calculate the fecal coliform density after exposure to UV light. The resulting correlation can be used to predict reasonably well the number of lamps necessary to meet the permit requirements for WWTPs.

6.5.7 Disinfection Standards

The level of disinfection required by the USEPA NPDES permit is commonly less than 200 fecal coliform counts/100 ml as a 30-day geometric mean. In general, a UV dose of 20 to 30 mW · s/cm^2 is required to achieve this level of disinfection in secondary treated wastewater with a 65% transmittance and TSS < 20 ppm. The UV dose requirement to meet specific limits depends on the nature of the particle with respect to numbers, size, and composition. Therefore, UV dose requirements will vary.

A more stringent limit of <2.2 total coliform/100 ml is required for water reuse in California and Hawaii. In such cases, filtered effluents with TSS 2 ppm or less and 65% transmittance may require UV dose as high as 120 mW · s/cm^2 to achieve this level of disinfection. The concentrations of solids, bacteria in the particles, and the PSD are the main limiting factors when attempting to meet stringent disinfection limits.

It appears that the UV dose required to meet the traditional coliform limits will achieve better virus inactivation results than the comparable chlorine dose. Figure 6.4 compares the relative doses of UV and chlorine required to inactivate selected organisms compared to fecal coliform indictor (Trojan Technologies Inc. 2000).

Figure 6.5 presents the monthly average values of fecal coliform after the UV exposure and TSS in effluent for 1998 and 1999 obtained from LOTT WWTP (Mhatre 2000). During the winter and rainy seasons (typically November through February in the Pacific Northwest), fecal coliform counts and TSS concentrations were found to be marginally higher than the other months of dry and warm seasons. At higher flow rates, efficiency of TSS removal in the secondary treatment system is lower, because of shorter retention times, which contribute to higher TSS and fecal coliform count. However, during dry and warmer seasons, the secondary treatment system removes a higher percentage of TSS, and as a result, fecal coliform counts and TSS levels are much lower.

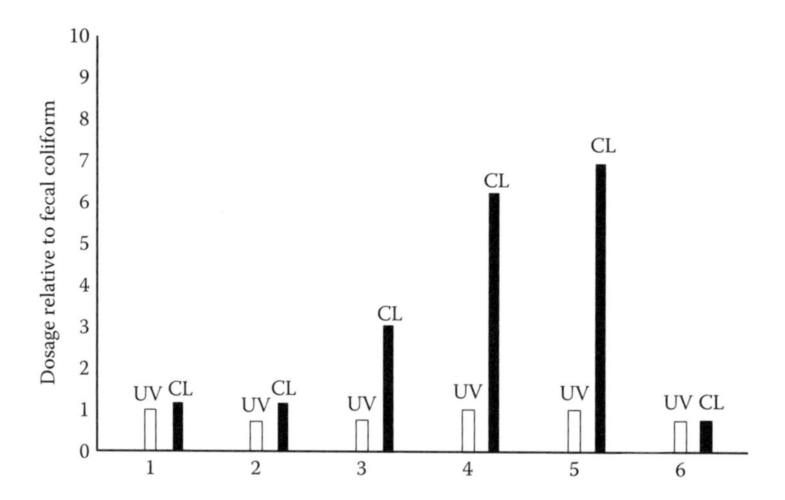

FIGURE 6.4 Comparison of the relative effectiveness of chlorine versus UV on bacteria and viruses: (1) *Escherichia coli*, (2) *Salmonella typhosa*, (3) *Staphylococcus aureus*, (4) *Polio virus type* 1, (5) *Coxsackie AZ virus*, and (6) *Adenovirus type*. (From Trojan Technologies, Inc. (2000). Overview of UV disinfection, Ontario, Canada.)

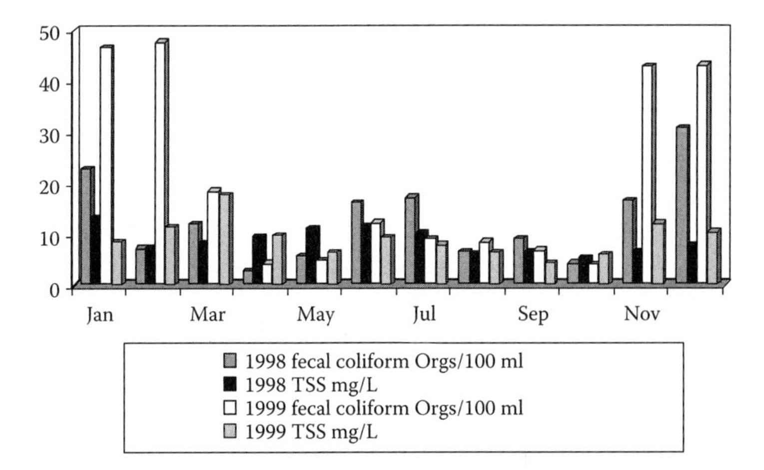

FIGURE 6.5 Results of TSS fecal coliform counts at the LOTT WWTP: monthly average 1998–1999. (From Mhatre, A. (2000). Personal communication. City of Olympia, LOTT wastewater treatment plant, Olympia, Washington.)

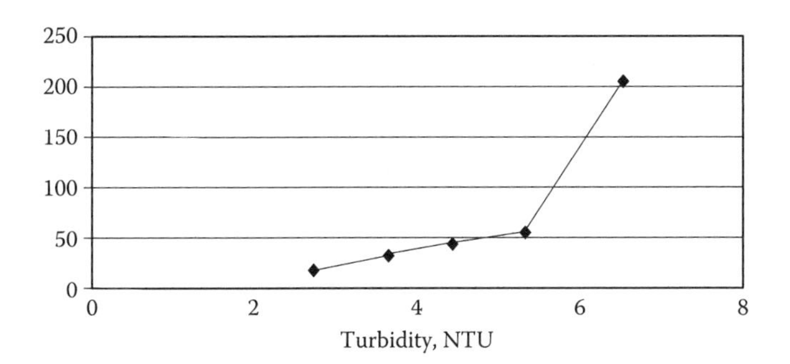

FIGURE 6.6 Turbidity in nephelometric turbidity units versus fecal coliform counts. (From Mhatre, A. (2000). Personal communication. City of Olympia, LOTT wastewater treatment plant, Olympia, Washington.)

Figure 6.6 presents values of effluent turbidity and fecal coliform counts for the LOTT facility. Higher effluent turbidity may indicate higher TSS level. Composed of bacteria-laden particles of varying number and sizes, suspended solids reduce the UV intensity in an effluent by absorbing and scattering UV light, with lower disinfection and higher coliform counts as results.

6.6 BRINE CONCENTRATORS FOR RECYCLING WASTEWATER

Brine concentrators are vapor compression evaporator systems that produce distilled water and a very small salt concentrate stream. These are ideal for water recycling because the concentrate stream is so low that wastewater can be treated economically with a very high recovery and with no liquid discharge.

Mini-Case Study: Saving the Colorado River

The market for brine concentrators initially arose because of federal clean water regulations. The Colorado River, a major source of drinking and irrigation water for the southwestern United States, had been growing increasingly saline as a result of human activities. It was to control such damage to the environment here and elsewhere that the US EPA, in the 1970s, promulgated regulations curtailing discharges to the Colorado River and forbidding construction of new plants that could not achieve zero discharge of water into the river.

To comply with the regulations, both new and existing power plants that were using river water had to recycle their water wastes. Although vapor compression was not new to the industry as an energy source, the technological fit with power plants was natural because it allowed these facilities to use electricity they were generating as the source of the mechanical energy needed in recycling water.

Federal law was not the only factor prompting the industry to turn to the use of brine concentrators for recycling wastewater. All over the country, there were local siting regulations, as well. Thus, with the advent of the private power industry in the early 1990s, entrepreneurs turned to zero-discharge water systems, which allowed them to use sites with a limited water supply, far away from discharge points. Similarly, the move to clean-burning natural gas diminished the importance of locating power plants near sources of fuel or water.

As of 2004, there were approximately 60 brine concentrators in the United States. They are sold as package plants, designed and constructed with energy conservation principles. All the original units were at coal-fired power plants, but as metal smelters, manufacturers of chemicals and semiconductors, and other enterprises began to recognize the need to eliminate water discharges, the use of brine concentration has spread.

The sections that follow discuss brine concentrators in detail, beginning with process basics.

6.6.1 Evaporator Basics

In a continuous evaporator process, the ratio of the feed to the waste (concentrate) flow rate is called the concentration factor (CF). By mass balance

$$F = D + W,$$

where F = feed flow rate (lb/h), D = distillate flow rate (lb/h), and W = waste flow rate (lb/h).

It follows that for a concentration factor of F/D, the ratio of distillate flow to feed flow would be the reciprocal of F/D:

$$\frac{D}{F} = \frac{CF - 1}{CF}.$$

This second quantity, *D/F*, is called the recovery ratio and has a maximum (although impossible) value of 1. This is the proportion of the treated water that is recovered as distilled water. For nonvaporizing species, the CF times the concentration of the species in the feed water becomes the concentration in the waste stream. The waste stream concentration becomes the determining factor in setting the CF, hence the evaporation rate for a given feed rate.

The following tabulation is helpful in providing a feel for these numbers:

CF	Recovery
2	0.5
5	0.8
10	0.9
100	0.99

In calculations of feed and concentrate streams, mass flow rates are frequently replaced by the volumetric flow rates. This equivalence is justified because most feed waters and distilled water are very close to each other in density, and the concentrate stream is usually quite small compared to the feed and distillate streams.

The brine concentrator is a single-step evaporator, and if steam is used, it is a single-effect (stage) evaporator. Multiple effects should be used if steam is the heat source used. In the multiple-effect (stages) evaporation technique, the steam evaporated out of one of the effects (stages) is used as the heating steam in the next effect (stage). The steam economy, defined as the ratio of the motive steam to the total energy required for evaporation, is approximately inversely proportional to the number of effects. For example, a 10-effect evaporator has a steam economy of 0.10.

Specific energy consumption is expressed as kilowatt-hours per thousand gallons of distillate. For high-CF systems, where the feed and distillate are approximately equal, specific energy consumption is essentially equal to kWh/1000 gallons of feed.

For brine concentrators, however, common specific energy consumption is 100 kWh/1000 gallons of distillate. Upon doing the necessary conversions, we see that this is equivalent to a 12-effect evaporator. Thus, the mechanical vapor compression technique is considerably more energy efficient than using steam. Vapor compression, however, relies on a consistently adequate supply of electricity.

6.6.1.1 Controlling the CF

The CF in any given brine concentrator is limited by the concentrations obtained in the concentrate, which, in turn, are reduced by the presence of scaling compounds. However, if the concentrate produces scaling compounds, the evaporator requires constant cleaning, an extremely unreliable operating mode. The prevalent scaling compounds in brine concentrators are calcium sulfate, calcium carbonate, magnesium sulfate, and silica, all of which appear in brine concentrator applications because BC feed waters are typically concentrates of groundwater or river water.

Calcium carbonate is eliminated as a precipitating compound by acidifying the feed, thus converting the carbonates to carbon dioxide. As shown in Figure 6.7, a pH

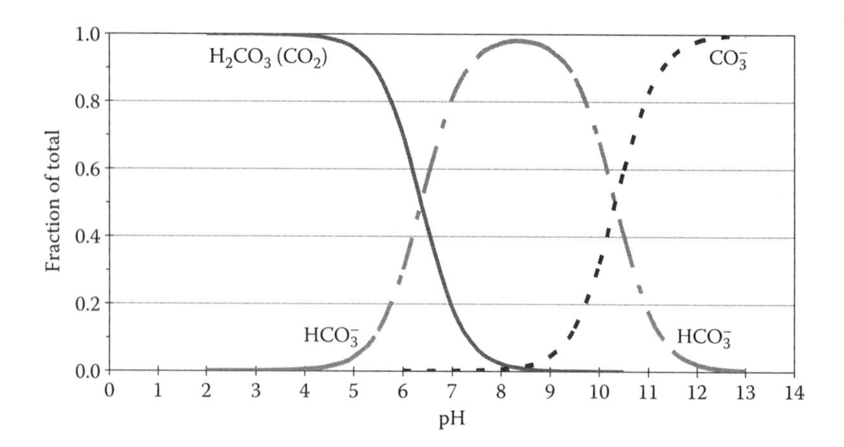

FIGURE 6.7 Alkalinity constituents with pH.

below 6.5 will achieve the conversion. If the mass flow rate of suspended calcium sulfate going out the waste steam exceeds the mass flow rate precipitated out of the feed water, then the suspended calcium sulfate must be recycled. The recycling is done with a hydrocyclone on the waste flow, with the concentrated suspended solids (bottoms of the cone) stream being rerouted back to the evaporator and the dilute suspension (top flow of cone) sent to the waste stream (Hodel 1993).

Calcium sulfate and silica saturations are such that high-CF operation is possible only on the dilute feeds. The saturation concentration of calcium sulfate is approximately 3000 mg/L in brines. The saturation concentration of silica is 150 mg/L. Thus, for a typical water supply with a silica concentration of 5 mg/L feeding a cooling tower, with a tower CF of 7, the allowable CF in an evaporator treating cooling tower blowdown is only approximately 2. With scale inhibitors, these solubility limits can probably be exceeded by a factor of 2, providing the scale inhibitor is stable at 212°F, the boiling point of water (Dalan 2000).

The seeded slurry method of scale control offers a way of overcoming the limitations with these two main scaling compounds. This technology, which has been around since the 1970s, uses a suspension of calcium sulfate to seed the circulating brine, which enables calcium sulfate, silica, and other sparingly soluble salts to precipitate on the slurry particles instead of the heat transfer tubes (Dalan 2000). The solubility limits of calcium sulfate (conservatively stated as 3000 mg/L in brines) and silica (conservatively stated as 150 mg/L) can be exceeded by a factor of 10 to 100.

If the feed to the evaporator is saturated at these quantities in a facility that does not have the seeded slurry system, the evaporator will immediately precipitate salts on the heat transfer tubes. In such cases, a crystallizer, which is a forced circulation evaporator, must be used instead of a regular brine concentrator-type evaporator (Dalan 2000).

6.6.1.2 Falling Film Evaporation

Brine concentrators use the heat transfer mechanism known as falling film evaporation. The water flow is vertical by gravity down a tube, and a film of water forms on the tubes' inner diameter. The vapor that is evaporated migrates to the center of the tube and travels downward with the water. To maintain the film over the whole length of the tube, only a small portion of the water is evaporated per length of tube and the rest is recirculated. The amount of evaporation per one tube travel, called the extraction per pass, is typically 3%–5%, indicating that for even thin film and long tube travels, a film is maintained for the whole length of the tube. The water evaporated is replaced by fresh feed to the recirculation loop.

Distribution devices that produce falling films in all the tubes include patented tube inserts or distribution plates. Distribution plates are perforated plates located above the top tube sheet with strategically sized and located holes to produce a film in the top of each tubes (Hodel 1993).

With a falling film, the required temperature difference between the heating vapor and brine is low, typically 5°F–10°F, thus making this mechanism compatible with economical (low pressure ratio) compressors.

Overall heat transfer coefficients (based on the inside tube area) for the falling film mechanisms range from 300 to 600 Btu-h^{-1} ft^{-2} $°F^{-1}$. When the seeded slurry mode is used, no fouling factor is necessary.

6.6.1.3 Typical Feed Waters

Two typical feed waters for a brine concentrator from a power plant are cooling tower blowdowns and regenerants from an ion exchange–type demineralizer. Cooling tower blowdown compositions can be calculated from the makeup water chemistry by multiplying the makeup water concentration by the cooling tower CF. Some zero-discharge facilities discharge their plant wastes, including the regenerant wastes, into their cooling tower basins (Dalan 2000).

Boiler blowdown from a steam cycle is basically steam condensate, a dilute water stream. This flow typically is part of the brine concentrator feed stream.

Plant wash downs are also typically included. This water is the service water contaminated with local minerals and oil that are part of a typical plant. For oil-fired plants, the main contaminant is oil; for coal-fired plants, it is coal and ash dust; and for gas-fired plants, it is typically pretty close to the composition of the local service water, but still with suspended solids.

In coal-fired plants, sulfur dioxide is removed from the flue gas by scrubbers. Scrubbers typically use solutions that convert the gases containing oxides of sulfur to ionic sulfate. Lime solutions therefore produce a scrubber blowdown saturated in calcium sulfate, which cannot be used as brine concentrator feed unless treated with the seeded slurry method of evaporation.

In newer installations, using natural gas as the power plant fuel, a frequent brine concentrator feed is the RO concentrate. RO is frequently used as a pretreatment of cooling tower blowdown, or as a first step in a demineralizer train (typically using service water as feed) and ending with an electrodeionization (EDI) demineralizing step.

6.6.2 Zero-Discharge Systems Using Brine Concentrators

For simple systems, such as the production of demineralized water for a steam cycle in a plant that needs to practice zero discharge, RO is used as a first step in the

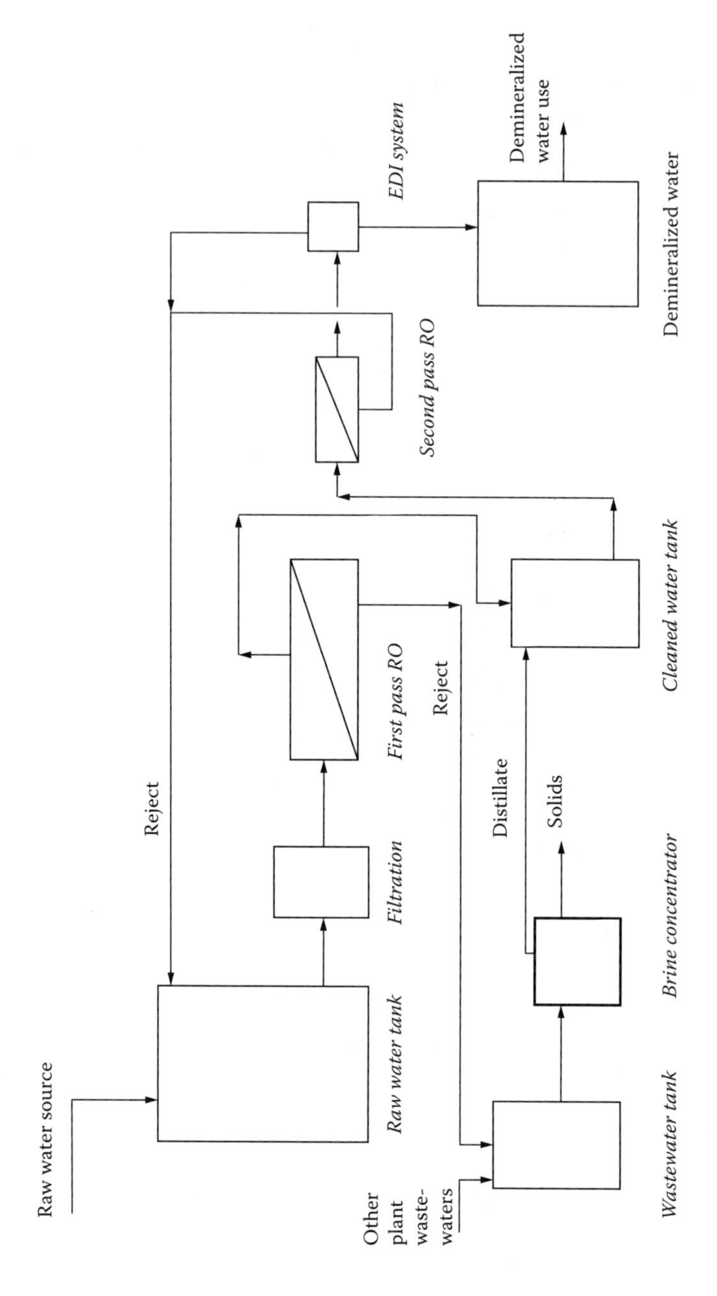

FIGURE 6.8 Schematic of the brine concentrator as part of demineralized water production.

demineralizing train. RO is frequently accomplished in two passes, with the permeate (clean stream) reprocessed in second RO system, thus producing water very low in TDS as feed to demineralizing equipment. Demineralization is done by ion exchange, or a newer process called electrodeionization.

For zero discharge to be possible, the concentrate from the first pass of the RO must be treated with a brine concentrator. Figure 6.8 shows a block diagram of a generic demineralizer converted to perform as a zero-discharge system.

For plants with cooling towers, almost exclusively power plants, cooling tower blowdown is sometimes treated with RO, assuming the cooling tower CF is not so high that the circulating cooling tower is saturated in either silica, calcium carbonate, or calcium sulfate. The RO product receives a second pass and is fed to a demineralizer, while the first pass RO concentrate is fed to a brine concentrator system, as shown in Figure 6.9.

6.6.3 Alternative Zero-Discharge Methods

We mention briefly staged cooling and high efficiency reverse osmosis (HERO) systems, two technologies that may, as time passes, supplant brine concentrators as ZLD systems. If, however, the waste stream flows are high enough, concentrating devices, including brine concentrators, will still be needed in true ZLD systems.

6.6.3.1 Staged Cooling

In power plants with water-cooled condensers, the effective cooling tower CF can be driven quite high if the scale formation chemicals in the cooling tower are removed by softening (removing calcium and magnesium) a slip stream of the cooling. A variation of this idea, called staged cooling, has been patented by Eau Tech Partners, and is in operation in at least three plants. Figure 6.10 is a system schematic (Sanderson and Lancaster 1988, 1989).

The system divides the plant cooling load into two condenser sets and two cooling towers, with the blowdown from the first cooling tower being sent to a chemical softener and then as the feed to the second cooling tower. The second cooling tower then has a slip stream softener, thus enabling a high cooling tower CF on the second tower. A final concentrating device, perhaps a pond, must be used. The final solid concentrate of the staged cooling system is reported as 100,000 mg/L TDS, approximately one-third of what is possible in an evaporator system (Sanderson and Lancaster 1988, 1989).

6.6.3.2 HERO Process

A recently patented system overcomes the silica limit in RO separation by softening the feed and operating the RO system at high pH, where silica becomes soluble. Softening the feed removes calcium and magnesium, so that calcium carbonate, calcium sulfate, and magnesium sulfate do not precipitate in the concentrate.

A schematic of the proprietary operating system is shown in Figure 6.11.

The RO system CF is limited by the osmotic pressure exerted by the RO concentrate. With the current state of RO membranes and pumps, a final TDS of 100,000 mg/L is possible. A final concentrating device is still needed. Figure 6.11 shows the use of an evaporation pond.

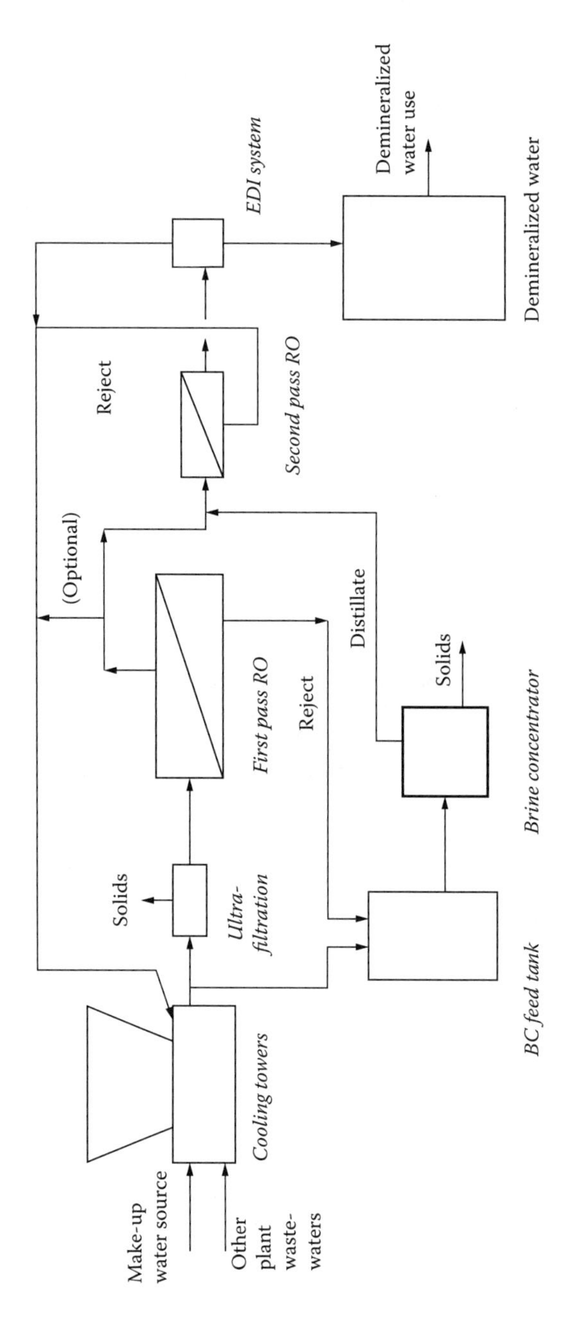

FIGURE 6.9 Schematic of brine concentration as part of cooling tower blowdown treatment.

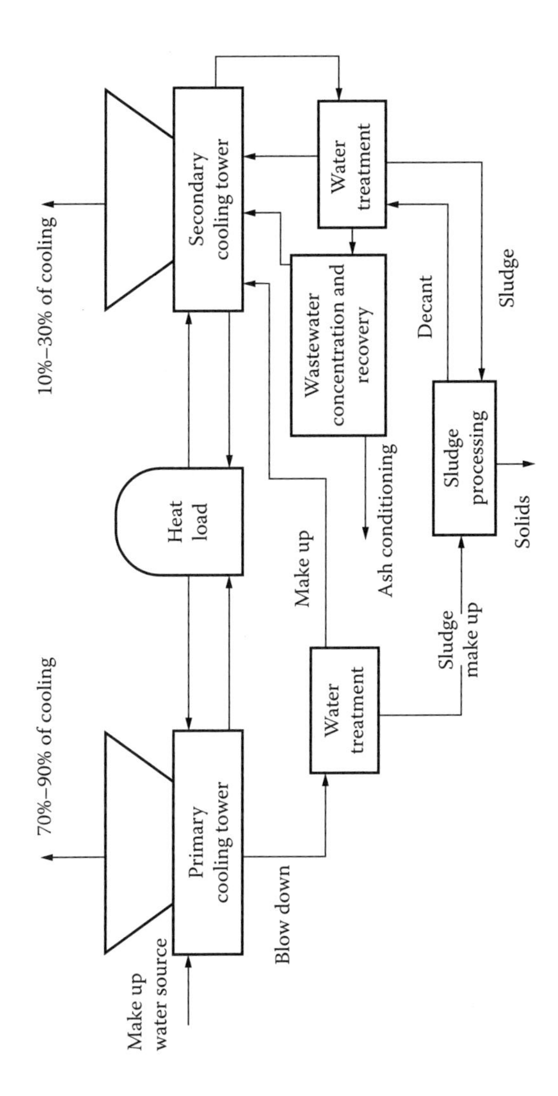

FIGURE 6.10 Schematic of a stage cooling system.

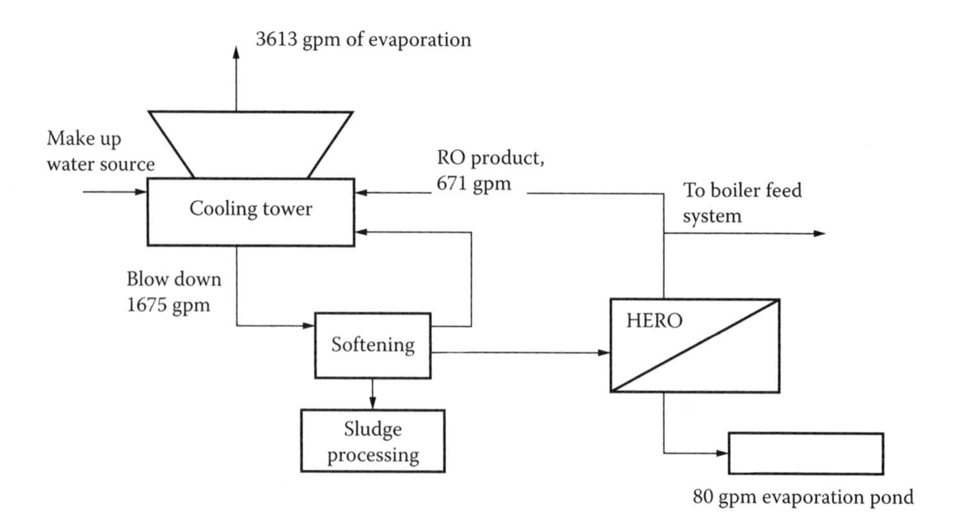

FIGURE 6.11 Schematic of the HERO process.

6.6.4 Economics of Brine Concentrator Systems

On the basis of the parametric cost information on brine concentrator systems given earlier, we will now overview some economic aspects of these systems.

Brine concentrators and, by extension, integrated ZLD systems only make sense in grassroots facilities when

- There is a shortage of surface or well water
- The facility needs water to operate
- There is no water discharge option, such as a source of potable water or a river
- A discharge permit is unobtainable

ZLD takes away the siting constraint; that is, it is no longer necessary to locate the plant near a large usable water source or suitable discharge point. The economic advantage of such an effect is hard to generalize, being very specific to the actual circumstance.

The net present value of a zero-discharge facility is negative. Dalan and Rosain (1992) found that a 265-gallons per minute (gpm) brine concentrator had operating costs of $8.94/1000 gallons of water treated (total of $1,252,600/year), while the avoided cost of extra demineralizer regeneration chemicals was $216,000/year. The avoided cost amounts to 1.54/1000 gallons. Since any dollar amount here is an operating expense, the cash flow is negative.

A typical specific operating expense is in the $5–$7/1000 gallons treated range. In this estimate is the price of the electricity at a retail price of 0.05/kWh. In grassroots power plant planning, the electricity is many times considered a parasitic load on the power plant (electricity needed to produce the power). In this accounting method, the cost of electricity is zero. The elimination of electricity as an operating cost brings the total treatment cost down to the $2–$3/1000 gallons range.

TABLE 6.6
Operating Cost Breakdown for a 265-gpm System Resulting in a Cost of $9.62/1000 Gallons of Feed

Item	Consumption	Unit Cost	Annual Cost
Operating labor	1 mh/h	$50/mh	$438,000
Maintenance (labor)	2 mh/h	$50/mh	$87,600
Maintenance (materials, including spare parts)			$80, 000
Electricity	1617 kWh	$0.05/kWh	$707,000
Chemicals			
Sulfuric acid	293 lb/day	$0.06/lb	$6416
Polymers	40 lb/day	$1.50/lb	$21,900
Total chemicals			$28,316
Total annual operating cost			$1,340,916
$/1000 gallons of feed			$9.62

Source: Dalan J. and Rosain, R. (1992). Zero discharge wastewater treatment facility for a 900 MW GCC power plant. Report No. EPRI TR-100375, Electric Power Research Institute, Palo Alto, CA (updated by Dalan to 2002).

The typical energy load for a brine concentrator is 100 kWh/1000 gallons of water produced.

The determination of water economics is very geography specific. For example, in western Washington, water and sewer bills typically are in the $1–$2/1000 gallon range, whereas in eastern Washington, water costs more, if indeed it is economically available.

For non–power plant applications, local high prices for electricity can be overcome by seeking out alternate energy sources. Compressors (the majority energy user) in brine concentrator plants that are steam driven or natural gas (via a natural gas engine) driven have been installed. Table 6.6 presents a breakdown of typical operating costs (Dalan and Rosain 1992; Haussman and Rosain 1996).

For existing facilities, installing a zero-discharge plant makes sense when

- The discharge permit conditions change, with the result that the existing facility can no longer be operated economically
- The cost of water rights plus treatment fees exceeds the operating costs of a zero-discharge facility

This last point is illustrated by the following mini-case study.

Mini-Case Study: Calculating Payback for a Zero-Discharge System

A power plant has a 300-gpm blowdown stream from the plant. Preliminary estimates show that a zero-discharge plant would cost $4.55/1000 gallons to operate. The local utility raises the water and disposal fees to $3.00/1000 gallons, at a total cost to the plant of $6.00/1000 gallons.

The capital cost of 300 gpm is approximately \$7.5 million. The annual savings is (\$1.45 × 300 × 1440 × 365)/1000, or \$228,636/year. This represents a project with a simple return on investment of 3% and a simple payback of 32 years.

6.7 PROGRESS TOWARD ZERO DISCHARGE IN PULP AND PAPER PROCESS TECHNOLOGIES

The US pulp and paper manufacturing industry is the country's fourth largest consumer of process water. The pulp and paper manufacturing industry has a long history of recycling and reuse. The kraft pulping process is unique in that most of the chemicals from spent liquor can be recovered for reuse in subsequent cooks, and therefore, this process can be a good example of a closed-loop or a zero-discharge manufacturing system. Having exhausted many gains achievable through end-of-pipe pollution control, the pulp and paper industry is inventing and implementing process changes, process modifications, and retrofits, to improve wastewater quality, lessen air pollution, and achieve better solid waste management by means of recycling and reuse.

Technological advances aimed at reducing the formation of dioxins and furans during pulp bleaching have led to a series of process changes, including (1) eliminating the use of certain defoamers that contained dioxin and furan precursors, (2) decreasing the use of elemental chlorine as bleaching chemical, (3) and increasing the use of chlorine dioxide for pulp bleaching. As the industry has implemented these changes, dioxin and furan concentrations in bleaching effluent have dropped well below detection limits established by the US EPA. Indeed, dioxin has been zeroed out. Moreover, the process changes have made recovery of energy, process water, and bleaching chemical a feasible approach to energy conservation, water conservation, water reuse, and pollution prevention.

The recycling and recovery of all pulping and bleaching process wastewater is termed *closed cycle*, and this chapter discusses the positive results obtained in the pulp and paper industry by doing just that.

The prevalent technologies used for ZLD systems are membrane processes, primarily RO, followed by evaporation, and then by crystallization. In the pulp and paper industry, to date, filtration followed by evaporation has been used (Dalan 2000). A zero-discharge system can produce from industrial wastewater a clean stream suitable for reuse in the plant and a concentrate stream that can be disposed of in an environmentally benign manner, or further reduced to a solid. For pulp and paper, the concentrate streams, if they are black liquor, are used as fuel in heat recovery boilers.

While there are several definitions of zero discharge, in practice, the term most commonly means that no water effluent stream will be discharged from the processing site. Zero-discharge systems have several advantages:

- Minimum consumption of freshwater
- Capability of recovering valuable resources
- Reduction in volume of sludge
- Better water quality
- Flexibility in facility site selection, since no receiving waterway is needed for wastewater treatment

Disadvantages include maintenance problems (e.g., scaling and corrosion), reduced plant reliability, and the presence of certain trace chemicals not found in wastes from the more traditional processes. The cost of installing a zero-discharge system can probably only be justified for grassroots mills. However, for closed-loop operations using existing equipment, lost is not prohibitive.

6.7.1 Two Case Studies

6.7.1.1 Louisiana-Pacific Corporation: Conversion to Totally Chlorine Free Processing

To take advantage of the benefits of a zero-discharge system, the Louisiana-Pacific Corporation's (L-P) Samoa pulp mill, located on the northern California coast, converted to totally chlorine free (TCF) pulp processing. The mill, constructed in 1964, produces an average of 650 tons of bleached kraft pulp per day from waste wood chips generated by local sawmills. In January 1994, the mill became the only North American kraft mill to replace chlorine or chlorine-containing compounds with hydrogen peroxide and oxygen in all its bleaching agents. The noncorrosive chemistry of these TCF bleaching chemicals and the similarity of pH and temperature conditions between bleaching stages make it easier to recycle TCF bleaching wastewaters. Other key technologies enabling high recycle rates are extended digester cooking, improved brown-stock washing, closed screening, oxygen delignification, a high-efficiency recovery boiler, and advanced green liquor filtration. These pollution prevention technologies, coupled with innovative process changes, have enabled L-P to push the technical limits of CC operation and dramatically reduced the environmental impact of the mill (Bicknell and Holdsworth 1995; Jaegel and Spengel 1996).

Data collected from a large number of mills across the United States and reported by the National Council of the Paper Industry for Air and Stream Improvement (NCASI 1997) and Das (1999) show that as a result of these changes, between 1988 and 1996, effluent concentrations of 2,3,7,8-tetrachlorodibenzo-*p*-dioxin and 2,3,7,8-tetrachlorodibenzo-*p*-furan were very significantly reduced, with most of the data clustered in the region of concentration.

6.7.1.2 The World's First Zero Effluent Pulp Mill at Meadow Lake: The Closed-Loop Concept

The $250 million Millar Western Meadow Lake Mill is located on a 247-acre site approximately 200 miles northwest of Saskatoon, Saskatchewan. It uses mechanical action supplemented by mild chemicals to turn aspen wood chips into bleached chemi-thermomechanical pulp (BCTMP), approximately 240,000 metric tons per year. More efficient than the kraft process, this approach uses half the trees to make the same amount of pulp, producing almost 1 ton of pulp for each ton of wood on a water-free basis. The Millar Western BCTMP process also eliminates chlorine compounds and odorous sulfur-based impregnation chemicals. This environmentally friendly mill uses hydrogen peroxide to increase the brightness of the pulp, making it suitable for printing and writing grades of paper as well as for tissue and paper towels.

The plant is the first pulp mill in the world to operate a successful ZLD system. Effluent from the thermomechanical pulping process is concentrated from 2% solids to 35% solids by three falling film vapor compression evaporators, followed by two steam-driven concentrators that further concentrate the effluent to approximately 70% solids. Of the 1760 gpm of effluent sent to the system, 1720 is recovered as high-purity water for reuse in the pulping process. Solids are burned in the boiler; the smelt is cast into ingots and stored on-site for future chemical recovery.

In the early 1990s, Millar Western Pulp (Meadow Lake) Ltd. announced plans to build a mill in northern Saskatchewan; the community was concerned about the pollution it would generate, especially effluent discharged to the Beaver River. Though a biological treatment system planned at the mill would have made the effluent cleaner than river water, Millar Western decided to go one step further and eliminate all effluent discharge from the pulp mill. The zero effluent system at Meadow Lake is the first of its kind in the world. The evaporator system, the key equipment in the water recovery process, was designed and supplied by the Resources Conservation Company (RCC) (Fosberg 1992). All effluent coming out of the mill is treated in the water recovery plant. As a result, the mill only needs approximately 300 gpm of makeup water to replace water lost to the atmosphere by evaporation. The same type of pulp mill without a water recovery plant would need approximately 2500 gpm of raw water makeup. The effluent treatment system started up in January 1992, when the mill went online.

Millar Western's Meadow Lake BCTMP mill is an example of successful closure of the water cycle in a mechanical pulp mill. An earlier attempt in Canada to close a kraft mill by recycling bleach plant effluents through the kraft chemical recovery process had failed on account of the buildup of corrosive materials (Smook 1992). Thus, it is useful to study in detail the advanced system in place at Meadow Lake.

The effluent produced by the BCTMP process at Meadow Lake is discharged at a rate of almost 1800 gpm. It has a temperature of 150°F and a pH of approximately 8 and contains approximately 20,000 ppm dissolved solids. Figure 6.12 shows a more detailed view of the water recovery portion of the system, consisting of five stages: clarification, evaporation, concentration, stripping, and incineration.

6.7.2 Clarification

The first unit operation to receive pulp mill wastewater is the floatation clarifiers. Since removal of fiber is very important to the performance of the evaporators, the mill decided to install two clarifiers instead of one. This allows for maximum removal efficiency and flexibility. Chemicals are added to aid in flocculation and floatation of the solids.

To ensure that upsets in the pulp mill do not directly affect the evaporators, an online meter measures suspended solids in the clarifier accepts stream. When the suspended solids exceeds 900 ppm, the clarifier accepts are directed to the settling ponds. Clarifier accepts normally go directly to the evaporators in the winter to conserve heat. In the summer, the accepts go preferentially to the settling pond to dump heat since the heat balance changes from season to season.

6.7.3 Evaporation

The heart of the zero effluent system is three vertical tube, falling-film vapor compression evaporators that operate as explained earlier (Section 6.6.1). At 100 ft tall,

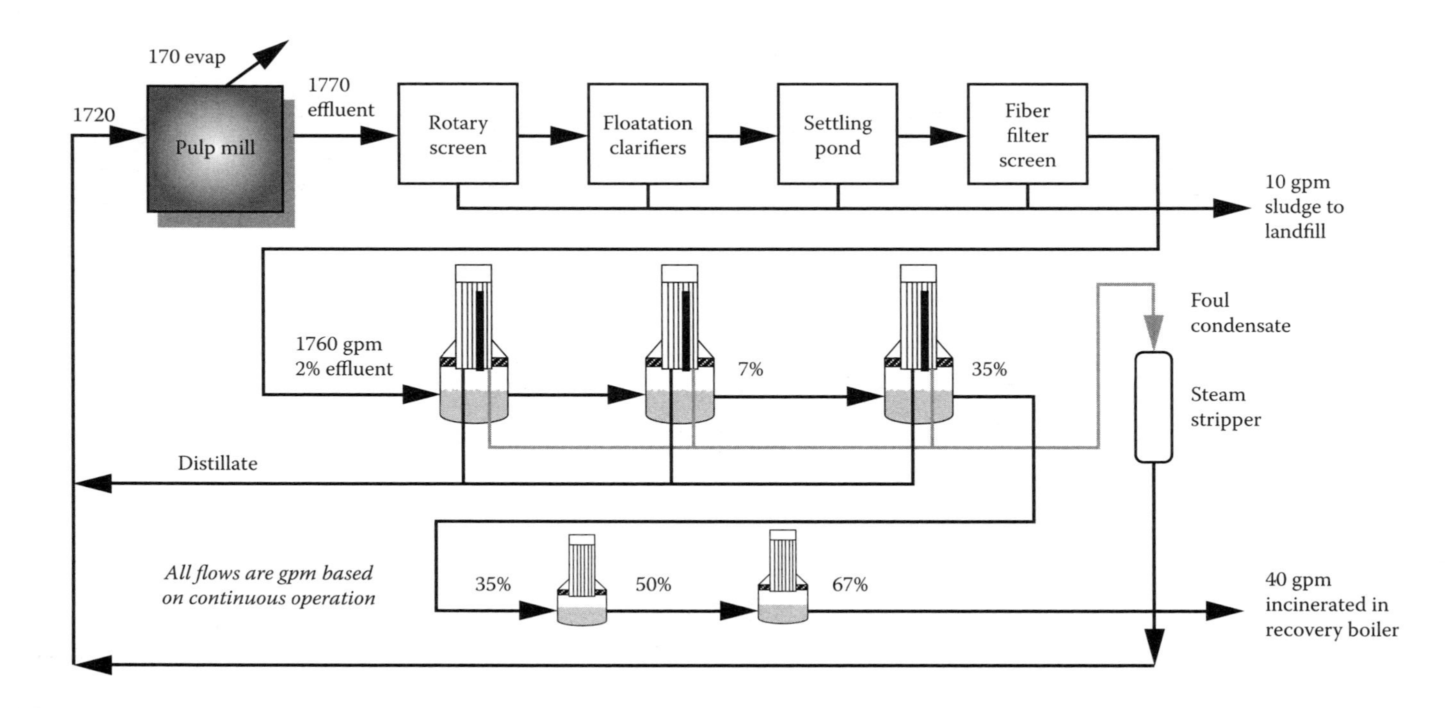

FIGURE 6.12 Effluent treatment system.

and with thousands of square feet of heat transfer surface, this is the largest train of mechanical vapor recompression evaporators in the world. The evaporators concentrate effluent from 2% solids to 35% solids, by means of an energy-efficient mechanical vapor compression process that recovers distilled water from the effluent. The evaporator consists principally of a heating element, a vapor body, a recirculation pump, and a vapor compressor.

The effluent is pumped from the vapor body sump to the top of the heating element (tube bundle). A distributor is installed on the top of each tube, causing the effluent to flow down the inside of each tube in a thin film. The distributor helps prevent fouling of the heat transfer tubes by keeping them evenly and constantly wet. It also allows the mill to operate at reduced capacity if desired, since the heating surfaces will remain wet regardless of the amount of effluent being processed. (The evaporators are also capable of handling 1.2 times more than design flow rates from the pulp mill, which gives the mill a significant amount of catch-up ability.) When the effluent reaches the bottom of the tubes, the recirculation pump sends it back to the top for further evaporation.

As the effluent flows through the heated tubes, a small portion evaporates. The vapor flows down with the liquid. When it reaches the bottom of the tube bundle, the vapor flows out of the vapor body through a mist eliminator and then to the compressor. The compressed steam (at a few pounds per square inch) is then ducted to the shell side of the tube bundle, where it condenses on the outside of the tubes. As it does so, it gives up heat to the tubes, resulting in further evaporation of the liquid inside. A large amount of heat transfer surface is provided, which minimizes the amount of energy consumed in the evaporation process. Operation of the vapor compression evaporator system requires only 65 kWh per 1000 gallons of feed.

As the vapor loses heat to the tubes, it condenses into distilled water, which flows down the outside of the tubes. Because the water that first condenses out of the steam is cleaner than water condensing later, baffles are provided within the heating element to create two separate regions for condensing. Steam flows first through the clean condensate region where most condenses. The remaining vapor, which is rich in volatile organics such as methanol, condenses in the foul condensate region of the heating element.

A major portion (70%) of the clean condensate is sent directly to the pulp mill for use as hot wash water at the back end of the mill. The balance of the clean condensate goes to the distillate equalization pond where it is combined with makeup water from Meadow Lake and serves as the cold water supply to the mill. The foul condensate, which contains the volatile organic materials, is reused after stripping in a steam stripper. The steam stripper top product (which contains the concentrated organics) is incinerated.

6.7.4 Concentration

Like the three evaporators, the two concentrators have a vertical tube, falling-film design. Rather than using a vapor compressor to drive the system, the concentrator is operated with steam generated by the recovery boiler. The evaporation process in the concentrators is essentially the same as in the evaporators, but the effluent is concentrated further, to approximately 67% solids. The concentrated effluent is incinerated in the recovery boiler. The lead concentrator takes the liquor from 35% to 50%, while the lag concentrator goes from 50% to 67% solids.

6.7.5 Stripping

The foul condensate, only approximately 10% of the total condensate, is stripped of volatile organic compounds in a packed column stripper. The VOCs are selectively concentrated in the foul condensate because of the condensate segregation features built into the evaporator heating elements. Process steam from the concentrator is sent to a reboiler, which generates stripping steam from a portion of the stripped condensate. The stripped condensate is combined with the clean condensate and reused in the mill. The concentrated VOCs are incinerated in the recovery boiler as a concentrated vapor.

6.7.6 Incineration

At the recovery boiler, the organic components of the effluent are incinerated, a process that also generates steam to operate the concentrators. Inorganic chemicals in the effluent are recovered in the smelt from the boiler, which is cast into ingots and stored on site. The mill is considering recovering the sodium carbonate, which would then be converted to sodium hydroxide, a major chemical used in the BCTMP process.

6.8 SUCCESSFUL IMPLEMENTATION OF A ZERO-DISCHARGE PROGRAM

In July 1996, a paper company located on the west bank of the Mississippi River undertook a program to eliminate the discharge of industrial wastewater to the river. A wastewater recycling system consisting of pumps, surge tank, and filtration system reduced discharges by 99%. The successful pollution prevention includes the annual elimination of 562 million gallons of wastewater, 149,000 lb of total suspended solids, and 57,000 lb of biochemical oxygen demand. The plant primarily manufactures colored construction-grade paper from a mixture of secondary fiber, stone-ground wood pulp, and kraft pulp.

The company initiated a zero-discharge program, described in detail by Klinker (1996), which has two goals:

- Eliminate the discharge of wastewater into the Mississippi River
- Improve the efficiency of water use in manufacturing to reduce the mill's dependence on river water

Recycling treated wastewater into the mill's water supply system would accomplish these goals.

Besides the regulatory motivation for reusing wastewater, the company had concerns about periodic interruptions in the flow of water to the mill. A local power company's hydroelectric plant located immediately upstream caused these interruptions. On occasion, the utility lowered the river level by halting water flow through a canal that also feeds water from the Mississippi River to the paper mill. When this occurred, the mill had to stop its manufacturing process until there was sufficient volume of water to run the mill. By reusing wastewater, the company could reduce its dependency on river water and avoid this disruption to production.

The nucleus of the zero-discharge program, a closed-loop wastewater recycling system in the mill, not only offers environmental benefits but also generates difficulties because of the increase in the volume of recycled wastewater used in manufacturing and the expenses associated with addressing these problems.

6.8.1 Closing the Loop

Figure 6.13 is a diagram of the company's closed-loop wastewater recycling system. Before the zero-discharge program began, the Mississippi River supplied all the process and cooling water for the mill. Freshwater from the river entered the mill, passed over a fine mesh screen, and entered a 2700-gallon freshwater tank. The house pump directed it to process and cooling water demand points in the mill.

The resulting process wastewater underwent treatment in the company-owned, activated sludge, WWTP. Discharge was through a process-wastewater outfall designated outfall 01. Cooling wastewater discharge was at outfall 02. Wastewater sludge underwent dewatering on a belt filter press followed by land application on a company-owned agricultural land.

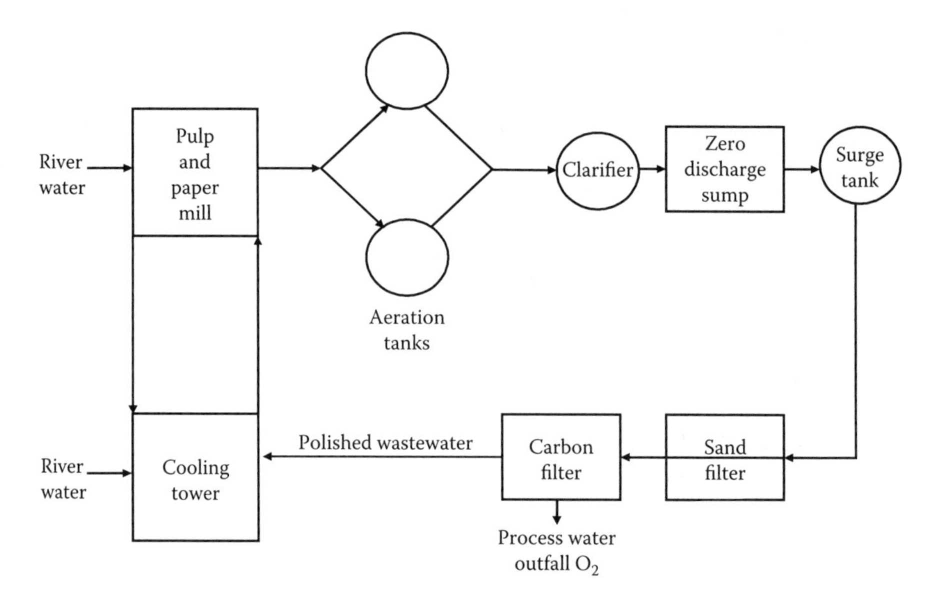

FIGURE 6.13 Schematic of a pulp and paper mill closed-loop recycling system for zero discharge system.

Before the zero-discharge program, the plant discharged an average of 607,000 gallons of process wastewater and 1.14 million gallons of cooling wastewater into the river each day. By using the closed-loop system to pump increased amounts of treated wastewater into the freshwater tank, the mill was able to state that its process and cooling water consisted of nearly 100% recycled wastewater. The total volume of wastewater discharged to the river decreased by 82% (Klinker 1996).

6.9 CONCLUSIONS

The US pulp and paper industry has made significant progress toward reducing water consumption and increased water recycling and reuse through innovative technologies and process modification. Some mills have implemented processes that "close the loop" and proved to be successful zero-discharge bleach plant systems. The effects of the EPA's Cluster Rule, which became applicable on April 15, 2001, have yet to be fully felt. It is reasonable to expect, however, that the new linkage of federal regulations aimed at reducing air and water pollution will result in an even higher level of processed water recycling and reuse within the mills, as well as greater pollution prevention and zero discharge in water, air, and solid waste areas the next decade or two. As technologies to reduce facility water, chemical, and energy use have advanced, other chemical industries, such as pulp and paper and power generating industries, have increasingly embraced the use of reclaimed water for a wide-ranging suite of purposes: from process water, boiler feedwater, and cooling tower use to finishing toilets and site irrigation. Current technologies produce reclaimed water that can provide the same performance as more expensive potable water. As water resources become increasingly valued around the world, industrial water reuse is expected to expand (Da Silva and Goodman 2014).

REFERENCES

Bicknell, B. and Holdsworth, T. (1995). Comparison of pollutant loadings from ECF, TCF and ozone/chlorine dioxide bleaching. International Non-Chlorine Bleaching Conference Proceedings.

Blackburn, J. and Ford, D. (1998). Wilson–Formosa discharge agreement—Summary.

Byers, W. (1995). Zero discharge: A systematic approach to water reuse. *Chemical Engineering*, **102**, 96.

Dalan, J. (2000). 9 things to know about zero liquid discharge. *Chem. Eng. Progress*, **98**(11), 71–76.

Dalan, J. and Rosain, R. (1992). Zero discharge wastewater treatment facility for a 900 MW GCC power plant. Report No. EPRI TR-100375, Electric Power Research Institute, Palo Alto, CA.

Das, T. K. (1999). Process technology advances in the pulp and paper industry. American Institute of Chemical Engineers Annual National Meeting, Paper No. 273e, Dallas.

Das, T. K. (2001). Ultraviolet disinfection application to a wastewater treatment plant. *Clean Products and Processes*, **3**(2), 69–80.

Das, T. K. (2004). *Disinfection*. In the Kirk-Othmer *Encyclopedia of Chemical Technology*, 5th Ed., Vol. 8, pp. 605–672, Hoboken, NJ: Wiley.

Das, T. K. (Ed). (2005). *Toward Zero Discharge: Innovative Methodology and Technologies for Process Pollution Prevention*, John Wiley and Sons, Hoboken, NJ.

Das, T. K. and Ekstrom, L. P. (1999). UV application to a major wastewater treatment plant on the west coast. Paper No. 100d, American Institute of Chemical Engineers Spring Annual Meeting, Houston.

Da Silva, A. and Goodman, A. (2014). Reduce water consumption through recycling. *Chemical Engineering Progress*, April 2014, 29–37.

Delaney, T., and Schiffman, R. I. (1997). Organizational issues associated with the implementation of ISO 14000. *Environmental Engineer*, **33**(1).

Dzombak, D. A. (2013). Use of treated municipal wastewater as power plant cooling system makeup water: Tertiary treatment versus expanded chemical regimen for recirculating water quality management. *Environmental Engineers and Scientists*, **49**(2), 30.

Ebara Corporation, Tokyo, Japan. (1997–2000). Progress on our zero emission challenge. Summary of Annual Reports.

Erkman, S. and Ramaswamy, R. (2001). *Industrial Ecology as a Tool for Development Planning. Case Studies in India*: Sterling Publishers, New Delhi and Paris.

Ford, D. (1996). Zero discharge and environmental regulations, the toxic release inventory, and natural laws. *Environmental Engineer*, **32**(4).

Ford, D., and Blackburn, J. (1997). Proceedings, Bio 1997, Ninth Annual Council of Biotechnology Centers Meeting, Houston, Texas, June B-12.

Ford, D. (1997). Formosa plastics—An industrial case study. International Conference on Global Development, Rice University, Houston Advanced Research Center, National Academy of Sciences, James A. Baker III Institute for Public Policy, The Woodlands, TX (March).

Ford, D., Blackburn, J., and Mounger, K. (1994a). Project summary prepared by the technical review commission, Wilson–Formosa agreement.

Ford, D., Blackburn, J., and Mounger, K. (1994b). Wilson-formosa zero discharge agreement. July.

Formosa Plastics Corporation. (1991). Application for Texas water commission wastewater discharge permit for the proposed Formosa Plastics Corporation expansion facility. Point Comfort, Texas.

Fosberg, T. (1992). Case study: Water recovery systems at the world's first zero effluent pulp mill. International Water Conference—51st Annual Meeting, Pittsburgh.

Goldblatt, M. E., Eble, K. S., and Feathers, J. E. (1993). Zero discharge: What, why, and how? *Chem. Eng. Progress*, **89**, 22.

Harm, W. (1980). *Biological Effects of Ultraviolet Radiation*. IUPAB Biophysics Series, Cambridge University Press, 29 pp.

Haussman, C., and Rosain, R. (1996). Power plant wastewater treatment technology review report. Report No. EPRI TR-10781, Electric Power Research Institute, Palo Alto, CA.

Hodel, A. (1993). Evaporators spawn zero discharge. *Chem. Processing*, **56**(9), 26–30.

Huff, C. B. (1965). Study of ultraviolet disinfection of water and factors in treatment efficiency. *Public Health Report*, **80**(8), 695–705.

Jaegel, A. and Spengel, D. (1996). Multimedia environmental performance of TCF closed bleach plant kraft pulp production. International Non-Chlorine Bleaching Conference Proceedings.

Klinker, R. (1996). Successful implementation of a zero discharge program. *TAPPI Journal*, **79**(1), 97–102.

Kawabata, T. and Harada, T. (1959). The disinfection of water by the germicidal lamp. *J. Illuminating Eng. Soc.*, **36**, 89.

Kiranmayee, V. and Manian, C. V. (2000). Zero discharge systems—A practical approach. Indian Chemical Engineering Congress, Science City, Kolkata, India, Paper # WTR21 (Dec. 18–21).

LOTT, Lacey, Olympia, Tumwater and Thurston. (1994). Pilot study report on UV disinfection of wastewater. The city of Olympia, Washington.

Lindenauer, K. G. and Darby, J. L. (1994). Ultraviolet disinfection of wastewater: Effects of doses on subsequent photoreactivation. *Water Resources*, **28**(4), 805.

Loge, F. J., Darby, J. D, and Tchobanoglous, G. (1996a). UV disinfection of wastewater: Probabilistic approach to design. *J. Environ. Eng.*, **1078**.

Loge, F. J., Emerick, R. W., Heath, M., Jacangelo, J., Tchobanoglous, G., and Darby, J. L. (1996b). Ultraviolet disinfection of secondary wastewater effluents: Prediction of performance and design. *Water Environmental Research*, **68**(5), 900.

Mhatre, A. (2000). Personal communication. City of Olympia, LOTT wastewater treatment plant, Olympia, Washington.

Morris, G. D. (1993). Formosa plastics labors to clean up its image. *Chemical Week*, June 23, 18–19.

NCASI, National Council of the Paper Industry for Air and Stream Improvement, Inc. (1997). Progress in Reducing the TCDD/TCDF Content of Effluents, Pulps, and Wastewater Treatment Sludges from the Manufacturing of Bleached Chemical Pulp: 1996 NCASI Dioxin Profile. Special Report No. 97-04. Research Triangle Park, NC, National Council of the Paper Industry for Air and Stream Improvement, Inc.

Parsons Engineering-Science. (1993). Outfall diffuser modeling studies for Formosa Plastics, Inc.

Rosain, R. M. (1993). Reusing water in CPI plants. *Chemical Engineering Progress*, **89**, 28.

Sanderson, W. G. and Lancaster, R. L. (1988). The staged high recycle cooling process—Acceptance test operations at the Wheelabrator/Shasta Power Plant. ASME/IEEE Power Generation Conference, 88 JPGC/PWR-52, Philadelphia, PA.

Sanderson, W. G. and Lancaster, R. L. (1989). The water concentrating cooling tower—An application of staged cooling as a final concentration step in a zero discharge system. 50th Annual Meeting International Water Conference, IWC-89-12, Pittsburgh, PA.

Scheible, O. K. (1987). Development of a rationally based design protocol for the ultraviolet light disinfection process. *J. Water Pollut. Control Fed.*, **59**(1), 25–31.

Scheible, O. K. and Bassell, C. D. (1981). Ultraviolet disinfection of secondary wastewater treatment plant. EPA Municipal Env. Res. Lab Report, EPA-600/S2-81-152, Cincinnati, Ohio.

Scheible, O. K., Casey, M. C., and Forndran, A. (1986). Ultraviolet disinfection of wastewaters from secondary effluent and combined sewer overflows. EPA-600/S2-86/005, USEPA, Cincinnati, Ohio.

Smook, G. A. (1992). *Handbook for Pulp and Paper Technologies*. 2nd Ed., Angus Wilde Publications, Bellingham, WA.

The University of Texas at Austin. (1993). Operating training course for Formosa Plastics, conducted by D. L. Ford and E. E. Gloyna, Jan. 19–21.

Trojan Technologies, Inc. (2000). Overview of UV disinfection. Ontario, Canada.

U.S. Environmental Protection Agency. (1985). Ambient aquatic life water quality criteria for chlorine. EPA 440/5-84-030, Environmental Research Laboratories, Duluth, MN, Gulf Breeze, FL, Narragansett, RI.

U.S. Environmental Protection Agency. (1992). Second draft user's manual for UVDIS, Version 3.1 UV disinfection process design manual, EPAG0703, Contract No. 68-C8-0023, USEPA, Cincinnati, OH.

U.S. Environmental Protection Agency. (1999). Wastewater technology fact sheet, ultraviolet disinfection.

Washington Department of Ecology (Ecology). (1998). Ecology's criteria for sewage works design. Chapter T5.

White G. C. (1999). *The Handbook of Chlorination and Alternative Disinfectants*, 4th Ed., Wiley, New York, pp. 1206–1236.

WHO (1996). World water project. World Health Organization.

Zacerkowny, O. (2002). Not a drop leaves the plant. Pollution Engineering, 19–22 (Feb.).

7 Wastewater Treatment, Reuse, and Disposal

Mohamed K. Mostafa, Ramesh C. Chawla, and Robert W. Peters

CONTENTS

7.1 INTRODUCTION AND BACKGROUND

Increasing industrialization has positively influenced living conditions in the past few decades but at a cost. Industries and homes produce large amounts of wastewater every day. Unfortunately, improper treatment and subsequent disposal of this wastewater into water resources can greatly impair the environment and surface water conditions. Clean water is central to the survival of any community (Reynolds and Richards 1996) and wastewater can severely compromise water quality. Although the content of wastewater varies largely with source, its contaminants are categorically pathogens, bacteria, organic content, inorganic particles, pharmaceuticals, toxins, animals, and so on (Reynolds and Richards 1996; Lin and Lee 2001).

The amount of waste generated in a small area can accumulate and lead to diseases if not treated or disposed of properly (Vesilind 2003). The natural environment uses some of the same treatment mechanisms applied at wastewater treatment plants (WWTPs) such as aeration from water agitation (waterfalls), sedimentation, filtration through sand, and so on; however, the natural environment takes a long time to degrade these substances to relatively trace quantities. The US Environmental Protection Agency (EPA) defines wastewater as the used or spent water from a home, community, industry, or farm that contains suspended or dissolved matter (EPA 2013). Densely populated and industrialized cities produce more waste than smaller cities. Therefore, natural attenuation cannot treat this volume of waste within a reasonable period. Outbreaks from waterborne illnesses began to reemerge with the industrial age. That is because the industrial age brought new contaminants into the ecosystem. As technology advances, new chemicals are being produced, which, in turn, get into the water supply from various sources.

7.1.1 Significance of Water Quality

The importance of water cannot be overemphasized. However, water contamination levels help determine what we can or cannot use it for. The quality of water depends on the physical, chemical, and biological parameters of water (Reynolds and Richards 1996). Government agencies including the US EPA have strict regulations on allowable maximum contaminant levels for different characteristics of water as they relate to human health and the ecosystem (Diersing 2009; Johnson et al. 1997). Impurities are accumulated from anthropogenic sources. The US EPA defines water pollution as the presence of objectionable or harmful material that adversely affects the water quality (EPA 2013). The quality of water is determined or evaluated on the basis of different criteria applicable to the beneficial uses. Parameters such as temperature, pH, dissolved oxygen, total nitrogen, total phosphorus, bacteria, turbidity, and organic and metal contaminants are monitored to help control pollution (Reynolds and Richards 1996). These characteristics are evaluated using the daily average or allowable maximum as applicable to the water body under consideration.

7.1.2 Removing Oxygen-Demanding Material

Wastewater discharged into the environment undergoes natural detoxification, decomposition and separation processes (Vesilind 2003). Activities of microorganisms, filtration through sediments, aeration (e.g., waterfalls/cascades), sedimentation, and ultraviolet (UV) radiation from sunlight are several natural processes involved in wastewater decontamination (Reynolds and Richards 1996). Microorganisms specifically use significant amounts of dissolved oxygen while breaking down the waste. This depletion in oxygen can have a devastating effect on the ecosystem. Both biochemical oxygen demand (BOD) and chemical oxygen demand (COD) are widely used criteria for water quality assessment. Analysis of wastewater is essential to adequately treat the wastewater to meet regulations for discharge (Vesilind 2003).

BOD is the amount of dissolved oxygen needed by the aerobic biological microorganisms in a body of water to break down organic material present in a given water sample at a certain temperature over a specified period. This quantity is usually expressed in milligrams per liter. Typically, a sample has a specific incubation period at a certain temperature (e.g., 5 days at 20°C) (Lin and Lee 2001). Two methods of BOD testing include the dilution method and manometric method. The dilution method includes measuring the dissolved oxygen concentrations in a sample before and after an incubation period. The analysis is performed using 300-ml incubation bottles in which buffered dilution water is dosed with seed microorganisms and stored for 5 days in a dark room at 20°C to prevent dissolved oxygen production via photosynthesis (Madaeni and Eslamifard 2010).

$$\text{Unseeded BOD} = (D_0 - D_5)/P$$

$$\text{Seeded BOD} = [(D_0 - D_5) - (B_0 - B_5)f]/P$$

where $(D_0 - D_5)$ = change in dissolved oxygen at day 5 from initial dissolved oxygen at time $t = 0$, $(B_0 - B_5)$ = change in dissolved oxygen of seed control at day 5 from initial dissolved oxygen at time $t = 0$, P = decimal volumetric fraction of wastewater utilized, and f = ratio of seed volume in dilution solution to seed volume in BOD test on the seed control.

COD is used to chemically determine the amount of oxidizable component of the wastewater. Both BOD and COD help determine the relative oxygen depletion for wastewater from different sources (Reynolds and Richards 1996). Like BOD, COD is expressed in milligrams per liter or parts per million, which indicates the mass of oxygen consumed per liter of solution. Therefore, industries can use test results to determine the strength of their waste to design and evaluate the appropriate treatment systems (Lin and Lee 2001).

7.1.3 Preventing Eutrophication

Eutrophication is the enrichment of water by nutrients. Eutrophication is a natural process that helps provide nutrients for aquatic plant species to grow (Grady et al. 2011). However, unnatural amounts of nutrients can be considered pollution. If there are excess nutrients in the water, these nutrients can cause algal blooms that affect waterways (Grady et al. 1999). If the concentration of nitrogen or phosphorus becomes significantly high, these blooms can block waterways by creating a thick layer of algal mass. This mass of algae can block out sunlight that is needed for certain aquatic species for photosynthesis. Also, these blooms can congest waterways. The death of the algae is followed by its decay, which not only alters the smell and taste of water but results in oxygen depletion because of oxidation (Grady et al. 2011).

Monitoring the dissolved nitrogen concentration of any waste stream is vital in the prevention of eutrophication. Regular analysis will help develop a model for nitrogen speciation in the waste stream, which enables the industry to determine the extent and kinetics of denitrification in its wastewater. Different methods are used depending on the effluents from applicable industries. Biological nutrient removal (BNR) is a process that removes nitrogen and phosphorus from wastewater (Grady et al. 2011). "BNR processes require the oxidation of ammonia-N to nitrate-N through nitrification and the reduction of nitrate-N to nitrogen gas N_2 through denitrification, thereby removing the nitrogen from the wastewater and transferring it to the atmosphere in an innocuous form" (Grady et al. 1999).

7.1.4 Removing Pathogens and Bacteria

The US EPA defines a pathogen as any disease-causing organism. The major pathogen groups are viruses, protozoa, and helminthes (worms). Bacteria are also disease-causing organisms but are generally regarded as nonpathogenic (Mara and Nigel 2003). These microorganisms are easily transmitted and can cause a wide variety of effects including respiratory infections, eye infections, diarrhea, aseptic meningitis, poliomyelitis, herpangina, myocarditis, gastroenteritis, pneumonia, typhoid fever, paratyphoid fever, bacillary dysentery, cholera, hookworm, and ascariasis, to name just a few (Mara and Nigel 2003). Thus, these must be detected and removed from wastewater before being discharged into water resources (Mara and Nigel 2003).

Wastewater effluents are not tested for pathogens to determine microbiological quality, because laboratory analyses are difficult to perform and are quantitatively unreliable. Instead, it is generally determined by induction using indicator organisms. Some indicator organisms are bacteria coliforms, fecal coliform, and *Escherichia coli* (Mara and Nigel 2003). Detection of these indicator bacteria in water suggests that pathogens and other organisms may be present. Generally, conventional wastewater treatment removes 99.0% to 99.9% of pathogenic microorganisms, but since they can be infectious in small amounts, outbreaks can still occur (Mara and Nigel 2003). This high removal rate seems adequate, but depending on the latency, persistence, and infectious dose, specific pathogens may be able to survive and infect a local population. In any case, the water quality standards by the EPA must be met before discharge of wastewater into the environment. Therefore, a random collection of different samples should be analyzed for the presence of indicator organisms and then the membrane filter test method should be used to count the number of organisms per sample. The geometric average of the collected data is compared to the standard setting for disposal decision (Mara and Nigel 2003).

$$\log \bar{a} = \frac{\sum (\log a)}{N},$$

where a is the number of organisms per sample volume, $\bar{a}$ is the geometric average, and N is the number of samples.

7.1.5 Emerging Contaminants

7.1.5.1 Xenobiotic Organic Chemicals

Xenobiotic organic chemicals (XOCs) are emerging types of contaminants that consist of synthetic products from the pharmaceutical and chemical industries (Grady et al. 1999). Xenobiotics are foreign chemicals that are not found in nature. The pharmaceutical and chemical industries have grown dramatically over the past century and their formulation of XOCs has created many new and critical issues for waste disposal. Because XOCs are new to the environment, it is imperative to understand the following: (1) their biodegradation mechanisms, (2) their effects on the environment and water quality, (3) how to detect trace amounts of these materials, and (4) most importantly, how to safely dispose of these materials (Grady et al. 1999). A major concern since the 1980s has been establishing new technologies and methodologies to deal with each new XOC as it is developed.

Since the molecular structures of XOCs are often very similar to normal biological materials, this makes their disposal more difficult (Grady et al. 1999). A clear problem is that XOCs must be transformed or metabolized by microorganisms into simple chemicals that are either not toxic or can be easily removed. For this to work, the microorganisms must be able to metabolize every type of XOC, despite the fact that the microorganism's enzymes have clearly never seen this type of compound before and did not evolve to handle this type of compound. Therefore, XOC biodegradation requires that a microorganism has an enzyme that has the capability of binding and metabolizing each different XOC (Grady et al. 1999). Mineralization of XOCs requires that the product of the first reaction of the XOC biodegradation will serve as the substrate of the second and so on, until a final biogenic product is formed. Because this is a multistep process, this requires a number of different types of microorganisms that have the capability of metabolizing each specific substrate as it becomes available. A population of microorganisms is therefore required to completely biodegrade each XOC (Grady et al. 1999).

For XOC degradation to be successful, the proper environmental conditions (temperature, pH, and nutrients) must be compatible with all of the microorganisms in the system. Solid retention times will typically be long and complicated because of this (Grady et al. 1999). Often, this process is broken down into different stages because of the different microorganisms required and the different environmental conditions required for each microorganism's growth. This type of sequential process makes the treatment of the wastewater time consuming (Grady et al. 1999).

The enzymes required for biodegradation are often inducible and are synthesized by the microorganisms usually only when the XOC is present. If the XOC is close in structure to a normal substrate, that is, a biogenic compound, then it is easier for the enzyme and microorganism to metabolize since the XOC fits into a common metabolic pathway (Grady et al. 1999). Conversely, the greater the difference the XOC is from any common biogenic compound, the more difficult it is to biodegrade and consequently greater numbers of different types of microorganisms are required. How XOCs inhibit microbial growth and substrate removal is not well understood, but one possible mechanism is that the XOC can act as a competitive inhibitor of an enzyme that is critical for the microorganism's growth (Grady et al. 1999).

7.1.5.2 Endocrine-Disrupting Compounds and Other Emerging Contaminants

Endocrine-disrupting compounds (EDCs) can be natural or synthetic chemicals that interfere with hormone regulation and can promote a number of serious medical complications in living systems. They interfere with the synthesis of a hormone, hormone metabolism, or hormone actions (Li et al. 2013). This can lead to alterations in development or reproduction (Li et al. 2013). EDCs often interfere with estrogen functions, which are essential for growth and differentiation, and can lead to a number of conditions that include feminization, fertility problems, and breast cancer (Li et al. 2013; Mnif et al. 2010). EDCs also have been known to have developmental effects in fish populations (Mnif et al. 2010). Fortunately, there are cell bioassays available to test the overall EDC potential of any sample (Mnif et al. 2010).

The unintended release of antibiotics has also been of significant interest. Antibiotics are used in several applications to fight disease by fighting bacterial infections. Antibiotics are used in livestock operations, pharmaceutical manufacturing, and hospitals (Pruden et al. 2013). This wide use of antibiotics in these categories has caused waterways to become contaminated with trace amounts of these chemicals. Because antibiotics have been present, some bacteria have become resistant, and there are also antibacterial-resistant genes (Pruden et al. 2013). The spread of antibiotics could be limited by reducing the use of antibiotics and carefully monitoring how antibiotics are disposed. A large portion of antibiotic use in the Unites States is for livestock production; therefore, promoting healthier environments could reduce this use by decreasing the density of animals in a given area and developing nutrient programs (Pruden et al. 2013). Also, preventing surface runoff and sediment erosion from areas using antibiotics could help reduce the amount of antibiotics getting into the environment (Pruden et al. 2013).

With the advancement of technology, new processing techniques have allowed the production of smaller and smaller particles. Nanoparticles are used in many consumer products (Chalew et al. 2013). Like XOCs, there is not much known about the degradation characteristics of nanoparticles. There are many different types of nanomaterials and their complex chemistry is associated with the treatment of contaminated waters (Chalew et al. 2013). Some effects derived from ingesting water contaminated with nanoparticles include the following: kidney damage, increased blood pressure, gastrointestinal inflammation, neurological damage, and cancer with metal nanoparticles causing DNA damage (Chalew et al. 2013). Chemical coagulation and flocculation can remove a large portion of nanoparticles that get enmeshed in the floc, but not all particles were removed (Chalew et al. 2013). Microfiltration and ultrafiltration were used to remove more nanoparticles with the microfilter removing large nanoparticle aggregates and ultrafiltration removing single nanoparticles but not dissolved ionic species (Chalew et al. 2013). While the amount was significantly reduced with filtration methods, other new more advanced methods of treatment need to be studied to determine the efficiency of removal of nanoparticles.

7.1.6 Reverse Osmosis (Hyperfiltration)

Reverse osmosis (RO) is a nonconventional separation process involving the use of pressure to force solvent through a semipermeable membrane. The pore size of a typical RO membrane filter is approximately 0.0001 μm, which prevents the passage of larger particles (materials) such as viruses, bacteria, large organic molecules, and minerals. Hence, monovalent and multivalent ions are removed by the filter, which means that it desalinates water (Cassano et al. 1997). Thus, when water passes through an RO membrane, very pure water is obtained.

The mechanism is similar to the principle of thermodynamics (entropy transfer) to initiate the flux of water across the membrane filter in the direction of greater decrease in solute concentration.

$$J = -\frac{dC}{dn},$$

where n is the direction perpendicular to the membrane boundary, J is the water flux, and C is the solute concentration (Bejan et al. 1996). The pressure exerted exceeds the osmotic pressure but acts opposite along the same line. Figure 7.1 illustrates the difference between the directions of water flux in RO as opposed to osmosis.

7.1.6.1 Applications

Wastewater reuse has been made possible and affordable by RO (Madaeni and Eslamifard 2010). Its application in the purification of recycled wastewater has been extended to winery (Ioannou et al. 2013; Tay and Jeyaseelan 1995), tanning (Cassano et al. 1997), and greasy (Cassano et al. 1997) wastewaters. The turbidity, COD, BOD, total dissolved solids, suspended solids (SS), SO_4, NH_4, CaH, and total hardness of the wastewater are maximally reduced (Ioannou et al. 2013) in each application. Figure 7.2 illustrates the wastewater treatment process using RO technology.

7.1.6.2 Advantages

1. RO provides the finest filtration mechanism because it has the smallest pores of available membranes today (Reynolds and Richards 1996).
2. Even with its high filtration capacity, RO membranes can last 3 to 4 years with proper maintenance (Reynolds and Richards 1996).

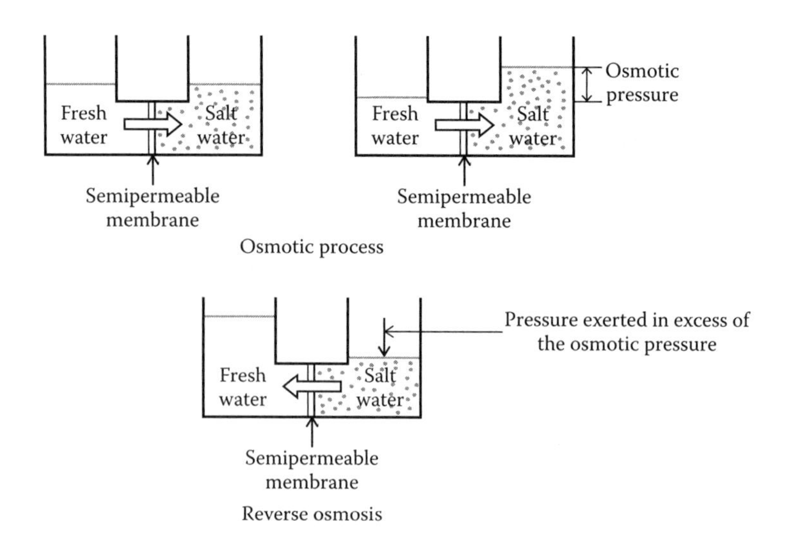

FIGURE 7.1 Osmotic and reverse osmosis processes. (Adapted from EPA 2005, *Membrane Filtration Guidance Manual*, EPA Report No.: EPA 815-R-06-009, Office of Water, Washington, D.C. [November].)

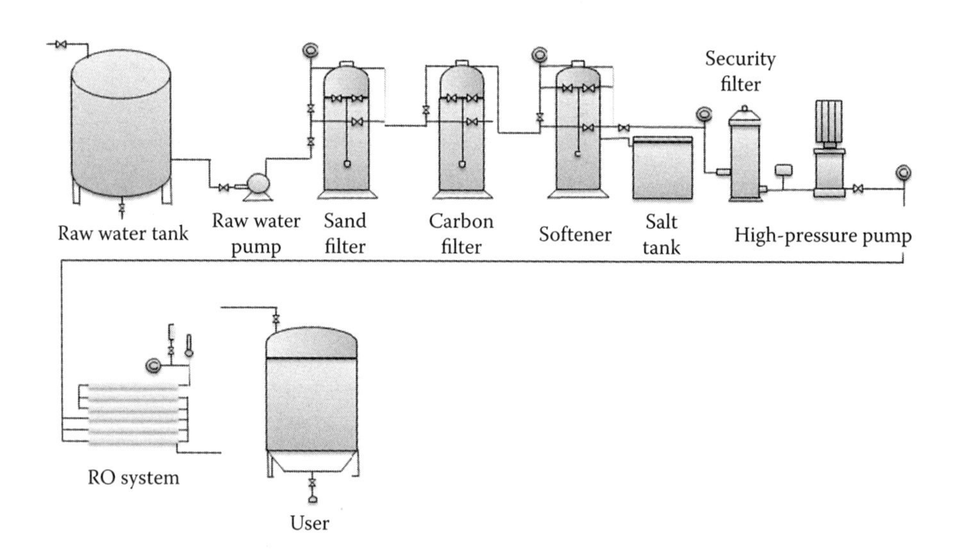

FIGURE 7.2 RO system. (Adapted from EPA 2005, *Membrane Filtration Guidance Manual*, EPA Report No.: EPA 815-R-06-009, Office of Water, Washington, D.C. [November].)

3. RO membranes have a cross-flow mechanism that allows the membrane to self-clean. It allows some of the fluids to flow across the membrane while the rest of the fluid flows downstream, thereby continuously backwashing contaminants from the membrane wall (Reynolds and Richards 1996).

7.1.6.3 Disadvantages

1. RO is not foolproof against all types of microorganisms or species of bacteria and viruses. As a result, RO processes should be combined with chlorination or a UV system if microbes are suspected to be present (Reynolds and Richards 1996).
2. Additional costs of maintenance could be incurred by combining RO processes with sedimentation processes and activated carbon prefilters for better performance and longevity (Reynolds and Richards 1996).
3. All RO membranes are not equally efficient; hence, they produce different water quality. The efficiency of each membrane depends on the quality of its material and constituents. Low-quality constituents could suffer from premature fouling (Rautenbach and Linn 1996).
4. Since RO involves a hyperfiltration membrane, it will remove just about anything including nutrients that may be useful to our body (Reynolds and Richards 1996).

7.1.6.4 Economics

The economics of setting up and operating an RO plant varies with location owing to source water content, electricity rate, and labor cost (Rautenbach and Linn 1996). Typical capital costs for setting up the plant includes the costs of pressure pumps, building, and pipes, whereas the cost of operation involves electricity cost depending on the power capacity of pumps, pretreatment cost, and maintenance (Rautenbach and Linn 1996; Reynolds and Richards 1996).

7.2 WASTEWATER QUANTITIES AND QUALITY

The most important data needs required for the design of any wastewater treatment plant (WWTP) include the quantity of the wastewater, the quality of raw wastewater, and the desired effluent quality. Knowing the flow rate helps in determining the suitable hydraulic design of the plant and the size of various treatment units, while knowing the quality of raw wastewater and the desired effluent quality helps in identifying the most suitable treatment method to be provided for the plant (Reynolds and Richards 1996). The quantity of wastewater produced varies widely from country to country and even between communities from the same country, where it depends on the lifestyle, climate, and water uses (Lin and Lee 2001). The average wastewater flow from residential areas in the United States is 265 L (70 gallons) per capita per day. In residential areas, wastewater accounts for approximately 60% to 85% of the potable water consumed (Lin and Lee 2001). Table 7.1 shows the amount of wastewater produced per person in different types of facilities. Wastewater flow rates for medium industrial developments range from

TABLE 7.1
Amount of Wastewater Produced per Person in Different Types of Facilities

Type of Facility	Gallons per Person per Day
Airports (per passenger)	5
Bathhouses and swimming pools	10
Camps:	
Campground with central comfort station	35
With flush toilets, no showers	25
Construction camps (semipermanent)	50
Day camps (no meals served)	15
Resort camps (night and day) with limited plumbing	50
Luxury camps	100
Cottages and small dwellings with seasonal occupancy	75
Country clubs (per resident member)	100
Country clubs (per nonresident member present)	25
Dwellings:	
Boarding houses	50
(additional for nonresident boarders)	10
Rooming houses	40
Factories (gallons per person, per shift, exclusive of industrial wastes)	35
Hospitals (per bed space)	250
Hotels with laundry (2 persons per room) per room	150
Institutions other than hospitals including nursing homes (per bed space)	125
Laundries—self-service (gallons per wash)	30
Motels (per bed) with laundry	50
Picnic parks (toilet wastes only per park user)	5
Picnic parks with bathhouses, showers, and flush toilets (per park user)	10
Restaurants (toilet and kitchen wastes per patron)	10
Restaurants (kitchen wastes per meal served)	3
Restaurants (additional for bars and cocktail lounges)	2
Schools:	
Boarding	100
Day (without gyms, cafeterias, or showers)	15
Day (with gyms, cafeterias, and showers)	25
Day (with cafeterias, but without gyms or showers)	20
Service stations (per vehicle served)	5
Swimming pools and bathhouses	10
Theaters:	
Movie (per auditorium set)	5
Drive-in (per car space)	10
Travel trailer parks without individual water and sewer hook-ups (per space)	50
Travel trailer parks with individual water and sewer hook-ups (per space)	100
Workers:	
Offices, schools, and business establishments (per shift)	15

Source: Adapted from Illinois EPA 1997, "Recommended Standards for Sewage Work," Part 370 of Chapter II, EPA, Subtitle C: Water Pollution, Title 35: Environmental Protection, Illinois Environmental Protection Agency (IEPA), Springfield, Illinois. ftp://www.ilga.gov/JCAR/AdminCode/035/03500370sections.html, accessed September 27, 2015.

14 to 28 m^3/(ha-day) (1500 to 3000 gal/(acre-day)), while for light industrial developments, the range is from 9 to 14 m^3/(ha-day) (1000 to 1500 gal/(acre-day)). For the commercial developments, wastewater flow rates normally range from 6.5 to 15 m^3/(ha-day) (800 to 1500 gal/(acre-day)) (Lin and Lee 2001).

The sewer network should be well designed to reduce the infiltration of groundwater to the network through defective or leaking pipe joints. The water may also enter the sewer system through manhole wells (Lin and Lee 2001). The amount of groundwater that enters the sewer system through infiltration ranges from 0.0094 to 0.94 m^3/(day-mm-ha) (100 to 10,000 gal/(day-inch-miles)) or more (Metcalf and Eddy, Inc. 1981). The maximum allowable infiltration rate is approximately 0.463 m^3/(day-km-cm) (500 gal/(day-mile-inch)) of pipe diameter (Lin and Lee 2001). The quantity of infiltration can be estimated as 10% of the average domestic daily flow, or 3.0% to 5.0% of the peak hourly wastewater flow (Lin and Lee 2001). The infiltration rate can be reduced to 0.1852 m^3/(day-mm-ha) (200 gal/(day-mile-inch)) of pipe diameter, if a better pipe joint material was chosen and a tight control of construction methods was applied (Lin and Lee 2001).

7.3 WASTEWATER FLOWS AND CHARACTERISTICS

The quality of raw and treated wastewater is identified by measuring the physical, biological, and chemical characteristics of the wastewater (Lin and Lee 2001). The main physical characteristics of municipal wastewater include color, total dissolved solids (TDS), total suspended solids (TSS), volatile SS, settleable solids, turbidity, odor, and temperature (Reynolds and Richards 1996). The dissolved solid and SS as well as the temperature are the most important parameters in wastewater treatment. Temperature controls the biological activities and chemical reaction, while the solids content controls the size and operation of the treatment units (Lin and Lee 2001). Turbidity in wastewater is mostly caused by suspended particles, which range in size from clay (1.0–4.0 μm) to coarse suspensions (0.5–1.0 mm) (Reynolds and Richards 1996). The wastewater should be transferred quickly to the nearest WWTP to avoid biochemical reactions in the sewer system. The biochemical reactions may cause corrosion in the sewer pipes. The fresh domestic wastewater is usually characterized by a light tan color. If the fresh wastewater takes more than 6 h to reach the WWTP, biochemical oxidation may occur and the wastewater color will change from light tan to black (Reynolds and Richards 1996). The black color indicates low dissolved oxygen concentration, which requires increased air flow rate during the aeration process. The odor of fresh domestic wastewater is usually not offensive and has a smell similar to oil and soap. The smell becomes unpleasant during biochemical oxidation, where offensive compounds such as mercaptans, hydrogen sulfide, skatol, and indol are produced. Hydrogen sulfide is produced by the decomposition of organic material, as well as by the biological reduction of sulfates. It is the main cause of the offensive odor in wastewater because it is characterized by the unpleasant rotten egg smell (Reynolds and Richards 1996). The dissolved solids refer to small solid particles that pass through a filter paper with a 2.0-μm average pore size, while the SS refer to solid particles that retained on the filter paper and do not pass through (EPA 1997a). The fixed solids refer to solid particles that remain after the filtered residue is ignited at 1020°F (550°C), while the volatile solids refer to solid particles that burn off after ignition (Reynolds and Richards 1996). The settleable solids refer to solid particles that can settle out by sedimentation using a 1.0-L Imhoff cone over a given period. In the usual municipal wastewater, the settleable solids represent approximately 60% to 65% of the SS (Reynolds and Richards 1996).

The main chemical characteristics of municipal wastewater include total organic carbon (TOC), COD, pH, alkalinity, sulfate ion $\left(SO_4^{-2}\right)$, chloride ion (Cl^-), hardness including calcium (Ca) and magnesium (Mg), heavy metal ions, grease content, various forms of phosphorus, and various forms of nitrogen, as well as priority pollutants and trace elements (Reynolds and Richards 1996). The TOC is the amount of organic compounds in wastewater and it is measured in terms of the carbon content in organic materials (Florescu et al. 2011). The COD is the amount of oxygen required for chemical decomposition of organic compounds (Reynolds and Richards 1996). Knowing the pH, the alkalinity, the sulfate ion, and the chloride ion in the treated wastewater is essential to examine the appropriateness of reusing the treated wastewater. pH and alkalinity are also very important parameters that control the performance of some treatment processes (Reynolds and Richards 1996). Knowing the concentrations of some heavy metals, such as zinc (Zn), nickel (Ni), mercury (Hg), copper (Cu), lead (Pb), cadmium (Cd), arsenic (As), silver (Ag), and chromium (Cr) in the raw wastewater is essential to assess the treatability of a wastewater (Reynolds and Richards 1996). Knowing the concentrations of the grease and the trace elements such as copper, zinc, iron, and cobalt in raw wastewater is also useful for the control of some biological processes (Lin and Lee 2001; Reynolds and Richards 1996). Phosphorus may be present in wastewater in the form of organic phosphorus, which include proteins and their breakdown products, or inorganic phosphorus, such as the phosphate ion $\left(PO_4^{-3}\right)$. Nitrogen can be present in wastewater in many chemical forms including ammonium $\left(NH_4^+\right)$, organic nitrogen, total Kjeldahl nitrogen, and nitrite $\left(NO_2^-\right)$ (Reynolds and Richards 1996). The priority pollutants are divided into toxic organic and inorganic chemicals. These chemicals are extremely toxic to humans even at low concentrations. If the concentrations of these chemicals at the plant effluent exceed the allowable limits, they must be reduced at their source of origin (Reynolds and Richards 1996). The toxic inorganic chemicals include nonmetals (such as selenium), metalloids (such as arsenic), and heavy metals (such as barium [Ba], Hg, Pb, Cr, and Cd). Examples of toxic organic chemicals include vinyl chloride, lindane, benzene, toxaphene, carbon tetrachloride, methoxychlor, trichloroethylene, xylenes, endrin, toluene, *para*-dichlorobenzene, 1,1,1-trichloroethane, 2,4,5-trichlorophenoxy, 2,4-dichlorophenoxyacetic acid, 1,2-dichloroethane, and 1,1-dichloroethylene.

The main biological characteristics of municipal wastewater include BOD, the microbial life in wastewater, and the nitrogenous oxygen demand (Lin and Lee 2001). The BOD is the amount of oxygen consumed by microbes, mainly bacteria, over a 5-day period at 20°C to oxidize organic compounds under aerobic conditions (Reynolds and Richards 1996). The nitrogenous oxygen demand is the amount of oxygen needed by nitrifying bacteria to convert ammonia nitrogen to nitrate nitrogen (Reynolds and Richards 1996). The microbial life in the wastewater includes bacteria, viruses, protozoa, algae, fungi, nematodes, and rotifers. The bacterial population is very important in biological treatment processes and represents the most dominant microbe with a number of cells ranging from 100×10^3 to 100×10^6 cells per milliliter. The fecal coliforms are harmful microbial contaminants, which are a species of *Aerobacter aerogenes* and *E. coli*, and live in the intestines of humans, soil, and warm-blooded animals (Reynolds and Richards 1996). The fecal coliform is the main cause of the spread of many waterborne diseases, such as dysentery, hepatitis A, typhoid fever, gastroenteritis, and cholera (Cabral 2010). Counting the number of colonies of coliform bacteria in the wastewater sample will help in evaluating the removal efficiency of coliform by certain treatment processes and in examining the appropriateness of discharging the treated wastewater to surface water bodies (Reynolds and Richards 1996).

7.4 TREATMENT ADVANTAGES AND LIMITATIONS

WWTPs are designed to remove contaminants from wastewater before releasing them into receiving waters and to accelerate the natural purification process, which occurs in natural receiving waters (Lin and Lee 2001). By treating the wastewater, the treated effluent can be a valuable resource that can be reused or recycled. Proper wastewater management is also very important to protect public health and prevent the spread of waterborne diseases, such as kidney failure, respiratory disease, hepatitis, congenital heart disease, typhoid, cholera, diarrhea, encephalitis, eye infections, pleurodynia, diabetes mellitus, fever, meningitis, myocarditis, rash, gastroenteritis, and paralysis (Safe Drinking Water Foundation 2009). All WWTPs worldwide are required to reduce the organic compounds and SS concentrations to acceptable limits. In addition, many treatment plants were designed to achieve high removal efficiency for nutrients and pathogenic microorganisms found in wastewater (Horan 1990). Several treatment units are required to efficiently and effectively achieve all of these requirements. The most common municipal WWTPs include primary and secondary treatment plants, physical–chemical treatment plants, and tertiary treatment plants (Reynolds and Richards 1996).

7.5 WASTEWATER COLLECTION SYSTEMS

Wastewater collection systems are underground conduits to collect and convey wastewater generated from residential, industrial, and commercial areas to the nearest WWTP (Lin and Lee 2001). These systems also include manholes, oil and grease traps, pump stations, and inverted siphons (Reynolds and Richards 1996). Wastewater from homes, industries, and businesses enters the collection system through service lines. The lateral and branch pipes are used to collect wastewater from different service lines and convey it to larger lines called main lines (Parcher 1998; WEF 2010). The branch or lateral lines can serve a small number of streets. The main lines transfer the wastewater flow to the largest lines in the sewer system, called trunk lines. Finally, the trunk lines transfer wastewater directly to the treatment plant (Parcher 1998; Ragsdale 2014).

Manholes are important parts of the collection system, where they provide access to the system for inspection, cleaning, and clearing stoppages. Manholes should be installed at junctions of conduits and at every change in sewer slope, size, and direction (Ragsdale 2014). Inverted siphons are smaller pipes used if the sewer line must pass under an obstacle such as streambeds, railroads, or a roadway (Sanders 2009; WEF 2010). During low-flow conditions, debris and grit accumulate inside the inverted siphons, which may cause a blockage and cause a severe flood risk. Inverted siphons rely primarily on high velocities to remove any blockage in the pipe (Ragsdale 2014).

Collection lines are installed with sufficient downhill slope to allow the wastewater to be transported by gravity, as well as reduce the number of pumps needed to convey the wastewater to the treatment plant. The flow velocity in the collection lines should not be less than 2.0 ft/s to prevent septic conditions (Ragsdale 2014; WEF 2010). Lift stations are built at lower elevations of the sewer network in order to lift the sewage up from lower elevations to higher elevations, so that it can flow by gravity (Sanders 2009). The capacity of a line can be determined through knowing the type of the pipe, the slope of the line, and the size of the pipe. The collection systems are designed to accommodate the peak flow conditions. Inflow and infiltration are also taken into account when designing a collection system (WEF 2010). Inflow may come from surface water runoff, which enters the sewer system directly through illegal connections that permit storm water to enter the system or through submerged manhole covers. Infiltration problems occur when groundwater enters the sewage system through leaking joints or broken pipes (Ragsdale 2014).

The most popular materials used for sewer pipes include vitrified clay pipe (VCP), reinforced concrete pipe (RCP), ductile iron pipe (DIP), cast iron pipe (CIP), acrylonitrile butadiene styrene (ABS), and polyvinyl chloride (PVC). VCP is the most common pipe material used in sewer applications with a pipe diameter ranging from 4 to 36 inches. It is constructed with bell and spigot rubber gasket joints to maintain a tight seal and to prevent sewage spills (Sanders 2009). VCP is strong enough to withstand heavy trench loading and provides high corrosion resistance. RCP is larger than VCP in size, where the pipe diameter ranges from 18 to 60 inches (Parcher 1998; Ragsdale 2014). Concrete pipes are constructed with mortise and tenon connections or bell and spigot joints, as well as bitumastic compounds or rubber gaskets to create a seal between the two pipes (Sanders 2009; WEF 2010). The potential drawback of using concrete pipes is the internal corrosion caused by sewer gases. DIP and CIP are mainly used in areas characterized by high trench loading such as under a dirt or gravel road that carries heavy equipment or under a railroad track. Iron pipes can also be used as an inverted siphon to convey water under roadbeds or streams (Parcher 1998). The internal surface of the iron pipe is also subject to corrosion caused by sewer gases. ABS and PVC pipes are made from plastic and they are used for small-diameter sanitary sewers (WEF 2010). These pipes are characterized as being lightweight, flexible, and easy to install, and their internal surface is impervious to corrosion caused by inorganic salts, sewer gases, and organic acids found in wastewater. The main drawback of using such pipes is their inability to withstand heavy trench loading (Ragsdale 2014; WEF 2010).

7.6 WASTEWATER PROCESSING/UNIT OPERATIONS

7.6.1 Preliminary Treatment

Wastewater usually contains large amounts of SS and grit that may interfere with treatment processes or cause mechanical wear and increase maintenance cost for treatment equipment. The main purpose of preliminary treatment at the WWTP is to prevent damage to pumps and subsequent treatment units and unit operations through clogging. Preliminary treatment operations typically include screening devices (coarse screens, bar racks, and fine screens), comminution devices (shredders, cutters, and grinders), grit removal chambers, and flow equalization. The screen devices are usually used to remove large floating and suspended materials such as plastic, wood, paper, cloth, garbage, and so on. Comminution devices are used to reduce the size of large particulates in the wastewater stream before beginning the treatment process (EPA 2000a; Tillman 1996). The grit chamber is constructed to collect heavy inorganic solids such as sand, gravel, small stones, cinders, and grit that have passed through screens and consequently reduce the volume of sediment in the sedimentation basins (Tillman 1996). Preliminary systems are also designed to remove large amounts of oil and grease. Flow equalization is the process of controlling the treatment plant inlet flow rate in order to improve the performance of treatment processes and reduce the cost and size of treatment units (Lin and Lee 2001; Vesilind 2003).

7.6.2 Primary Treatment

The main objective of primary treatment is to reduce the flow velocity of the wastewater to approximately 1 to 2 ft/min (0.3 to 0.6 m/min) to allow low-density materials to float and SS to settle out. Primary settling tanks remove and collect the settled sludge solids for further treatment or the solids can be transferred directly for final disposal (Sperling 2007). Floating materials are removed by skimming. Approximately 50% to 60% of the incoming TSS, 65% of the oil and grease, and 25% to 35% of BOD_5 are removed during primary treatment (Lin and Lee 2001).

7.6.2.1 Sedimentation/Clarification

Sedimentation, or clarification, is the process of allowing suspended particles heavier than water (i.e., with a specific gravity greater than 1.0) to settle out under the influence of gravity (Lin and Lee 2001). Sedimentation tanks can be circular, rectangular, or square. Circular and rectangular settling tanks are the most common systems used to treat wastewater (Sperling 2007). Rectangular settling tanks are constructed with lengths ranging from 50 to 300 ft (15 to 90 m) and widths ranging from 10 to 80 ft (3 to 24 m). The tank depth should exceed 7 ft (2 m) (Vesilind 2003). The inlet of the rectangular basin is located at one end and typically consists of small pipes and baffles used to dissipate the inlet velocity in order to prevent short-circuiting and to diffuse the flow evenly across the cross section of the tank. The outlet is located at the other end of the tank and consists of weirs (Lin and Lee 2001). Figure 7.3 illustrates a typical rectangular primary settling tank.

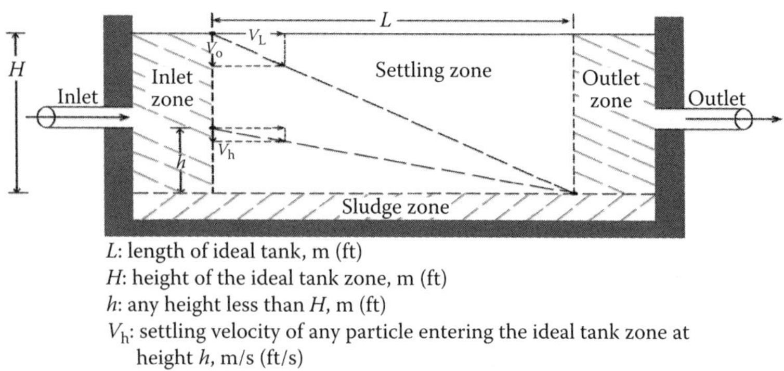

FIGURE 7.3 Typical rectangular primary settling tank. (Adapted from EPA 1997b, *Wastewater Treatment Manual: Primary, Secondary, and Tertiary Treatment*, Ardcavan, Wexford, Ireland.)

The settled sludge is collected in a hopper located at the end of the basin, either by single bottom scrapers mounted on a travelling bridge or by flight scrapers mounted on parallel chains (Sperling 2007; Tillman 1996). Circular settling tanks range from 10 ft (3 m) to more than 300 ft (90 m) in diameter and 8 to 13 ft (2.4 to 4 m) in depth (Vesilind 2003).

The inlet of the circular basin is typically located at the center of the tank and the wastewater flows outward. The outlet is in the form of an overflow weir and extends around the perimeter of the basin with baffles extending 8 to 12 inches (20 to 30 cm) beneath the wastewater surface to retain grease and floating material. The settled sludge is collected in a hopper located in the middle of the tank bottom (Lin and Lee 2001; Tillman 1996). Figure 7.4 depicts a typical circular primary clarifier/settling tank. The construction cost is higher for rectangular tanks than circular tanks, where thinner walls are constructed in circular tanks that act as tension rings. Rectangular tanks require fewer pipes than circular tanks (Vesilind 2003).

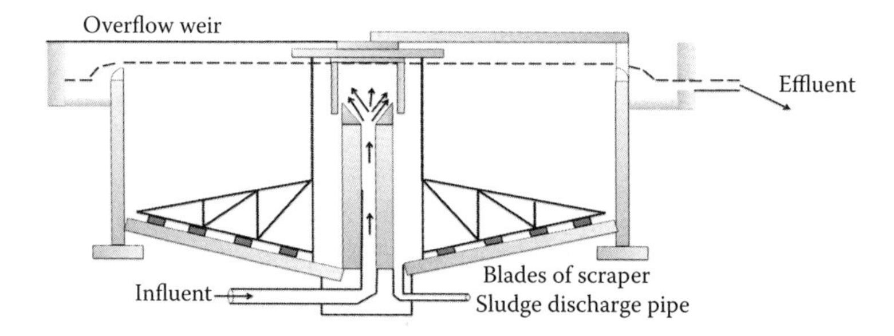

FIGURE 7.4 Typical circular primary settling tank. (Adapted from EPA 1997b, *Wastewater Treatment Manual: Primary, Secondary, and Tertiary Treatment*, Ardcavan, Wexford, Ireland.)

Square settling tanks remove floatable materials and settleable solids in the same manner as circular tanks (Sperling 2007). Square units are not as cost-effective as circular units because of their high construction cost; square tanks require thicker walls than circular tanks. Corner sweeps are used to remove settled solids from corners. Mechanical and operational problems can occur as a result of the use of corner sweeps (Vesilind 2003).

The typical operating conditions of the primary settling tank include the following: (1) pH, should be in the range of 6.5 to 9; (2) temperature, varies with climate; (3) influent SS ranging from 5 to 15 mg/L; (4) effluent SS ranging from 0.3 to 5 mg/L; (5) dissolved oxygen < 1.0 mg/L; (6) influent BOD ranging from 150 to 400 mg/L; (7) effluent BOD ranging from 50 to 150 mg/L; (8) the percentage of solids ranging from 4% to 8%; and (9) the percentage of volatile matter ranging from 40% to 70% (Spellman 2009). The performance of the primary clarifier depends on many factors, including the performance of the preliminary treatment processes, the flow rate through the tanks, and other wastewater characteristics, such as temperature, nature and quantity of industrial wastes, strength, and the density, shape, and size of the solid particles (Tillman 1996). Key factors in primary treatment include the following (Tillman 1996):

- Weir overflow rate, [(m^3/day)/(lineal meter)] = flow [m^3/day]/weir length [lineal meter]
- Solids loading rate, [(g/day)/(m^2)] = solids into clarifier [g/day]/surface area [m^2]
- Retention time, [h] = [volume of settling zone (m^3) × 24(h/day)]/[flow (m^3/day)]
- Surface loading rate, [(m^3/day)/(m^2)] = flow [m^3/day]/surface area [m^2]

7.6.2.2 Coagulation/Flocculation

Coagulation and flocculation processes involve adding a chemical reagent to wastewater to combine with slow-settling SS and nonsettleable colloidal solids to produce a rapid-settling floc. The coagulation process requires rapid mixing while adding coagulant to destabilize very fine SS and colloidal particles in order to enhance the agglomeration of the destabilized particles. The flocculation process requires slow mixing in order to promote agglomeration and form rapid settling flocs (Reynolds and Richards 1996). Figure 7.5 shows a typical coagulation and flocculation process.

The removal efficiency of COD, phosphorus, BOD, SS, and pathogens when applying coagulation and flocculation processes is typically 30% to 60%, 70% to 90%, 40% to 70%, 60% to 90%, and 80% to 90%, respectively. In comparison, settling without adding coagulant may remove only 25% to 40% of the BOD_5, 5% to 10% of the phosphorus loadings, 40% to 70% of the SS, and 50% to 60% of the pathogens (Vesilind 2003). In wastewater treatment, the most commonly used coagulants are aluminum salts, iron salts, lime, and polyelectrolytes. Key factors that influence the coagulation and flocculation of wastewater include pH, SS, turbidity, temperature, duration and degree of agitation, coagulant aid, anionic and cationic composition and concentration, and nature and dosage of the coagulant (Reynolds and Richards 1996).

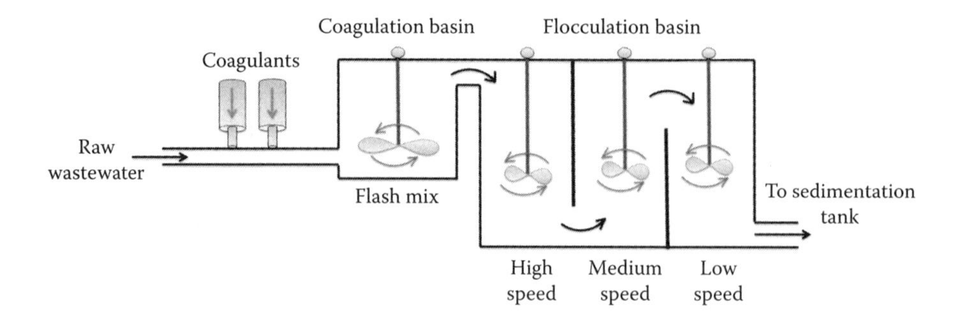

FIGURE 7.5 Coagulation and flocculation processes. (Adapted from EPA 2002, *Wastewater Treatment Manuals*: *Coagulation, Flocculation, and Clarification*, Wexford, Ireland; and EPA 2015a, *Conventional Treatment*, Drinking Water Treatability Database, Washington, D.C. http://iaspub.epa.gov/tdb/pages/treatment/treatmentOverview.do?treatmentProcessId=1934681921, accessed September 27, 2015.)

The typical operating conditions of coagulation and flocculation processes include (1) coagulant dose (dependent on the quantity and quality of the wastewater) and (2) pH (the coagulation process should operate at a pH value suitable for the wastewater temperature) (Pizzi 2010). Advantages of coagulation include increased efficiency of solids removal, enhanced precipitation of phosphate ion, and an increased ability of the primary sedimentation tanks to operate at higher overflow rates. Disadvantages of the coagulation process include high chemical cost, an increase in primary sludge mass in the sludge handling units, and the production of complex solids that are sometimes difficult to dewater and thicken (Reynolds and Richards 1996; Vesilind 2003).

The selection of a coagulant for wastewater treatment requires the use of laboratory studies and often pilot studies. The jar test is an excellent laboratory technique used to select the proper coagulant type and dosage required for coagulation. In this test, equal amounts of the wastewater sample are poured into a series of beakers and then each beaker is treated with a different dose of the coagulant. The contents are rapidly stirred to allow the coagulant to be fully mixed with the wastewater sample in each beaker and then the contents are slowly stirred to simulate flocculation. After a certain period, typically between 30 and 60 min, the stirrers are turned off to allow the floc formed to settle to the bottom of the beaker. Preliminary results from the test include the percent color and turbidity (SS) removed, the floc size and formation time, and the pH of the coagulated wastewater (Reynolds and Richards 1996).

7.6.2.3 Filtration

Filtration is a process of passing wastewater through a porous medium to remove SS and colloidal material. It is used in wastewater treatment to produce a high-quality effluent, where it can efficiently filter (1) chemically treated raw wastewaters, (2) chemically treated secondary effluents, and (3) untreated secondary effluents (Reynolds and Richards 1996; Tillman 1996). Filtration processes can effectively improve disinfection and reduce turbidity. Filtration is typically applied after coagulation, flocculation, and sedimentation processes or after secondary biological treatment (Tillman 1996). Filters are also classified according to the filter bed media, which include single-medium filters, dual-media filters, and multimedia filters (see Figure 7.6). Single-medium filters have one type of media, typically crushed anthracite coal or sand. Dual-media filters have two types of media, typically sand and crushed anthracite. These processes allow more depth filtration rather than just surface filtration. These filters offer more storage capacity for solids in the bed and thus increase the interval between backwashes. Multimedia filters have three types of medium, typically sand, crushed anthracite, and garnet or ilmenite. These filters minimize head loss buildup and thus increase the interval between backwashes and permit longer filter runs (Reynolds and Richards 1996; Rowe and Isam 1995). The dual and multimedia filters are the most widely used in tertiary and advanced wastewater treatment (Reynolds and Richards 1996). Filtration may be provided by enclosed pressure filters or open gravity filters (Tillman 1996). The typical operating conditions of filtration process include the following: (1) water is applied at a rate 1.5 to 2 gallons/min/ft^2 of filter media surface, (2) turbidity level at the raw wastewater should not exceed 1000 turbidity units (TU), (3) the run time ranges from 12 to 72 h dependent on the raw wastewater quality, and (4) backwash should start once the head loss reaches 8 ft (Spellman and Joanne 2001).

According to the loading rate, filters are classified as high-rate sand filters, rapid sand filters, and slow sand filters. The loading rate (flow velocity) is the flow rate of wastewater applied over a surface area of the filter and can be determined by (Lin and Lee 2001)

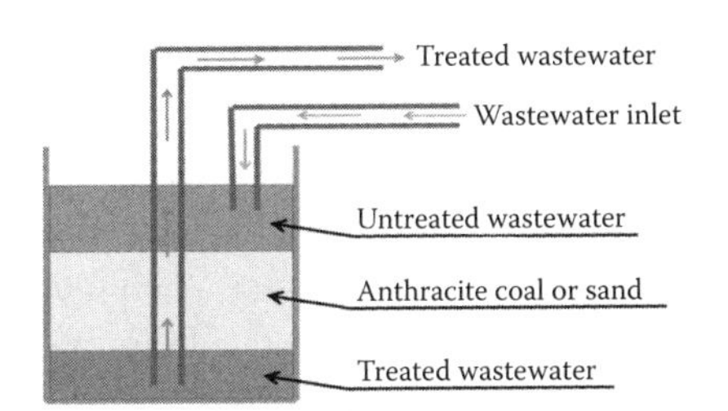

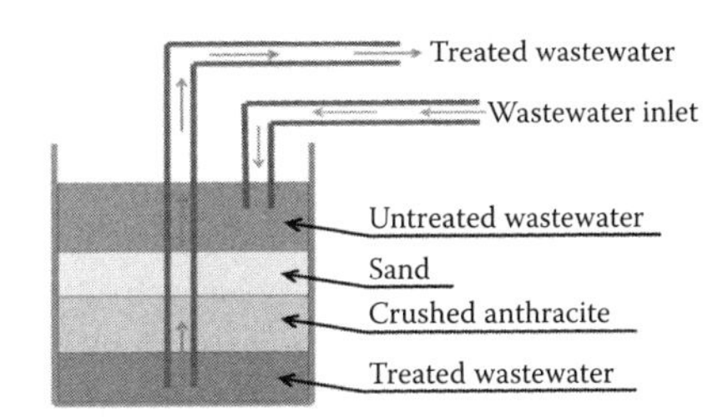

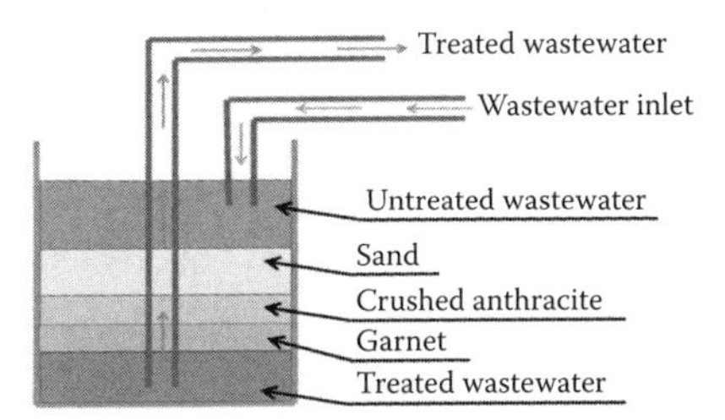

FIGURE 7.6 Single-medium, dual-media, and multimedia filters. (Adapted from EPA 1984a, *Tertiary Granular Filtration: Problems and Remedies*, EPA Report No.: 832-R-84-113, Office of Water Program Operations, Washington, D.C. [August].)

$$U = Q/A$$

where U = loading rate (gpm/ft^2), Q = flow rate (gpm), and A = surface area of filter (ft^2).

Slow sand filters have been used in early times and still prove to be efficient. They have a very high effectiveness in removing protozoa, such as *Cryptosporidium* and *Giardia*. Rapid sand filters are the most commonly used filters over the past several decades (Lin and Lee 2001). Over time, the filter media become clogged with particulate matter removed from the wastewater. The onset of clogging is usually detected by an increased head loss or decreased filtration rates typically resulting in an increased contaminant concentration (i.e., breakthrough). The filter media should be backwashed (reversing the flow) if the head loss reaches a maximum preset value (Tillman 1996).

7.6.3 Secondary Treatment/Biological Treatment

Secondary treatment processes are responsible for removing the colloidal and dissolved organic and inorganic solids, which remain after primary treatment (Horan 1990). The majority of the SS found in wastewaters can also be removed by secondary treatment. Phosphorus and nitrogen removal can also be achieved through biological treatment (Lin and Lee 2001). Secondary treatment processes can remove up to 85% of suspended matter and BOD_5, but are not effective in removing heavy metals, viruses, and dissolved minerals (Lin and Lee 2001; Nadakavukaren 2011). In biological treatment, microorganisms feed on organic matter found in wastewater, converting them into simpler compounds. Favorable environmental conditions, such as nutrients, dissolved oxygen, and temperature must be provided for microbial growth and proliferation (Lin and Lee 2001). The most common biological treatment systems are attached growth systems (rotating biological contactor [RBC] and trickling filters), suspended growth systems (activated sludge), and dual-process systems (combination of suspended and attached growth treatment) (Horan 1990; Qasim 1999). Other biological treatment systems include the oxidation ditches, aerated lagoons, phosphorus removal units, stabilization ponds, contaminant ponds, biological nitrification, denitrification, and high-purity oxygen–activated sludge (Lin and Lee 2001).

7.6.3.1 Activated Sludge

In the suspended growth process, the living microorganisms are mixed with the wastewater under a continuous supply of air in order to survive and feed on the organics found in the wastewater (Lin and Lee 2001). The mixture of wastewater and microorganism population (activated sludge) is called "mixed liquor." The air can be supplied to the system either by static aerators located at the surface of the tank or by air diffusers located near the bottom of the tank (Tillman 1996). As living microorganisms grow, they clump together (flocculate) to form an active mass of microorganisms (biological floc) (Grady et al. 1999). The produced floc will easily settle in the secondary clarifier and then returned to the aeration basin or wasted to a sludge handling unit for treatment and disposal (Lin and Lee 2001; Vesilind 2003). Approximately 40% to 60% of the settled sludge is returned to the aeration tanks and the rest is wasted (Tillman 1996). Figure 7.7 shows a conventional activated sludge process.

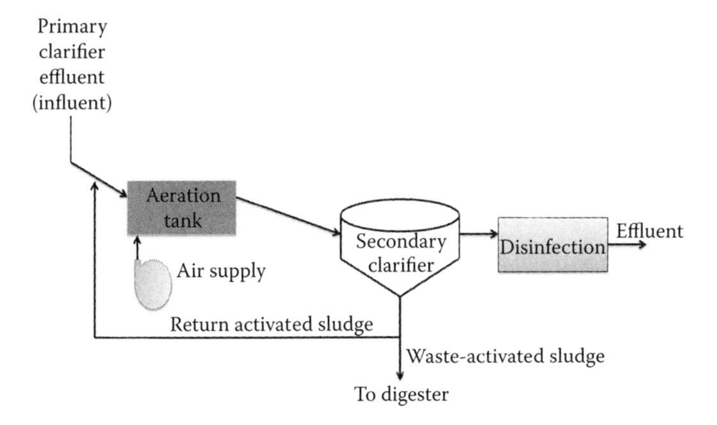

FIGURE 7.7 Conventional activated sludge process. (Adapted from EPA 1997b, *Wastewater Treatment Manual: Primary, Secondary, and Tertiary Treatment*, Ardcavan, Wexford, Ireland.)

The typical operating conditions of coagulation and flocculation processes include the following: (1) aeration—sufficient aeration is required to satisfy the organism oxygen requirements and to prevent the loss of activated sludge, (2) alkalinity—sufficient alkalinity is required to keep the pH in the range of 6.5 to 9 and to support the nitrification process, (3) nutrients—activated sludge process requires sufficient nutrients to perform well, (4) pH (should range from 6.5 to 9.0), and (5) temperature—warm temperature is preferred for the denitrification process (Spellman 2009). The main parameters in designing and operating the activated sludge process are the sludge volume index (SVI), mean cell residence time (MCRT), and food-to-microorganism (F/M) ratio. These parameters are calculated by using the following relations (Tillman 1996):

- Sludge volume index (SVI), ml/g = [30 min settled volume, (ml/L) × 1000]/[mixed liquor SS, (mg/L)]
- Mean cell residence time (MCRT), days = [solids in total system, (lb)]/[solids wasted + lost, (lb/day)]
- Food-to-microorganism (F/M), ratio = [BOD or COD in primary effluent, (lb/day)]/[lb volatile SS in aeration tanks]

7.6.3.2 Trickling Filters and RBCs

The trickling filter is the most common fixed or attached growth process used for municipal wastewater treatment (Grady et al. 2011). A trickling filter may be rectangular, square, or circular and consists of a fixed bed of coarse material, usually plastic media or stone slates, covered with microorganisms (Tillman 1996). A fixed nozzle or a rotating distributor arm is used to spray wastewater from the primary effluent to the filter media at a controlled rate. An underdrain system is used to carry the treated wastewater to the subsequent treatment units (Lin and Lee 2001; Reynolds and Richards 1996; Tillman 1996). Figure 7.8 shows a typical trickling filter process.

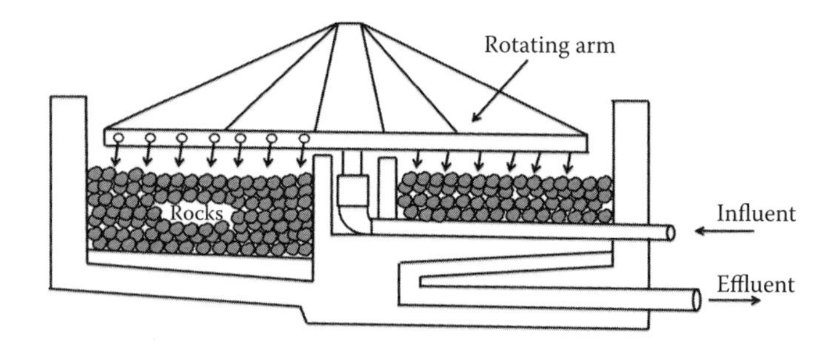

FIGURE 7.8 Schematics of a typical trickling filter. (Adapted from EPA 2000b, *Trickling Filters*, EPA Report No.: 832-F-00-014, Wastewater Technology Fact Sheet, Washington, D.C. [September].)

The advantages of the trickling filter are its simplicity of operation, low biomass yield, low power requirements, and resistance to shock loads. Its main disadvantages are low BOD removal (85%) and high SS concentration in the effluent (Lin and Lee 2001; Reynolds and Richards 1996; Tillman 1996). Application of the trickling filter process includes providing a large surface area for the microorganisms to grow on the surface of the support media (forming a biological slime) and feed on organic matters found in wastewater. Air diffusers are used to supply oxygen through the void in the filter media to trickle the wastewater downward through the bed media (Lin and Lee 2001). The slime layer periodically sloughs off and settles in the secondary sedimentation basin (Tillman 1996).

The typical operating conditions of coagulation and flocculation processes include the following: (1) liquid retention time, ranging from 8 to 20 min; (2) BOD loading, ranging from 5 to 500 lb/day/1000 ft^3 (0.08 to 8.0 kg BOD/m^3/day); and (3) ventilating area, 1 ft^2 (0.1 m^2) for each 10 to 15 ft (3 to 4.6 m) of the tower periphery, and 10.7 to 21.5 ft^2 (1 to 2 m^2) in the underdrain area per 1000 m^3 of media (Vesilind 2003). The main parameters in designing and operating the trickling filter process are the hydraulic loading rate, the recirculation ratio, and the organic loading rate. These parameters are calculated by using the following relations (Tillman 1996):

- Hydraulic loading rate, [(gal/day)/ft^2] = [Flow, gal/day (including recirculation)]/[Media top surface, sq. ft]
- Recirculation, ratio = [Recirculation flow, (MGD)]/[Average influent flow, (MGD)]
- Organic loading rate, [(lb/day)/(1000 cu. ft)] = [BOD into filter, (lb/day)]/[Media volume, (1000 cu. ft)]

An RBC consists of a horizontal shaft covered with a large-diameter circular plastic media (see Figure 7.9). The shaft is partially submerged in the incoming wastewater to allow the microbes that stuck on the surface of the shaft to get oxygen from the air as the shaft is rotated (Lin and Lee 2001). A thin layer of biological slime will cover the disk surface as the microbes in the incoming wastewater stick and grow on its surface (Tillman 1996). The excess growth periodically sloughs off

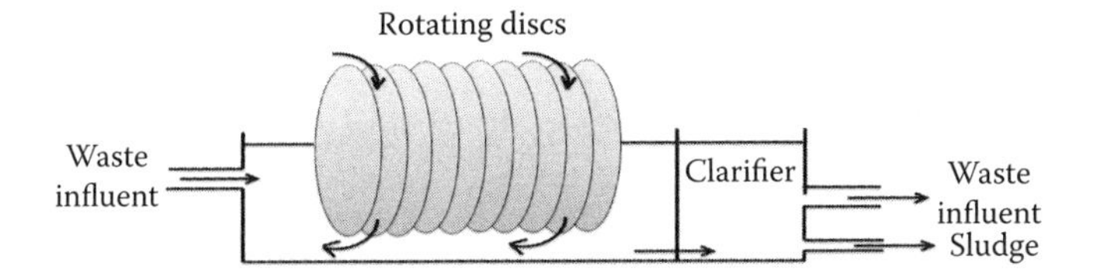

FIGURE 7.9 Rotating biological contactor. (Adapted from EPA 1984b, *Summary of Design Information on Rotating Biological Contactors*, EPA Report No.: 430/9-84-008, Office of Water Program Operations, Washington, D.C. [September].)

from the disks and settles in the secondary sedimentation basin (Tillman 1996). The advantages of the RBC system are the ability of handling a wide range of flows, low operating costs, short retention time, low power requirements, and low sludge production. The main disadvantages are that it requires frequent maintenance and covering to protect against freezing (Lin and Lee 2001; Reynolds and Richards 1996; Tillman 1996). The typical operating conditions of an RBC system include the following: (1) organic loading, exceeds 6.4 lb BOD/day/1000 ft^2 (0.031 kg BOD/m^2/day); (2) temperature, should be above 13°C (55°F); (3) biofilm control, the film thickness should range from 0.07 to 4.0 mm; and (4) dissolved oxygen, minimum acceptable dissolved oxygen level is 2 mg/L (Vesilind 2003).

7.6.3.3 Membrane Operations

The membrane process is an innovative technique used to separate specific compounds from an aqueous solution containing numerous compounds using a selective permeable ultrafiltration membrane. The traditional membrane processes are (1) RO, (2) dialysis, and (3) electrodialysis (Reynolds and Richards 1996). The driving force is the most important factor in the membrane process, which is used to transfer the solute across the membrane. The driving force can be expressed as the difference in pressure, concentration, or electric potential for the case of using RO, dialysis, and electrodialysis processes, respectively. The main drawback of using membrane processes is the low mass transfer rate per unit area of the membrane (Reynolds and Richards 1996). In the RO process, a hydrostatic pressure and a semipermeable membrane are used to separate a solvent (usually water) from a saline solution. The word *osmosis* refers to the transfer of the solvent water from the solvent side to the saline side. The typical operating conditions of the RO process include the following: (1) system pressure should range from 30 to 100 psi; (2) temperature should range from 40°F to 100°F (4°C to 38°C), pH may range from 3 to 11; (3) TDS level should be less than 2000 mg/L; and (4) turbidity must not exceed 1.0 NTU (Lin and Lee 2001; Reynolds and Richards 1996; Tillman 1996).

In the dialysis process, a semipermeable membrane is used to transfer solutes of different molecular or ionic size from the solution side to the solvent side (see Figure 7.10). This membrane is characterized by very small pore openings, which allow smaller molecules or ions to pass, but not larger molecules or ions (Reynolds and Richards 1996).

In the electrodialysis process, the selectively permeable membrane is used to separate inorganic electrolytes from an aqueous solution, and an electromotive force

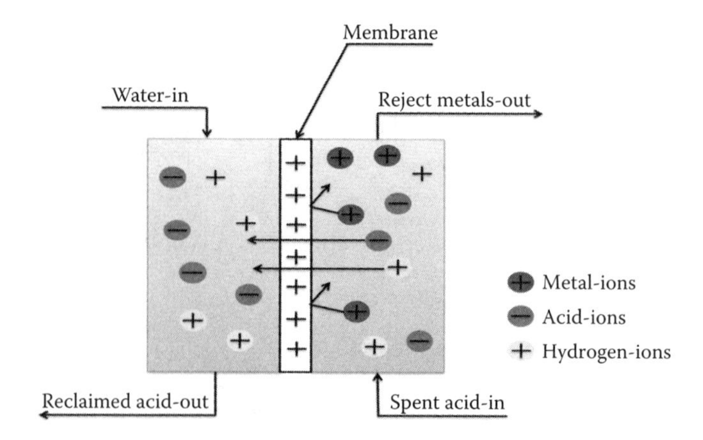

FIGURE 7.10 Dialysis membrane process. (Adapted from Fumatech 2014, "Diffusion Dialysis," Functional Membranes and Plant Technology, Bietigheim-Bissingen, Germany. http://www.fumatech.com/EN/Membrane-technology/Membrane-processes/Diffusion-dialysis/, accessed March 5, 2015.)

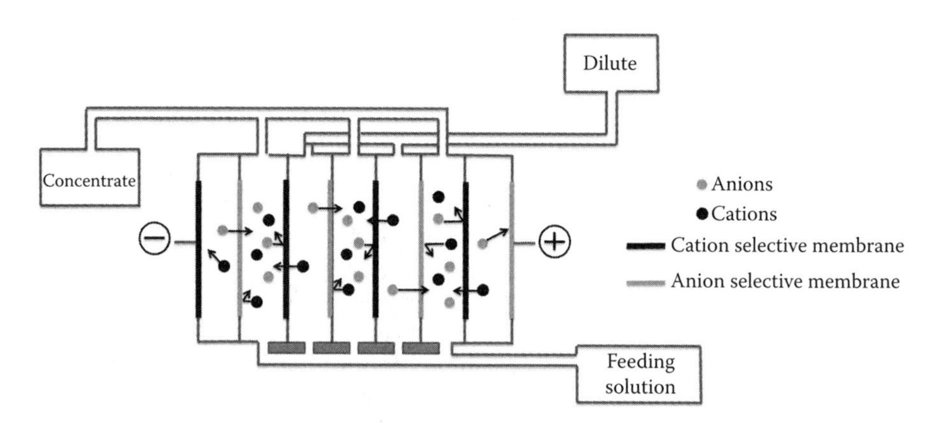

FIGURE 7.11 Electrodialysis membrane process. (Adapted from United States Department of the Interior Bureau of Reclamation 2010, *Reclamation: Managing Water in the West*, Washington, D.C.)

is applied to increase the rate of mass transfer. Electrodialysis can be used for desalinating of brackish water and seawater, as well as for demineralizing effluents in tertiary treatment (Reynolds and Richards 1996). Figure 7.11 shows a typical electrodialysis process.

7.6.4 Tertiary Treatment

Tertiary treatment, also called effluent polishing, directly follows secondary treatment and is responsible for improving the effluent quality before it is reused or discharged into the environment (i.e., receiving body of water). The effluent polishing improves the removal efficiency of nutrients and SS (Vesilind 2003). Most of the WWTPs contain at least one tertiary treatment process to remove contaminants, which remain after secondary treatment (Lin and Lee 2001). Tertiary treatment processes include lagooning, disinfection, filtration, and nutrient removal (Vesilind 2003).

7.6.4.1 Stabilization Ponds and Aerated Lagoons

In stabilization ponds or lagoons, wastewater is treated by using wind, sunlight, oxygen, and algae (Tillman 1996). Wastewater enters the pond using a pipe mounted at the edge or the center of the pond (see Figure 7.12). Bacteria grow and proliferate in the pond by consuming the oxygen released by the algae. Sunlight is the main source of energy for the algae to grow. The bacteria also release inorganics and carbon dioxide for use by the algae (Lin and Lee 2001). The ponds are usually shallower than the lagoons.

The lagoons are divided into aerated lagoons and unaerated or facultative lagoons (Vesilind 2003). The aerated lagoons depend on wind action plus aeration equipment to provide the oxygen required by bacteria to feed on organic material found in the wastewater. The air can be supplied to the lagoon either by mechanical aerators located at the surface of the lagoon or by air diffusers located in the bottom (Tillman 1996). An unaerated lagoon depends on wind action and algae to provide the oxygen needed by bacteria. This lagoon usually has an anaerobic zone at the bottom, an anaerobic and aerobic zone in the middle depth, and an aerobic zone near the surface (Reynolds and Richards 1996). Facultative lagoons are usually shallower and larger than the aerated ponds (Tillman 1996). The typical operating conditions of aerated lagoons include the following: (1) the pond must be kept shallow (depth, 3 to 4 ft) to maintain suitable aerobic conditions that provide adequate mixing, (2) the detention time could reach 30 days, and (3) BOD loading ranges from 15 to 50 lb/acre/day (EPA 1992). The main parameters in designing and operating the lagoon/pond are the flow rate, detention time, and organic loading rate. These parameters are calculated by using the following relations (Tillman 1996):

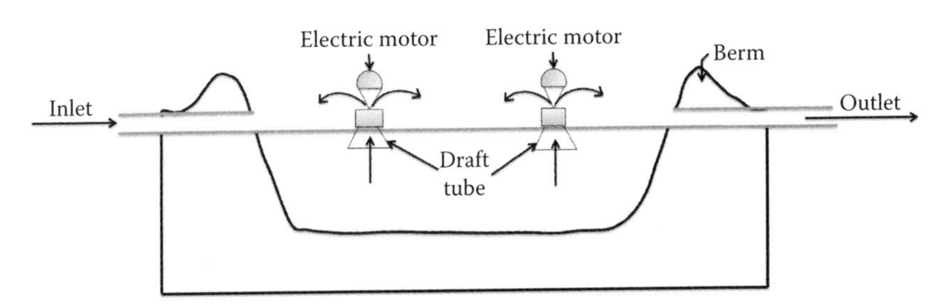

FIGURE 7.12 Wastewater stabilization ponds. (Adapted from EPA 2011, *Principles of Design and Operations of Wastewater Treatment Pond Systems for Plant Operators, Engineers, and Managers*, EPA Report No.: EPA/600/R-11/088, Office of Research and Development, Washington, D.C. [August].)

- Flow rate, (acres-ft/day) = [Flow into pond, (gal/day)]/[(7.48 gal/cu. ft) (43,560 sq. ft/acre)]
- Detention time, days = [Volume, (acre-ft)] × [flow, (acre-ft/day)]
- Organic loading rate, [(lb/days)/acre] = [BOD into pond, (lb/day)]/[pond area, (acres)]

7.6.4.2 Ammonia Removal

The ammonia concentration in the final effluent should be less than 3 mg/L to prevent toxicity to fish that live in the receiving water body (Reynolds and Richards 1996). Releasing high concentrations of ammonia decreases the dissolved oxygen concentration in the receiving water body because the nitrifying microorganisms consume dissolved oxygen to bio-oxidize the ammonia to nitrates (Crites et al. 2006). Nitrates stimulate the growth of undesirable algae and aquatic plants. Ammonia can be removed from the treated wastewater by physical means, such as air stripping processes, biological means (such as the nitrification process), or chemical means (such as ion exchange technology) (Reynolds and Richards 1996).

7.6.4.2.1 Physical Process

Air stripping is the most common physical method for removing ammonia from wastewater. Ammonia should be in a dissolved gas form (NH_3) to be easily stripped from wastewater (Reynolds and Richards 1996). Caustic soda or lime is added to the wastewater to increase the pH between 10.8 and 11.5, which helps convert ammonium hydroxide ions to ammonia gas (EPA 2000a). Ammonia is stripped from the wastewater in a stripping tower, which consists of a fan at the top of the tower to draw the air containing ammonia gas from the tower, packing media to provide air–wastewater contact, a tray to uniformly distribute the wastewater, a grid to distribute the incoming air and support the packing, a drift eliminator (demister) to control water loss from the tower, a collection basin at the bottom of the tower to collect the stripped wastewater, and the tower structure (see Figure 7.13) (Negulescu 1985; Reynolds and Richards 1996). The typical operating conditions of air stripping system include the following: (1) pH value should be 11 or higher, (2) temperature should not be lower than 40°C, and (3) the quantity of air should be at least 3000 m^3 of air per cubic meter of water (Sorensen and Jorgensen 1993).

7.6.4.2.2 Biological Processes

In biological processes, ammonia nitrogen is oxidized in a two-stage process, first to nitrite nitrogen and then to nitrate nitrogen (Tillman 1996). Proper aerobic conditions are required to convert nitrogen to nitrate (see Figure 7.14). Ammonia removal by nitrification processes can be achieved in either a one-stage or a two-stage process (Lin and Lee 2001). In single-stage nitrification, the oxidation of the carbon and the nitrogen is carried out in a single unit, while in the two-stage nitrification, the carbonaceous oxidation and the nitrification steps are carried out in different units (Tillman 1996). Using two-stage nitrification results in poorer settling, greater alkalinity consumption, and much higher oxygen demand (Lin and Lee 2001).

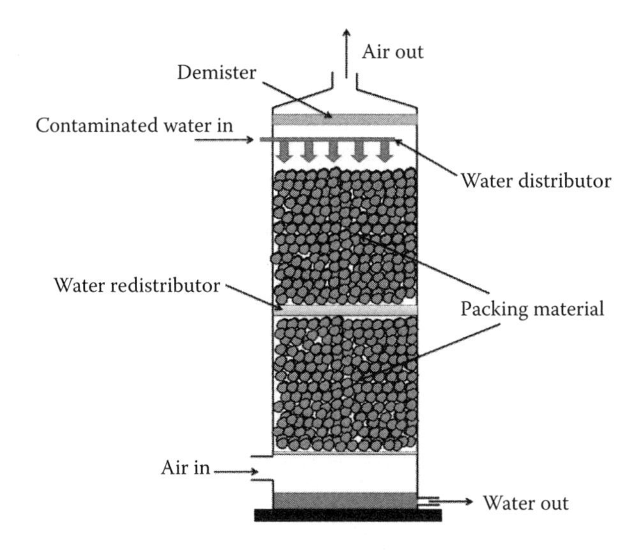

FIGURE 7.13 Air stripping process. (Adapted from EPA 2012, *A Citizen's Guide to Air Stripping*, EPA Report No.: EPA 542-F-12-002, Office of Solid Waste and Emergency Response, Washington, D.C. [September].)

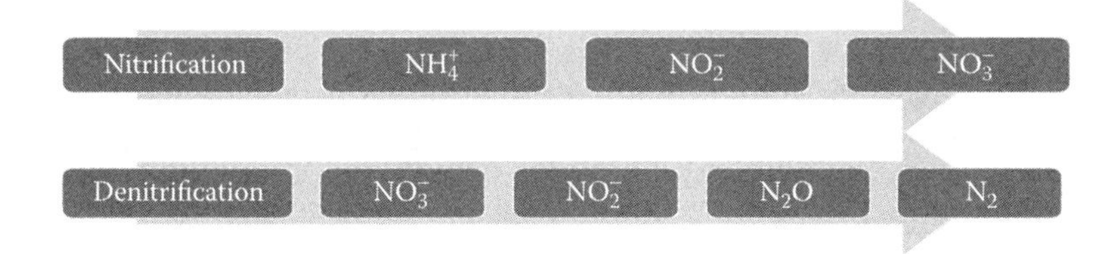

FIGURE 7.14 Nitrification and denitrification process steps.

The typical operating conditions of the nitrification process include the following: (1) temperature should range from 0°C to 20°C, (2) pH should range from 8 to 9, and (3) the loading rate should not exceed 0.003 kg NH_4-N/m^2/day (Vesilind 2003). Denitrification processes should be applied directly after the nitrification process in order to remove the nitrogen completely from the treated wastewater (Vesilind 2003). In this process, the denitrifying bacteria remove oxygen molecules from the nitrate compounds in order to convert nitrates (NO_3^-) to nitrogen gas (N_2) (see Figure 7.14) (Tillman 1996). The efficient nitrification process requires the water temperature to exceed 10°C (Lazarova et al. 2012).

7.6.4.2.3 Chemical Processes

Ion exchange technology is the most common chemical method for removing ammonia from wastewater. It depends on the chemical reaction between ions in the solid phase and ions in the liquid phase (Reynolds and Richards 1996). Wastewater is passed through a column of natural zeolites in order to separate the ammonia and ammonium ions from the wastewater (see Figure 7.15). The natural zeolites are more efficient for columnar use than synthetic zeolites because the surface of the natural

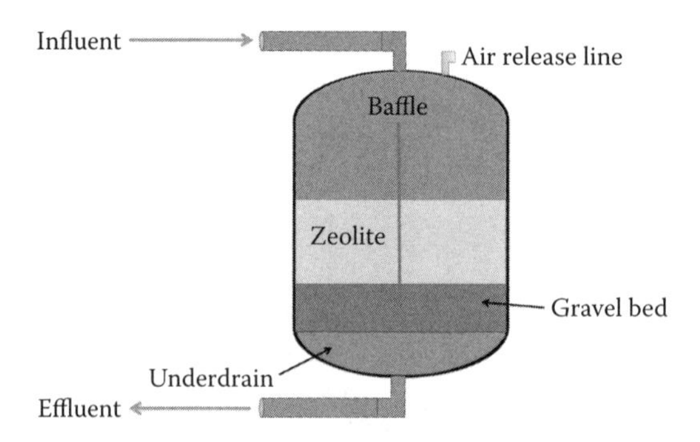

FIGURE 7.15 Zeolite softening unit. (Adapted from EPA 2015b, *Ion Exchange*, Drinking Water Treatability Database, Washington, D.C. http://iaspub.epa.gov/tdb/pages/treatment/treatmentOverview.do?treatmentProcessId=263654386, accessed September 27, 2015.)

zeolites is more granular than the surface of the synthetic zeolites (Reynolds and Richards 1996). Wastewater is passed through the column until the zeolites become saturated with ammonium ions. The saturated zeolites can be regenerated through elution of ammonium ions from the zeolite using a competing ion, or used as fertilizer, where the ammonium ions are slowly released from the zeolites to the soil. The regeneration process produces a regenerant solution that is difficult to treat (Flanigen et al. 1991). For instance, during the conversion of ammonium ions into ammonia by raising the pH of the solution, air in the counterflow mode should be applied to strip off the ammonia from the column, which is subsequently oxidized to nitrogen in the presence of a catalytic agent, such as chlorine (Flanigen et al. 1991). The typical operating conditions of ion exchange technology include the following: (1) the linear velocity at the inlets of desalting cells is approximately 10 cm/s, (2) the linear velocity at the inlets of concentrating cells is approximately 1 cm/s, (3) temperature is approximately 25°C, and (4) the distance between spacer rods is approximately 0.3 cm (Tanaka 2015).

7.6.4.3 Disinfection

Effluent disinfection is the final step in the treatment process, which aims to improve the effluent quality by killing or inactivating microorganisms present in the water, such as viruses, bacteria, and protozoans before release to the environment (Tillman 1996). The most common methods of disinfection are chlorination, UV disinfection, and ozonation (Lin and Lee 2001). Chlorination is the process of adding chlorine to the treated wastewater to destroy disease-causing microorganisms. Chlorine can be fed automatically or manually into a chlorine contact tank containing the treated wastewater (Tillman 1996). Chlorine can be applied in many forms, which include solid, gas, and liquid (hypochlorite form). Chlorination is an efficient and a low-cost technology used for disinfecting wastewater as well as for controlling odor and activated sludge bulking (Vesilind 2003). Chlorine can also be used for removing nitrogen compounds from the treated wastewater. The process of removing nitrogen by chlorine is called breakpoint chlorination. In this process, chlorine is added to the chlorine contact tank for at least 30 min to convert ammonium nitrogen into nitrogen gas; approximately 10 mg/L of chlorine is added to wastewater for every 1 mg/L of ammonium nitrogen. The pH is normally in the range of 9.5 to 11.0 (Tillman 1996). Figure 7.16 illustrates a typical chlorine disinfection unit. The dechlorination process should be applied directly after the chlorination process in order to remove all traces of residual chlorine from the treated wastewater before discharge to surface water bodies (Vesilind 2003). The common chemicals used for dechlorination include sodium sulfite, sulfur dioxide, and sodium metabisulfite (Tillman 1996).

Ozone (O_3) is a derivative of oxygen and considered the most powerful oxidant used to oxidize most waterborne microorganisms such as viruses, bacteria, mold, and yeast (Vesilind 2003). Ozone is an unstable gas, having a short half-life between 20 and 30 min in distilled water at a temperature of 20°C (Reynolds and Richards 1996). The half-life of ozone can be reduced to less than 20 min, if oxidant-demanding materials are found in the solution (Reynolds and Richards 1996; Rice et al. 1979). Ozone cannot be stored because of its short half-life; therefore, it is produced on-site by passing air between oppositely charged plates employing electric discharges. This method is used to break down oxygen molecules, and then the ozone molecules are produced by combining three oxygen atoms together. The ozone disinfection process is shown schematically in Figure 7.17. Air is refrigerated to a temperature below the dew point to remove the moisture and then is passed through twin-tower dryers with media, such as activated alumina or silica gel, to dry the air stream to a dew point of −104°F (−40°C) to −140°F (−60°C) (Reynolds and Richards 1996). The use of clean and dry air improves the production of ozone per unit of power used, increases the life of the units, and reduces unscheduled ozone generator maintenance (Jolley 1975).

UV disinfection is a good alternative to ozonation and chlorination because it is economical and effective in pathogen inactivation (EPA 1986; WPCF 1986). The UV light penetrates the cell walls of the microorganisms and disrupts the genetic

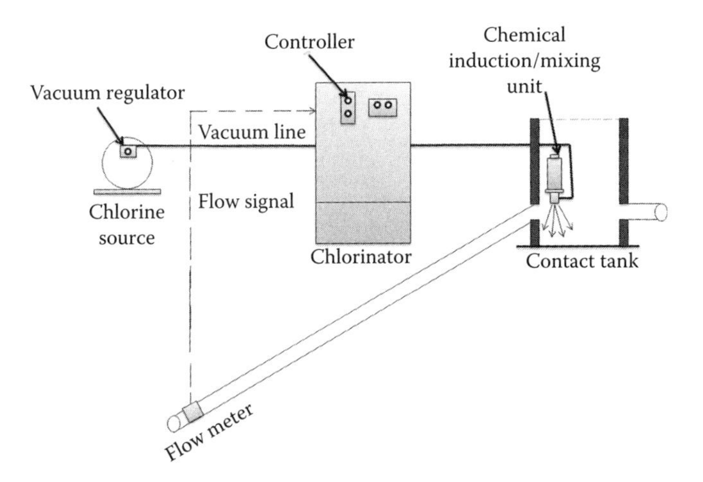

FIGURE 7.16 Chlorination unit. (Adapted from EPA 1999a, *Wastewater Technology Fact Sheet*: *Chlorine Disinfection*, EPA Report No.: 832-F-99-062, Office of Water, Washington, D.C. [September].)

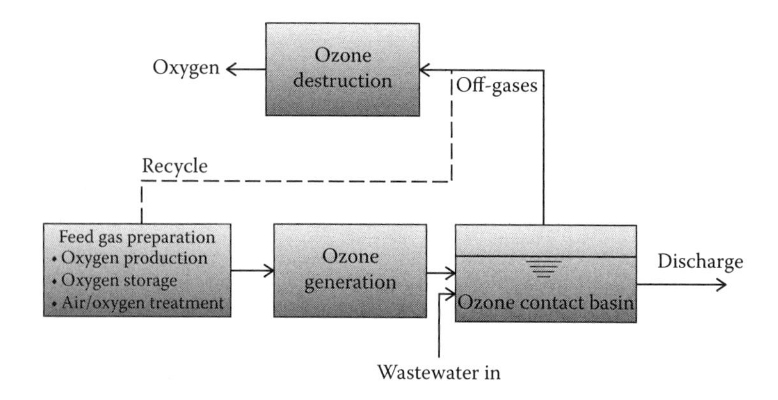

FIGURE 7.17 Ozone process schematic diagram. (Adapted from EPA 1999b, *Wastewater Technology Fact Sheet: Ozone Disinfection*, EPA Report No.: EPA-F-99-063, Office of Water, Washington, D.C. [September].)

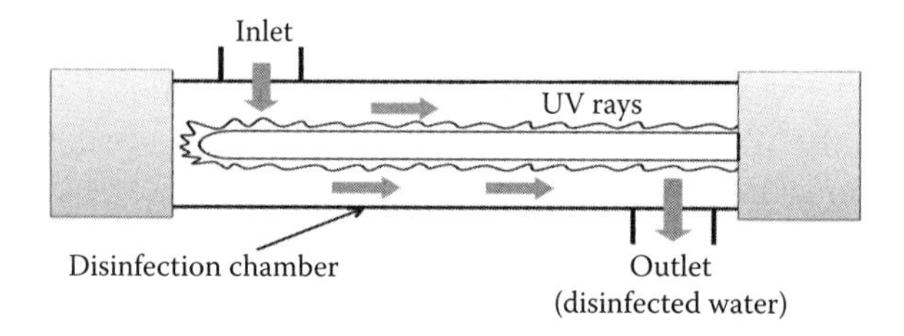

FIGURE 7.18 Ultraviolet disinfection process. (Adapted from EPA 2006, *Ultraviolet Disinfection Guidance Manual for the Final Long Term 2 Enhanced Surface Water Treatment Rule*, EPA Report No.: EPA 816-F-06-005, Office of Water, Washington, D.C. [February].)

material of the cells, which leads to the unlikelihood of occurrence of reproduction (Lahlou 2008). A special lamp is used with an optimum wavelength ranging from 250 to 270 nm used to destroy bacteria and other pathogens (Lahlou 2008). The UV unit can range in size to treat from less than 380 m^3/day (0.1 MGD) to greater than 190,000 m^3/day (50 MGD) (Reynolds and Richards 1996). Figure 7.18 illustrates a typical UV unit. The contact time normally ranges from 20 to 30 s. The lamp wall temperature ranges from 95°F to 122°F (EPA 1998).

7.6.5 Anaerobic Digestion

In the anaerobic digestion process, microorganisms play a major role in the decomposition of the organic matter found in sludge (Vesilind 2003). The anaerobic digestion unit can be operated at a low rate, which refers to one-stage digestion, or at a high rate, which involves one or two stages of digestion (Reynolds and Richards 1996). In low-rate digestion, fresh sludge is added to the digester three times daily. Three layers are formed inside the tanks as a result of digestion. A sludge layer is formed at the bottom of the digester, a supernatant layer is formed above the sludge, and a scum layer is formed at the top (see Figure 7.19) (Tillman 1996). In the two-stage high-rate digestion system, the first stage is used for sludge stabilization and the second stage is used for thickening the decomposed sludge. In a single-stage high-rate digestion system, a different thickening process replaces the system used at a second stage (Duggall 1966).

The detention time of anaerobic digesters typically ranges between 10 and 20 days. Nearly 50% of the organic solids are converted to liquid and gas. The liquid is typically returned to the inlet of the plant, the gas can be used in power generation, and the residual sludge is transferred to the following unit or disposed (Tillman 1996). The typical operating conditions of the anaerobic digestion process include the following: (1) the pH value should be in the range of 6.8 to 7.4, (2) mesophilic temperature ranges from 86°F to 95°F and thermophilic temperature ranges from 122°F to 132°F, (3) the hydraulic detention time ranges from 10 to 15 days, and (4) alkalinity concentration should be in the range of 1500 to 3000 mg/L (Lue-Hing et al. 1998). The main parameters in designing and operating the anaerobic digestion process are volatile solids loading, percent volatile solids reduction, detention time, and gas production. These parameters are determined using the following relations (Tillman 1996):

- Volatile solids loading, [(lb VS/day)/cu. ft] = [feed sludge VS, (lb/day)]/[digester volume, (cu. ft)]
- Volatile solids reduced, percent = [(lb VS_{in} – lb VS_{out})/(lb VS_{in})] × 100
- Detention time, days = [digester volume, (gal)]/[sludge feed, (gal/day)]
- Gas production, [cu. ft gas/lb VS_{fed}] = [gas produced, cu. ft per day]/[(VS_{fed}, (lb/day)) (reduction/100)]

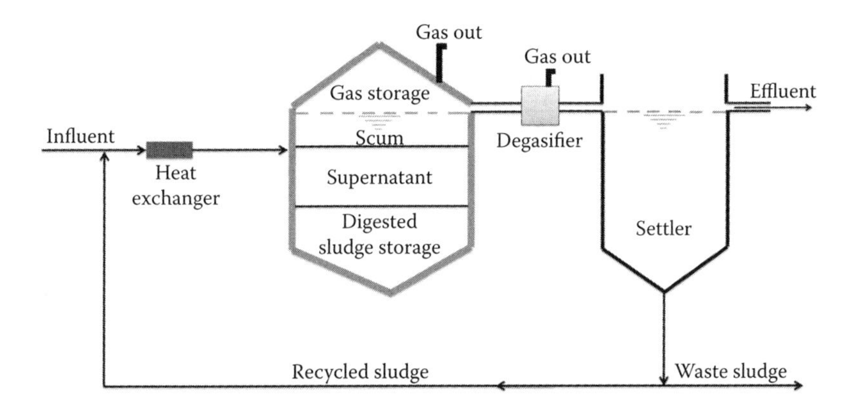

FIGURE 7.19 Anaerobic digestion process. (Adapted from EPA 1991, *Evaluating Sludge Treatment Processes*, Office of Wastewater Enforcement and Compliance, Office of Water, Washington, D.C.)

7.6.6 Solids Handling

The solids handling process aims to reduce the volume of the sludge and the pathogenic microbial content in the sludge before ultimate disposal to the environment (Reynolds and Richards 1996). The most common solids handling systems are gravity thickening and dissolved air flotation (DAF) thickening (Tillman 1996). The process selection depends on the ultimate method of disposal and the characteristics of the waste (Vesilind 2003). In conventional WWTPs, the sludges are mainly of an organic nature, such as raw or primary sludge, trickling filter humus, or excess activated sludge. In advanced WWTPs, the sludges may contain some natural organic matter but mostly have a chemical nature because they are produced from coagulation or precipitation processes (Reynolds and Richards 1996). Mixing the chemical and organic sludges usually produces a mixture that is difficult to manage and process. Thickening is a process that reduces the volume of sludge by increasing the solids content. For instance, if a sludge containing 3% solids is thickened to 6% solids, the sludge volume leaving the thickener is expected to be approximately half the volume of the feed sludge (Reynolds and Richards 1996). The optimal hydraulic loading rate ranges from 4 to 8 gal/ft^2/h (Wang et al. 2007). The gravity thickener is similar to a circular sedimentation tank, but the bottom of the thickener tank has a greater slope (Tillman 1996). Sludge enters at the middle of the tank, while a sludge blanket is comprised as a result of solids settling. A rake mechanism is used to stir the thickened sludge blanket in order to release gas bubbles and move the sludge toward a center sump for removal (Tillman 1996). The thickened sludge is pumped either to a surge holding tank or to a dewatering unit (Vesilind 2003). The supernatant flow leaves the thickener through the effluent weir located at the outer edge of the tank and returns to either the primary or the secondary treatment unit (Tillman 1996). Figure 7.20 illustrates a typical gravity thickener unit.

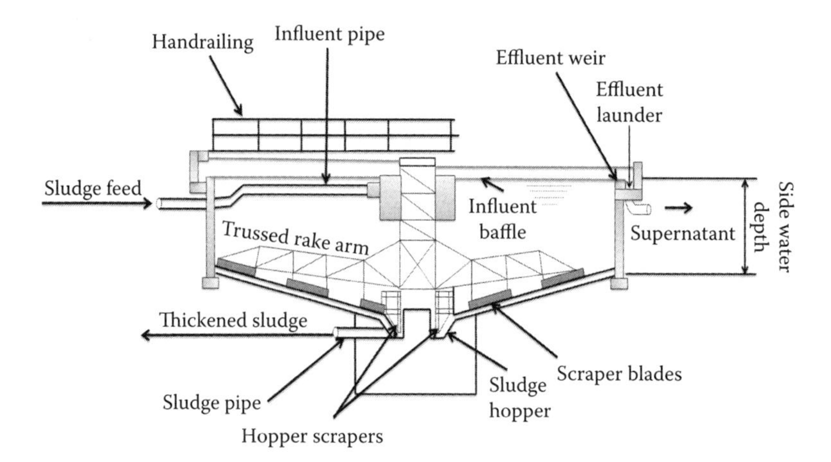

FIGURE 7.20 Typical gravity thickener. (Adapted from EPA 1979, *Automatic Sludge Blanket Control in an Operating Gravity Thickener*, EPA Report No.: EPA-600/2-79-159, Municipal Environmental Research Laboratory, Cincinnati, OH [November].)

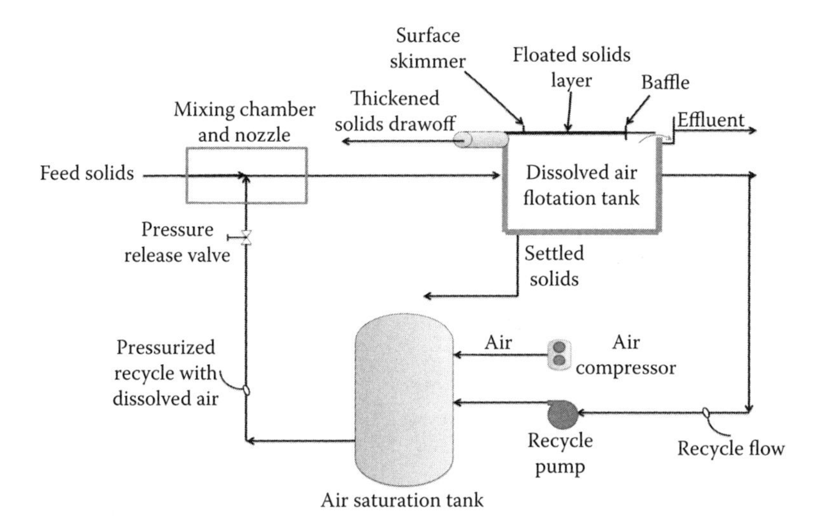

FIGURE 7.21 Schematic of DAF thickener. (Adapted from EPA 1991, *Evaluating Sludge Treatment Processes*, Office of Wastewater Enforcement and Compliance, Office of Water, Washington, D.C.; and EPA 2002, *Wastewater Treatment Manuals*: *Coagulation, Flocculation, and Clarification*, Wexford, Ireland.)

In the DAF thickening process, the sludge is thickened under an air pressure, normally between 40 and 70 psi (between 275.8 and 482.6 kPa) (Mines 2014). An air compressor is used to inject air into the wastewater. This wastewater/air mixture is mixed with the sludge and then flows into the flotation tank (Mines 2014). Under atmospheric pressure, air bubbles are released from the solution and carry the sludge particles to the surface of the tank to form a sludge blanket (Tillman 1996). This thickened sludge is skimmed out of the surface of the DAF tank using skimmers (Mines 2014). A recycle system is used to recycle part of the DAF effluent and mix it with the feed sludge (Tillman 1996). Figure 7.21 illustrates a typical DAF thickening process. The hydraulic loading rate should range from 0.5 to 2 gpm/ft^2, while the solids loading rate ranges from 0.80 to 2.8 lb/ft^2-h (Turovskiy and Mathai 2006).

7.7 TREATMENT PERFORMANCE

Domestic wastewater should be subjected to several types of treatment before disposal or reuse. The complete treatment of wastewater requires a sequential combination of physical, chemical, and biological unit processes (Kumar et al. 2010). The primary criterion for assessing the performance of the WWTP involves the degree of reduction of SS and BOD, which constitute organic pollution (Kumar et al. 2010). The efficiency of a WWTP depends mainly on the raw wastewater quantity and quality, as well as the proper design, construction, operation, and maintenance of the treatment units (Kapur et al. 1999; Qasim 1999). Poor performance of WWTPs may also be attributed to an increase in population and water use, which requires exceeding the plant design capacity (Dakers and Cockburn 1990). Performance evaluation of an existing WWTP is necessary to assess the effluent quality and to check the ability of the plant to handle higher organic and hydraulic loadings. The plant performance evaluation also helps in gathering additional data that can be used to improve the design efficiency of different treatment units (EPA 1971). An accurate sampling and laboratory analysis is essential for proper process control (Kaul et al. 1993).

7.8 WASTEWATER REUSE

The reuse of wastewater has been an important concept in our life and became an integral component of water resource management as a result of increasing water scarcity, which puts pressure on water resources (Asano 1998). Potential benefits of wastewater reuse include (1) reducing high-quality drinking water consumption by substituting drinking water with treated wastewater in applications that do not require high water quality, (2) providing an alternative water supply to meet the needs of the present and future generations, and (3) protecting the environment by reducing the amount of contaminants and any hazardous substances entering waterways or otherwise released into the environment (Asano et al. 2007; Tchobanoglous and Franklin 1991). Water reuse is particularly essential in communities characterized by rapid population growth and that have limited water resources (Asano et al. 2007). Major wastewater reuse applications include irrigation, industrial uses, and aquaculture.

Treated wastewater can be reused for irrigation of nonedible crops, gardens, and parks (Tchobanoglous and Franklin 1991). For these purposes, the treated wastewater should meet the secondary treatment standards (BOD < 20 mg/L and SS < 30 mg/L) (UNEP 2000). The pathogens and organic matter must be effectively removed to eliminate odors and protect public health. Using treated wastewater in irrigation purposes can significantly reduce the total treatment cost as a result of eliminating the cost of nutrient removal (WERF 2010). Treated wastewater is not appropriate to be used in sites with steep slopes because of the high runoff potential (WERF 2010).

The second application, reuse of treated wastewater in industrial purposes, is applicable if suitable industries exist near the WWTPs (UNEP 2000). Each industry has its own requirements for water quality; for instance, industries having cooling systems may require low water quality, while industries having boiling water systems for electricity generation may require very pure water (Tchobanoglous and Franklin 1991; UNEP 2000). The low water quality can be produced by applying only the secondary treatment. Since high-cost tertiary treatment is necessary to provide high water quality, the treated wastewater is better used in industries that require low water quality (e.g., water for cooling towers) (UNEP 2000). Additional treatment may be required to enable reusing the treated wastewater for industrial purposes. For example, the softening process is essential for reusing the treated wastewater for cooling purposes. The softening will help in protecting the heat-transfer surfaces against erosion and corrosion (WERF 2010).

Treated wastewater can also be used for aquaculture. This technique has been applied in many areas, such as Latin America, Peru, and the Middle East, for a considerable period. The wastewater-fed aquaculture is suitable to be applied in developing countries that cannot afford expensive wastewater treatment, and in arid and semiarid countries that suffer from water shortages (Tchobanoglous and Franklin 1991; UNEP 2000). Six constraints must be considered before applying the wastewater-fed aquaculture technique:

(1) lack of specific knowledge about aquaculture; (2) cultural and social acceptance of wastewater-fed practices; (3) the availability of suitable areas where treated wastewater is available for reuse; (4) level of urbanization, where the wastewater-fed systems are normally applied in countries that have low levels of urbanization; (5) mixing of industrial and domestic wastewater contaminates nutrient-rich sewage; and (6) a suitable climate for the rapid growth of aquaculture species and their food organisms (Asano 1998; UNEP 2000). The aquaculture system should be effectively managed to limit public health risks and wastewater should not be reused without adequate treatment.

7.9 SUMMARY AND CONCLUSIONS

Wastewater collection systems are installed to collect and convey wastewater generated from the source to the nearest WWTP. Collection lines should be installed with sufficient downhill slope to allow the wastewater to move via gravity and to reduce the number of pumps needed to convey the wastewater to the treatment plant, thus making the WWTP more energy efficient. The VCP is the most common pipe material used in sewer applications with pipe diameter ranging from 4 to 36 inches. Proper wastewater management is very important to protect public health and prevent the spread of waterborne diseases. WWTPs are designed to accelerate the natural purification process and to remove contaminants from wastewater before releasing it into receiving waters. The most important data required for the design of any WWTP include the quantity and quality of raw wastewater and the desired effluent quality. The quality of raw and treated wastewater is identified by measuring the physical, biological, and chemical characteristics of the wastewater.

The most common steps of wastewater treatment include preliminary treatment, primary treatment, secondary treatment, and tertiary treatment. The preliminary treatment units remove large amounts of SS and grit to prevent damage to pumps and subsequent treatment units. In primary treatment, the flow velocity is reduced to allow low-density materials to float and SS to settle out. Coagulation and flocculation are chemical processes that use chemical reagents to enhance solids settling. Secondary treatment processes are responsible for removing the colloidal and dissolved organic and inorganic solids, which remain after primary treatment. Tertiary treatment is responsible for improving the effluent quality before it is reused or discharged into the environment. For the sludge, the anaerobic digestion process is used to reduce the high organic loading of primary sludge. The solids handling process is also used to reduce the volume of the sludge and the pathogenic microbial content in the sludge before ultimate disposal to the environment. Performance evaluation of an existing WWTP is necessary to assess the effluent quality and to check the ability of the plant to handle higher organic and hydraulic loadings. Wastewater reuse is important to reduce pressure on water resources. The most common applications of wastewater reuse include irrigation, industrial uses, and aquaculture.

REFERENCES

Asano, T., (1998). *Wastewater Reclamation and Reuse*, Technomic Publishing Company, Inc., Lancaster, Pennsylvania.

Asano, T., L.B. Franklin, L.L. Harold, T. Ryujiro, and T. George, (2007). *Water Reuse: Issues, Technologies, and Applications*, Metcalf and Eddy, Inc., New York.

Bejan, A., G. Tsatsaronis, and M. Moran, (1996). *Thermal Design and Optimization*, John Wiley Sons, Inc., New York.

Cabral, J.P.S., (2010). Water microbiology: Bacterial pathogens and water, *International Journal of Environmental Research and Public Health*, ***7***(10): 3657–3703.

Cassano, A., E. Drioli, and R. Molinari, (1997). Recovery and reuse of chemicals in unhearing degreasing and chromium tanning processes by membranes, *Desalination*, ***13***: 251–261.

Chalew, T.E.A., S.A. Gauray, H. Haiou, and J.S. Kellogg, (2013). Evaluating nanoparticle breakthrough during drinking water treatment, *Environ. Health Perspect.*, ***121***(10): 1161–1166.

Crites, R.W., E.J. Middlebrooks, and C.R. Sherwood, (2006). *Natural Wastewater Treatment Systems*, CRC Press LLC, Boca Raton, Florida.

Dakers, J.L., and A.G. Cockburn, (1990). Raising the standard of operation of small sewage works, *Water Science and Technology*, ***22***(3–4): 261–266.

Diersing, N., (2009). "Water Quality: Frequently Asked Questions." Florida Brooks National Marine Sanctuary, Key West, Florida.

Duggall, K.N., (1966). *Elements of Environmental Engineering*, 1st edition, S. Chand and Company Ltd., Ram Nagar, New Delhi, India.

Environmental Protection Agency (EPA), (1971). *Process Design Manual for Upgrading Existing Wastewater Treatment Plants*, EPA Project No.: 17090 GNQ, Contract No.: 14-12-933, Technology Transfer, West Chester, Pennsylvania.

Environmental Protection Agency (EPA), (1979). *Automatic Sludge Blanket Control in an Operating Gravity Thickener*, EPA Report No.: EPA-600/2-79-159, Municipal Environmental Research Laboratory, Cincinnati, OH (November).

Environmental Protection Agency (EPA), (1984a). *Tertiary Granular Filtration: Problems and Remedies*, EPA Report No.: 832-R-84-113, Office of Water Program Operations, Washington, D.C. (August).

Environmental Protection Agency (EPA), (1984b). *Summary of Design Information on Rotating Biological Contactors*, EPA Report No.: 430/9-84-008, Office of Water Program Operations, Washington, D.C. (September).

Environmental Protection Agency (EPA), (1986). *Municipal Wastewater Disinfection*, EPA Report No.: EPA/625/1-86/024, EPA Design Manual, Washington, D.C. (October).

Environmental Protection Agency (EPA), (1991). *Evaluating Sludge Treatment Processes*, Office of Wastewater Enforcement and Compliance, Office of Water, Washington, D.C.

Environmental Protection Agency (EPA), (1992). *Summary Report: Small Community Water and Wastewater Treatment*, EPA Report No.: EPA/625/R-92/010, EPA Summary Report, Washington, D.C. (September).

Environmental Protection Agency (EPA), (1997a). *Volunteer Stream Monitoring: A Methods Manual*, EPA Report No.: EPA 841-B-97-003, Office of Water, Washington, D.C. (November).

Environmental Protection Agency (EPA), (1997b). *Wastewater Treatment Manual: Primary, Secondary, and Tertiary Treatment*, Ardcavan, Wexford, Ireland.

Environmental Protection Agency (EPA), (1998). *Ultraviolet Disinfection*, Fact sheet, Morgantown, West Virginia.

Environmental Protection Agency (EPA), (1999a). *Wastewater Technology Fact Sheet: Chlorine Disinfection*, EPA Report No.: 832-F-99-062, Office of Water, Washington, D.C. (September).

Environmental Protection Agency (EPA), (1999b). *Wastewater Technology Fact Sheet: Ozone Disinfection*, EPA Report No.: EPA-F-99-063, Office of Water, Washington, D.C. (September).

Environmental Protection Agency (EPA), (2000a). *Ammonia Stripping*, EPA Report No.: EPA 832-F-00-019, Wastewater Technology Fact Sheet, Washington, D.C. (September).

Environmental Protection Agency (EPA), (2000b). *Trickling Filters*, EPA Report No.: 832-F-00-014, Wastewater Technology Fact Sheet, Washington, D.C. (September).

Environmental Protection Agency (EPA), (2002). *Wastewater Treatment Manuals: Coagulation, Flocculation, and Clarification*, Wexford, Ireland.

Environmental Protection Agency (EPA), (2005). *Membrane Filtration Guidance Manual*, EPA Report No.: EPA 815-R-06-009, Office of Water, Washington, D.C. (November).

Environmental Protection Agency (EPA), (2006). *Ultraviolet Disinfection Guidance Manual for the Final Long Term 2 Enhanced Surface Water Treatment Rule*, EPA Report No.: EPA 816-F-06-005, Office of Water, Washington, D.C. (February).

Environmental Protection Agency (EPA), (2011). *Principles of Design and Operations of Wastewater Treatment Pond Systems for Plant Operators, Engineers, and Managers*, EPA Report No.: EPA/600/R-11/088, Office of Research and Development, Washington, D.C. (August).

Environmental Protection Agency (EPA), (2012). *A Citizen's Guide to Air Stripping*, EPA Report No.: EPA 542-F-12-002, Office of Solid Waste and Emergency Response, Washington, D.C. (September).

Environmental Protection Agency (EPA), (2013). *Green Communities*, Public Facilities Inventory and Evaluation, Washington, D.C.

Environmental Protection Agency (EPA), (2015a). *Conventional Treatment*, Drinking Water Treatability Database, Washington, D.C. http://iaspub.epa.gov/tdb/pages/treatment/treatmentOverview.do?treatmentProcessId=1934681921, accessed September 27, 2015.

Environmental Protection Agency (EPA), (2015b). *Ion Exchange*, Drinking Water Treatability Database, Washington, D.C. http://iaspub.epa.gov/tdb/pages/treatment/treatmentOverview.do?treatmentProcessId=263654386, accessed September 27, 2015.

Flanigen, E.M., J.C. Jansen, and V.B. Herman, (1991). *Introduction to Zeolite Science and Practice*, 2nd edition, Elsevier Science, Amsterdam, Netherlands.

Florescu, D., M.I. Andreea, C. Diana, H. Elena, E.I. Roxana, and C. Monica, (2011). *Validation Procedure for Assessing the Total Organic Carbon in Water Samples*, 12th International Balkan Workshop on Applied Physics, Constanta, Romania, (July 6–8).

Fumatech, (2014). "Diffusion Dialysis," Functional Membranes and Plant Technology, Bietigheim-Bissingen, Germany. http://www.fumatech.com/EN/Membrane-technology/Membrane-processes/Diffusion-dialysis/, accessed March 5, 2015.

Grady, C.P.L., T.D. Glen, and C.L. Henry, (1999). *Biological Wastewater Treatment*, 2nd edition, Marcel Dekker, Inc., New York.

Grady, C.P.L., T.D. Glen, G.L. Nancy, and D.M.F. Carlos, (2011). *Biological Wastewater Treatment*, 3rd edition, CRC Press, Boca Raton, Florida.

Horan, N.J., (1990). *Biological Wastewater Treatment Systems: Theory and Operation*, John Wiley and Sons Ltd., New York.

Illinois Environmental Protection Agency (EPA), (1997). "Recommended Standards for Sewage Work," Part 370 of Chapter II, EPA, Subtitle C: Water Pollution, Title 35: Environmental Protection, Illinois Environmental Protection Agency (IEPA), Springfield, Illinois. ftp://www.ilga.gov/JCAR/AdminCode/035/03500370sections.html, accessed September 27, 2015.

Ioannou, L.A., C. Michael, N. Vakondios, K. Drosou, N.P. Xekoukoulotakis, E. Diamadopoulos, and D. Fatta-Kassinos, (2013). Winery wastewater purification by reverse osmosis and oxidation of the concentrate by solar photo-Fenton. *Separation and Purification Technology*, ***118***: 659–669.

Johnson, D.L., S.H. Ambrose, T.J. Bassett, M.L. Bowen, D.E. Crummey, J.S. Isaacson, D.N. Johnson, P. Lamb, M. Saul, and A.E. Winter-Nelson, (1997). Meanings of environmental terms, *Journal of Environmental Quality*, ***26***(3): 581–589.

Jolley, R.L., (1975). Chlorine-containing organic constituents in chlorinated effluents, *J. WPCF*, ***47***(3): 601–618.

Kapur, A., A. Kansal, R.K. Prasad, and S. Gupta, (1999). Performance evaluation of sewage treatment plant and sludge bio-methanation, *Indian Journal of Environmental Protection*, ***19***: 96–100.

Kaul, S. N., P.K. Mukherjee, T.A. Sirowala, H. Kulkarni, and T. Nandy, (1993). Performance evaluation of full scale waste water treatment facility for finished leather industry, *Journal of Environmental Science and Health*, ***28***(6): 1277–1286.

Kumar, K.S., P.S. Kumar, and M.J.R. Babu, (2010). Performance evaluation of wastewater treatment plant, *International Journal of Engineering Science and Technology*, ***2***(12): 7785–7796.

Lahlou, Z.M., (2008). "Tech Brief: Ultraviolet Disinfection," A National Drinking Water Clearinghouse Fact Sheet, Morgantown, West Virginia.

Lazarova V, C. Kwand-Ho, and C. Peter, (2012). *Water-Energy Interactions in Water Reuse*. IWA Publishing, London, UK.

Li, Y., J.L. Colin, A.B. Katherine, A. Yukitomo, J. Zhongliang, T.T. Christina, R.T. Raymond, and S.K. Kenneth, (2013). Endocrine-disrupting chemicals (EDCs): In vitro mechanism of estrogenic activation and differential effects on ER target genes, *Environ Health Perspect.*, ***121***(4): 459–466.

Lin, S.D., and C.C. Lee, (2001). *Water and Wastewater Calculations Manual*, McGraw-Hill Publisher, New York.

Lue-Hing, C., R.Z. David, T. Prakasam, K. Richard, F.M. Joseph, and S. Bemard, (1998). *Municipal Sewage Sludge Management: A Reference Text on Processing*, Technomic Publishing Company, Inc., Lancaster, Pennsylvania.

Madaeni, S.S., and M.R. Eslamifard, (2010). Recycle unit wastewater treatment in petrochemical complex using reverse osmosis process. *Journal of Hazardous Materials*, ***174***(1–3): 404–409.

Mara, D., and J.H. Nigel, (2003). *Handbook of Water and Wastewater Microbiology*, Elsevier Science, London, UK.

Metcalf and Eddy, Inc., (1981). *Wastewater Engineering: Collection and Pumping*, McGraw-Hill, New York.

Mines, R.O., (2014). *Environmental Engineering: Principles and Practice*, 7th edition, Waveland Press Inc., Long Grove, Illinois.

Mnif, W., S. Dagnino, A. Escande, A. Pillon, H. Fenet, E. Gomez, C. Casellas, M.J. Duchesne, G. Hernandez-Raquet, V. Cavailles, P. Balaguer, and A. Bartegi, (2010). Biological analysis of endocrine-disrupting compounds in Tunisian sewage treatment plants, *Archives of Environmental Contamination and Toxicology*, ***59***(1): 1–12.

Nadakavukaren, A., (2011). *Our Global Environment: A Health Perspective*, 7th edition, Waveland Press Inc., Long Grove, Illinois.

Negulescu, M., (1985). *Developments in Water Science: Municipal Waste Water Treatment*, Elsevier Science, New York.

Parcher, M., (1998). *Wastewater Collection System Maintenance*, Technomic Publishing Company, Inc., Lancaster, Pennsylvania.

Pizzi, N., (2010). *Water Treatment*, TIPS Technical Publishing, Inc., Carrboro, North Carolina.

Pruden, A., D.G. Larsson, A. Amezquita, P. Collignon, K.K. Brandt, D.W. Graham, J.M. Lazorchak, S. Suzuki, P. Silley, J.R. Snape, E. Topp, T. Zhang, and Y.G. Zhu, (2013). Management options for reducing the release of antibiotics and antibiotic resistance genes to the environment, *Environ Health Perspect.*, ***121***(8): 878–885.

Qasim, S.R., (1999). *Wastewater Treatment Plants: Planning, Design, and Operation*, 2nd edition, CRC Press LLC, Boca Raton, FL.

Ragsdale, F., (2014). "Wastewater System Operator's Manual: Wastewater Collection Systems," Ragsdale and Associates Training Specialists, LLC, Albuquerque, New Mexico.

Rautenbach, R., and T. Linn, (1996). High pressure reverse osmosis and nanofiltration, a zero discharge process combination for the treatment of wastewater with severe fouling/scaling potential, *Desalination*, ***105***(1–2): 63–70.

Reynolds, T.D., and P.A. Richards, (1996). *Unit Operations and Processes in Environmental Engineering*, Cengage Learning, Independence, Kentucky.

Rice, R.G., G.W. Miller, C.M. Robson, and A.G. Hill, (1979). "Ozone Utilization in Europe," Proc. 8th Annual AIChE Meeting, Houston, Texas.

Rowe, D.R., and M.A. Isam, (1995). *Handbook of Wastewater Reclamation and Reuse*, 1st edition, Lewis Publishers, Boca Raton, Florida.

Safe Drinking Water Foundation (SDWF), (2009). "Drinking Water Quality and Health," Registered Canadian Charitable Organization, Saskatoon, Saskatchewan, Canada. http://www.safewater.org/PDFS/resourceswaterqualityinfo/Resource_Drink_Wat_Qual.pdf, accessed July 13, 2014.

Sanders, D., (2009). "Pipe Joints and Critical Performance Requirements by System Application," Professional Development Advertising Section—CONTECH Construction Products Inc., West Chester, Ohio.

Sorensen, B.H., and S.E. Jorgensen, (1993). *The Removal of Nitrogen Compounds from Wastewater*, Elsevier Science Publishers, Amsterdam, Netherlands.

Spellman, F.R., and D. Joanne, (2001). *Handbook for Waterworks Operator Certification: Advanced Level*, Manual of Practice, Technomic Publishing Company, Inc., Lancaster, Pennsylvania.

Spellman, F., (2009). *Water and Wastewater Treatment Plant Operations*, CRC Press LLC, Boca Raton, Florida.

Sperling, M.V., (2007). *Basic Principles of Wastewater Treatment*, Volume II, IWA Publishing, London, UK.

Tay, J.H., and S. Jeyaseelan, (1995). Membrane filtration for reuse of wastewater from beverage industry, *Resources, Conservation and Recycling*, ***15***(1): 33–40.

Tanaka, Y., (2015). *Ion Exchange Membranes: Fundamentals and Applications*, Elsevier, Waltham, Massachusetts.

Tchobanoglous, T., and L.B. Franklin, (1991). *Wastewater Engineering: Treatment, Disposal, and Reuse*, 3rd Edition, Metcalf and Eddy, Inc., New York.

Tillman, G.M., (1996). *Wastewater Treatment: Troubleshooting and Problem Solving*, CRC Press LLC, Boca Raton, Florida.

Turovskiy, I.S., and P.K. Mathai, (2006). *Wastewater Sludge Processing*, John Wiley Sons, Inc., Hoboken, New Jersey.

United Nations Environment Programme (UNEP), (2000). *Environmentally Sound Technologies in Wastewater Treatment for the Implementation of the UNEP Global Performance of Action: Guidance on Municipal Wastewater*, Division of Technology, Industry and Economics, Osaka, Japan.

United States Department of the Interior Bureau of Reclamation, (2010). *Reclamation: Managing Water in the West*, Washington, D.C.

Vesilind, P.A., (2003). *Wastewater Treatment Plant Design*, IWA Publishing, London, UK.

Wang, L.K., K.S. Nazih, and T. Yung, (2007). *Biosolids Treatment Processes*, Humana Press Inc., Totowa, New Jersey.

Water Environment Federation (WEF), (2010). *Wastewater Collection Systems Management*, Manual of Practice, 6th Edition, McGraw-Hill Publisher, New York.

Water Environment Research Foundation (WERF), (2010). *Wastewater Reuse*, Fact sheet, Alexandria, Virginia.

Water Pollution Control Federation (WPCF), (1986). *Wastewater Disinfection*, WPCF Manual of Practice FD-10. Alexandria, Virginia.

8 Wastewater Treatment and Disposal for Unconventional Oil and Gas Development

Liwen Chen, Peyton C. Richmond, and Ross Tomson

CONTENTS

8.1 INTRODUCTION

Novel technologies such as horizontal drilling and hydraulic fracturing have enabled the economic exploration of natural gas from unconventional sources, such as shale gas, tight gas, tight oil, and coal seam gas (coalbed methane) [1]. Horizontal drilling increases the lateral exposed section length of the underground reservoir and allows more wellheads to be clustered into one surface location, which makes it easier and cheaper to complete and produce the wells. Hydraulic fracturing, also known as fracking, is a well-stimulation technology using hydraulically pressurized fluid or fracking fluid that usually contains proppants and chemical additives (discussed in detail later in this section). In fracking, the high-pressure fracking fluid is first injected into the wellbore, creating fissures in the deep-rock formations by the hydraulic pressure. Then, the pressure is reduced, allowing water and natural gas to flow back through the fissures (Figure 8.1 [2]). Thanks to these technologies, shale oil and gas production has dramatically increased in several regions of the world, including the United States, the United Kingdom, Poland, Ukraine, Australia, and Brazil.

On the other hand, the abovementioned technologies demand much more water than traditional technologies, partially attributed to the increased formation contact volume. For example, the quantity of water needed for drilling and fracking a horizontal well ranges from 2 million to 7 million gallons of water in the Marcellus Shale region. By contrast, only 1 million gallons of water is needed for traditional vertical wells [3]. In the Woodford Shale region, the water requirement increases 315% from 80,000 barrels (2.52 million gallons) to more than 250,000 barrels (7.88 million gallons) per completion after the transition from vertical to horizontal wells [4]. To reduce the local freshwater demand, wastewater reuse and recycle are common practices among producers, and 90% of the water is reused for subsequent fracking jobs in the Marcellus Shale region [3].

Overall, the water life cycle at the well site can be divided into the following major steps: source water acquisition; chemical mixing; well injection; flowback and produced water generation; water reuse or recycle; and wastewater transportation, discharge, or disposal (Figure 8.1). Thus far, much of the water used has been withdrawn from surface or ground sources. Shale plays in humid regions may rely on surface water supplies, whereas those in arid regions require mostly groundwater withdrawal. For instance, surface water from different river basins, for example, Susquehanna and Delaware basins, is used predominantly at the Marcellus Shale play, whereas groundwater from the Carrizo-Wilcox aquifer is heavily relied upon at the semiarid Eagle Ford play [5]. Major shale basins and saline aquifer locations in the United States are shown in Figure 8.2 [6].

Although water demand for shale fracking does not represent a significant fraction of the total water consumption today, the availability of freshwater may be threatened by competition and supply shortages in certain arid or semiarid areas. For example, overextraction of groundwater for agricultural irrigation in the arid Winter Garden region at Eagle Ford accounts for a 60-m water-level decline over a 6500-km^2 area and disappearance of several large springs, and this could impede current and future shale gas production activities [5]. Other shale regions, such as the

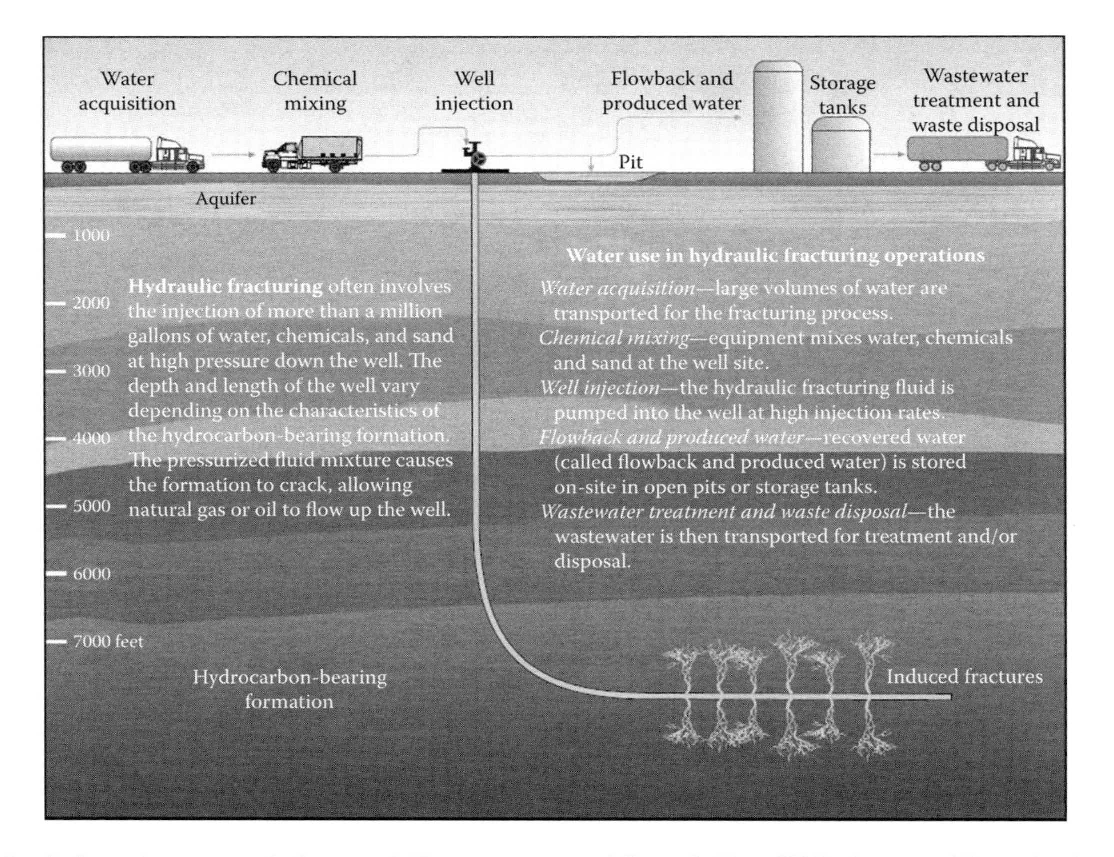

FIGURE 8.1 Hydraulic fracturing water cycle from acquisition to treatment and disposal. (From US Environmental Protection Agency, *Plan to Study the Potential Impacts of Hydraulic Fracturing on Drinking Water Resources*. 2011, Office of Research and Development.)

FIGURE 8.2 Shale basins and saline aquifer locations in the United States. (Reprinted with permission from Zhu H. and R. Tomson, *Exploring Water Treatment, Reuse and Alternative Sources in Shale Production*. 2013 [cited May 9, 2015]; Available from: http://www.shaleplaywatermanagement.com/2013/11/exploring-water-treatment-reuse-and-alternative-sources-in-shale-production/.)

Monterrey Basin in California, Denver-Julesburg Basin in Colorado, and basins in Argentina and Australia, also have similar problems where the fracking water need surpasses the sustainable groundwater withdrawal rate [7].

On the other hand, some researchers have argued that the total water use for unconventional gas development overall is relatively small compared with the water withdrawal for other energy production, such as cooling water for thermoelectric-power generation. The total water volume consumed by fracking in the last decade (2.5×10^8 to 3.0×10^8 m^3) accounts for merely 1% of the annual water loss from cooling thermoelectric-power generation [8]. In addition, the amount of water consumed per unit of energy could be 80% less than that needed by a conventional pulverized coal power plant if fueled by shale gas in a combined-cycle power plant [9].

The composition of fracking fluid generally contains 90% water, 9.5% proppant (i.e., sand or ceramic particles), and 0.5% chemical additives [10], although it varies among different companies. Proppant is granular material such as silica sand, ceramic media, or bauxite. It is used to support the fissures created by fracking and allow the shale gas to flow freely back to the wellbore when pressure is decreased. Chemical additives have diverse ingredients for various purposes. For instance, hydrochloride acid (HCl) is used after perforation to dissolve soluble minerals (such as limestone and dolomite) in the surrounding formation to improve porosity. Organic polymers are added to reduce friction between injection fluids and the wellbore so as to lower the energy costs for pumping. Gels or gelling agents based on water-soluble natural polymers are added during the fracking to increase the viscosity of the fracking fluid for better suspension of proppant in the fluid. Breakers such as ammonium $\left(NH_4^+\right)$ and peroxydisulfate $\left(S_2O_8^{2-}\right)$ are added after the fracking to bring down the fluid viscosity so as to facilitate the flowback. Anti-scalants are used to prevent scale precipitation in the well and formation. Biocides are needed to prevent polymer degradation caused by bacteria. Common proppant species and chemical ingredients with their usages are summarized in Table 8.1 [11].

Notably, this table is not exhaustive and oil and gas companies have their proprietary ingredients that are not disclosed to the public. Compatibility of the applied chemicals is usually investigated to guarantee successful fracking and production. Also, ingredients such as biocides, friction reducers (FRs), and scale inhibitors are harmful to human health and local environment, and are under regulation from state or federal legislations. For example, approximately 29 out of 650 chemical additives used in the fracking are identified as having known or possible human carcinogens, and they are either regulated under the Safe Drinking Water Act (SDWA) or listed as hazardous air pollutants under the Clean Air Act [12].

After completion of the fracking operation, wastewater begins to flow back through the wellbore when the hydraulic pressure is reduced. In the initial weeks of completion, the wastewater is mainly composed of the spent fracturing fluid and is normally referred to as the *flowback*. During the production, more and more underground natural formation water returns to the surface with oil and natural gas. The wastewater during this period is commonly known as *produced water*, which lasts throughout the well life. The composition of produced water more closely resembles

TABLE 8.1
Typical Fracking Fluid Additives and Their Purposes

Additive Type	Main Compound(s)	Purposes
Diluted acid (15%)	Hydrochloric acid or muriatic acid	Help dissolve minerals and initiate cracks in the rock
Biocide	Glutaraldehyde	Eliminates bacteria in the water that produce corrosive by-products
Breaker	Ammonium persulfate	Allows a delayed breakdown of the gel polymer chains
Corrosion inhibitor	*N,N*-dimethyl formamide	Prevents the corrosion of the pipe
Cross-linker	Borate salts	Maintains fluid viscosity as temperature increases
Friction reducer	Polyacrylamide, mineral oil	Minimizes friction between the fluid and the pipe
Gel (gelling agent)	Guar gum or hydroxyethyl cellulose	Thickens the water in order to suspend the sand
Iron control	Citric acid	Prevents precipitation of metal oxides
KCl	Potassium chloride	Creates a brine carrier fluid
Oxygen scavenger	Ammonium bisulfite	Removes oxygen from the water to protect the pipe from corrosion
pH adjusting agent	Sodium or potassium carbonate	Maintains the effectiveness of other components, such as cross-linkers
Proppant	Silica, quartz sand	Allows the fractures to remain open so the gas can escape
Scale inhibitor	Ethylene glycol	Prevents scale deposits in the pipe
Surfactant	Isopropanol	Used to increase the viscosity of the fracture fluid

Source: Ground Water Protection Council and ALL Consulting, *Modern shale gas development in the United States: A primer*. 2009, U.S. Department of Energy.

the underground formation water than the original fracking fluid, typically with much higher salt content and heavy metals.

The amount of total produced water returned to the surface is around 15 to 20 billion bbl per year in the United States (1 bbl = 42 US gallons), and the national average water–oil ratio and water–gas ratio, that is, volume of produced water generated from production activities per unit volume of oil and gas, are around 7.6 bbl/bbl and 260 bbl/MMcf (million cubic feet), respectively [13]. Considering the huge amount of wastewater generated and its potential health and environmental impact to the local society and other reasons such as water stress in certain arid areas mentioned above, the systematic management of wastewater is indispensable for sustainable oil and gas production.

On the basis of important factors such as water characteristics, regulatory standards, and economics, there are four major strategies [14] for shale gas and oil wastewater management: (1) deep well injection, (2) discharge to surface water, (3) disposal at commercial or municipal wastewater treatment facilities, and (4) reuse and recycle for future fracking jobs. Each of the options involves a certain level of treatment, and these topics will be elaborated in the following sections.

8.2 CHARACTERISTICS OF WASTEWATER IN UNCONVENTIONAL OIL AND GAS FIELD

The quantity and quality of the wastewater are of fundamental importance when choosing adequate management strategies. The quantity is usually represented by mass/volume per time or per well. However, the quality can be represented by the measurements of its major components, for example, total suspended solids (TSS), total dissolved solids (TDS), oil and grease content, biological oxygen demand (BOD), naturally occurring radioactive material (NORM) content, and so on.

The quantity of wastewater generated throughout the well life ranges from 15% to more than 200% of the initial volume of injected fracking fluid [15]. For example, assuming that the fracking fluid volume is 0.5 million gallons per well (normally between 0.2 and 0.8 million gallons per well according to literature [16]), the total volume of wastewater could be 1 million gallons per well. In addition, the quantity is influenced mainly by the type of hydrocarbon for production, the geological location of the well, and the age of the well. Barnett Shale formation in Texas produces three to four times the wastewater of the Marcellus Shale formation in Pennsylvania [17], and older oil wells could produce more than five times the volume of younger wells.

The quality of wastewater is affected by factors similar to those mentioned above. Most shale wastewater is weakly acidic to neutral with a pH ranging from 5 to 8. The salt content generally increases with the depth of the shale formation. For example, produced water from shale gas wells drilled at depths ranging from 5000 to 8000 ft have salt and mineral levels 20 times higher than those from coalbed methane wells drilled at depths of 1000 to 2000 ft [17]. In addition, since oil is more difficult than gas to remove from water, the hydrocarbon content of wastewater from oil wells is four to five times higher than that from gas wells. The TDS also varies in a wide range among various geological locations. For instance, the average TDS value in Bakken Shale in North Dakota was 271,485 mg/L, while the average TDS value was 22,504 mg/L in Eagle Ford Shale in Texas. In addition, even at a single location, such as Barnett Shale in Texas, the TDS ranged from 599 to 174,692 mg/L from different wells during February 2012 to April 2013 [18].

Generally speaking, the wastewater flow rate decreases while its TDS increases over the production time as shown in Figure 8.3 [19]. As we can see, the flowback usually returns earlier to the surface with less TDS and other components than produced water that comes out later.

Major compositions and their range of content in wastewater from shale oil and gas are categorized as follows and shown in Table 8.2 [19]:

- Suspended solids: that is, clay, sand, and silt represented by TSS
- Oil and grease: various organic compounds associated with hydrocarbons in the formation
- NORM: that is, ^{226}Ra and ^{228}Ra
- TDS: cations ($Fe^{2+/3+}$, Ca^{2+}, Mg^{2+}, Ba^{2+}, Sr^{2+}, etc.) and anions $\left(Cl^-, HCO_3^-, SO_4^{2-}, \text{etc.}\right)$

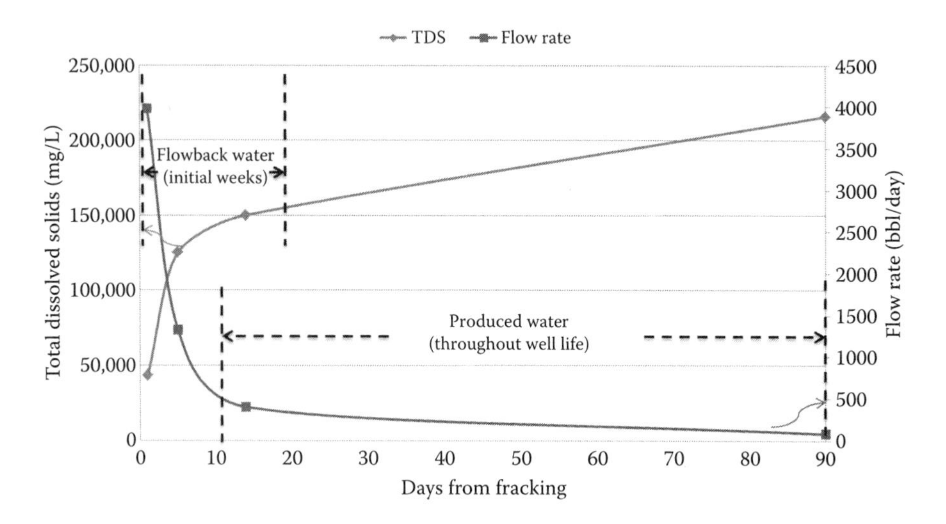

FIGURE 8.3 Flow rate and TDS of wastewater after fracking. (Data with permission from Hayes, T., *Sampling and analysis of water streams associated with the development of Marcellus Shale gas*. 2009, Gas Technology Institute: Des Plaines, IL.)

- Bacteria: sulfate-reducing bacteria (SRB)
- Residual chemicals from fracking fluid: polymers, gel, scale and corrosion inhibitors, and so on

TSS and scale-forming ions $\left(Fe^{2+/3+}, Ca^{2+}, Mg^{2+}, Ba^{2+}, Sr^{2+}, CO_3^{2-}, \text{and } SO_4^{2-}\right)$ in the wastewater are likely to form scale on the wellbore that gradually reduces the flow rate or even clogs the wellbore if not removed [20]. Common scales include calcium carbonate ($CaCO_3$), calcium sulfate ($CaSO_4$), magnesium carbonate ($MgCO_3$), barium sulfate ($BaSO_4$), strontium sulfate ($SrSO_4$), and iron sulfate ($FeSO_4$). In particular, barium sulfate ($BaSO_4$) formation needs to be prevented since the process is virtually irreversible and interferes with proppants in the fracking fluid. Oxidation and deposition of iron can reduce the permeability of the formation, thus reducing oil and gas production. NORMs are normally at very low concentrations but can concentrate in the wastewater if it is continuously reused without removal. Furthermore, since NORMs in the form of radium isotopes (^{226}Ra, ^{228}Ra) are normally coprecipitated with $BaSO_4$, the concentration of barium (Ba^{2+}) in the produced water could be a strong indicator of the presence of NORM [21].

High concentrations of TDS can decrease the efficiency of FRs and increase the energy costs for pumping. Produced water from Marcellus and Bakken Shale plays is known for high TDS (>100,000 mg/L) and especially high divalent cation content, and needs to be treated before reuse accordingly [22]. In addition, SRB that generate hydrogen sulfide (H_2S) can create iron sulfide (FeS) and result in microbiologically induced corrosion.

TABLE 8.2
Selected Components of Wastewater from Marcellus Shale Gas Wells

Parameter	Range	Median	Units
pH	4.9–6.8	6.2	No unit
Acidity	<5–473	NC	mg/L
Total alkalinity	26.1–121	85.2	mg/L
Hardness as $CaCO_3$	630–95,000	34,000	mg/L
Total suspended solids (TSS)	17–1150	209	mg/L
Chloride	1670–181,000	78,100	mg/L
Total dissolved solids (TDS)	3010–261,000	120,000	mg/L
Specific conductance	6800–710,000	256,000	umhos/cm
Ammonia nitrogen	3.7–359	124.5	mg/L
Nitrate–nitrite	<0.1–0.92	NC	mg/L
Nitrite as N	<2.5–77.4	NC	mg/L
Nitrate as N	<0.5–<5	NC	mg/L
Biochemical oxygen demand (BOD)	2.8–2070	39.8	mg/L
Chemical oxygen demand (COD)	228–21,900	8530	mg/L
Total organic carbon (TOC)	1.2–509	38.7	mg/L
Dissolved organic carbon	5–695	43	mg/L
Oil and grease (HEM)	<4.6–103	NC	mg/L
Bromide	15.8–1600	704	mg/L
Fluoride	<0.05–<50	NC	mg/L
Total sulfide	<3.0–3.2	NC	mg/L
Sulfite	7.2–73.6	13.8	mg/L
Sulfate	<10–89.3	NC	mg/L
Total phosphorus	<0.1–2.2	NC	mg/L

Source: Data with permission from Hayes, T., *Sampling and Analysis of Water Streams Associated with the Development of Marcellus Shale Gas*. 2009, Gas Technology Institute: Des Plaines, IL.

8.3 ECONOMICS OF WASTEWATER MANAGEMENT

Wastewater management (for both flowback and produced water) is primarily an economic decision mainly based on the quality of the source water and regulations of the state and federal government [23]. Wastewater with poor quality, that is, with high TSS, TDS, oil, and grease content, usually has fewer direct reuse opportunities and requires treatment, increasing the total cost for reuse. Meanwhile, the economic return for wastewater treatment normally diminishes over time because of its deteriorating water quality [24].

There are four primary options for wastewater management, i.e., *reuse*, *recycle*, *discharge*, and *disposal*, depending on the source water quality. These dedicated terminologies have intrinsic differences that merit explanation. *Reuse* generally involves little or no treatment, while *recycle* suggests involved on-site or centralized treatment to different degrees. *Discharge* normally refers to the release of wastewater to a surface water body and *disposal* means storage of wastewater through deep well injection. In addition, *reuse* and *recycle* are intermediate processes in the water-use flowchart used to reduce water acquisition; *discharge* and *disposal* are final steps in the flowchart seeking either harmless return of wastewater to nature or long-term, safe storage places for wastewater. Furthermore, a temporal combination of them, such as reuse and disposal, or recycle and discharge, is a common practice to maximize water use and the corresponding economic benefits.

Reuse is the easiest and cheapest management option especially if no treatment is involved. The cost of on-site reuse without treatment is typically less than $1/bbl [25]. More than 90% of flowback and produced water from the Marcellus Shale region were reused for new drilling activities in 2013, which reduced the total water management costs by up to 89% and miles trucked by 93% [14]. The opportunities for reuse are largely determined by the wastewater quality. Laboratory and field studies show that untreated produced water with TDS up to 23,000 mg/L can still be used as base fluid for cross-linked gel-based fracking operations [26]. Nevertheless, reuse is only a temporary solution and continuous reuse leads to the increased level of contaminants that eventually requires disposal. Hence, operating companies have been seeking ways, such as reducing the number of chemical additives and fluid chemistry modifications, to prolong the life of reuse for fracking fluids [27].

Recycling wastewater normally involves specific treatment technologies that remove TSS, acid-producing bacteria, and TDS. Effective treatment and recycling of the flowback fluids offer economic advantages associated with recapturing and reusing chemicals, reduced cost of water for subsequent treatments, and disposal. The cost is estimated at $3/bbl for pretreatment, that is, removal of TSS ($1/bbl) and bacteria and heavy metal ($2/bbl) through filtration and electric coagulation, and $6–8/bbl for TDS removal through reverse osmosis (RO), distillation, evaporation, and crystallization [28]. Since it is very costly to totally remove TDS and pure water is not required for fracking, TDS is commonly reduced to the level tolerable for the subsequent reuse and recycle operations.

Discharge to surface water requires the removal of TDS to below a certain level (≤500 mg/L) in Pennsylvania [29]. This requirement was triggered by a series of high-profile pollution incidents that threatened the safety of drinking water [23]. Since the corresponding treatment cost is exorbitant, most oil and gas producers have shifted to other disposal alternatives. Among them, disposal by deep well injection is currently the dominant management option regulated by an underground injection control (UIC) program from the Environmental Protection Agency (EPA). In 2007, 98% of produced water was injected underground from onshore wells, in which 59% was injected into production wells for enhanced oil and gas recovery and 40% was injected for nonproduction disposal in the United States [13]. The estimated cost of injection well disposal varied widely from $0.75 to $3 per barrel from different reports [25,30]. If injection wells are not locally available, transportation cost to the nearest Class II well locations becomes necessary. The transportation and disposal costs can be as high as $15–$18 per barrel if long-distance transportation is needed [16].

The quality and quantity of produced water and current management options among several typical shale plays are listed below [30]:

Barnett Shale play
- High water-to-gas ratio (>1000 gal/MMcf)
- Significant increase of TDS over time (50,000–140,000 mg/L) and low TSS
- Deep well injection for disposal owing to numerous available injection wells

Fayetteville Shale play
- Moderate water-to-gas ratio (200–1000 gal/MMcf)
- Low TDS (15,000 mg/L) and low scaling tendency
- Deep well injection for disposal owing to numerous available injection wells, excellent future potential for reuse because of good quality

Haynesville Shale play
- Moderate water-to-gas ratio (200–1000 gal/MMcf)
- Very high TDS and TSS (350 mg/L)
- Deep well injection, unattractive for reuse because of poor quality

Marcellus Shale play
- Low water-to-gas ratio (<200 gal/MMcf)
- High TDS (40,000–90,000 mg/L with long term >120,000 mg/L), yet low TSS (160 mg/L)
- Reuse is necessary owing to manageable quality and very few local deep well injection opportunities (only eight disposal wells in Pennsylvania in 2008 [31])

On the well pad, freshwater and wastewater are stored in different places. The former is stored in open impoundments with a cost of around \$0.46/bbl (\$3.86/m^3) for the lifetime of the well, while the latter is stored in frac tanks with costs of \$0.07–0.12/bbl/day (\$0.59–1.00/m^3/day) [9]. The material stream of water and their relative cost on the well pad are shown in Figure 8.4 [20].

8.4 FEDERAL AND STATE REGULATIONS

Environmental regulation of wastewater associated with shale gas production is addressed at the federal and state levels and reinforced by EPA. A state may gain "primacy" or the primary responsibility from EPA in implementing their state-level regulation programs by demonstrating that the adopted regulation is at least as stringent as the federal requirements [32]. The federal requirements are described in section 402 (b) of the CWA and 40 CFR Part 123. Wastewater discharged to surface waters (streams, rivers, lakes, etc.) is regulated under the National Pollutant Discharge Elimination System (NPDES) according to the Clean Water Act (CWA). Meanwhile, wastewater disposal through underground injection is regulated under the UIC program of the Federal SDWA that establishes health-based drinking water standards aimed at the prevention of contamination of underground sources of drinking water (USDW).

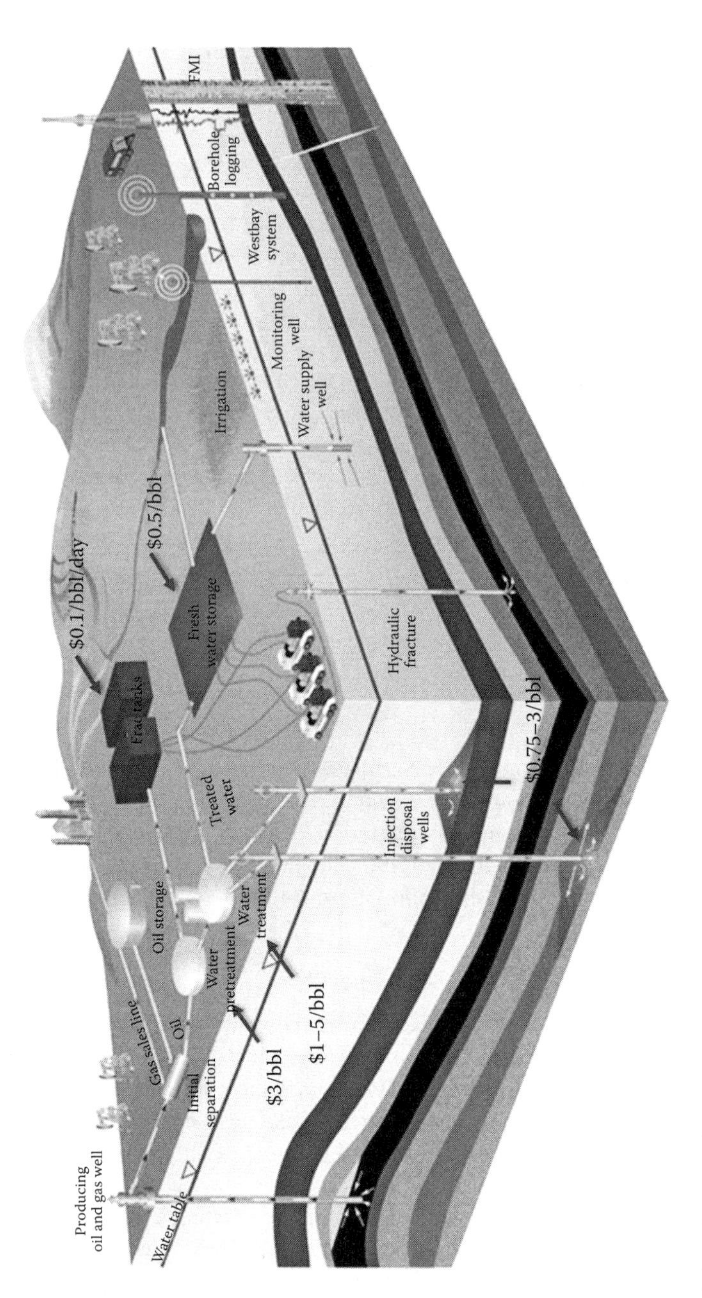

FIGURE 8.4 The material stream of water on the well pad. (Adapted with permission from Probert, T., *Water and Wastewater International*, 2012. 27(2): 22–24.)

The CWA requires that all discharges of wastewater to surface waters be permitted under the NPDES, which is usually administered by the states. NPDES permits are typically licenses for facilities to discharge certain amount of wastewater into the receiving water body under required conditions. For oil and grease in the produced water, all of the NPDES general permits require a monthly average limit of 29 mg/L and a daily maximum limit of 42 mg/L [13]. Limits for TDS, however, are varied among different states. Pennsylvania is the first of its kind to set maximum monthly average discharge standard for TDS (≤500 mg/L), total chlorides (≤250 mg/L), barium (≤10 mg/L), and strontium (≤10 mg/L) [29]. No federal regulations are yet available to specifically address the handling and disposal of NORM. However, the World Health Organization has a drinking water standard for NORM [33], and the Occupational Safety and Health Administration requires monitoring, protection, and education of exposed workers if radiation could exceed 5 millirem (mrem) in 1 h or 100 mrem in any five consecutive days [34].

The disposal of wastewater is regulated under the UIC program, in which six classes (I–VI) of injection wells are defined on the basis of the type of operation and nature of the fluid. More details on the classification of the wells can be found on EPA's website (http://water.epa.gov/type/groundwater/uic/wells_drawings.cfm). Produced water from oil and gas activities is injected into Class II wells for disposal, and Class II wells can further be subdivided into II-R (for enhanced recovery, approximately 80%), II-D (for disposal, approximately 20%), and II-H (for hydrocarbon storage) based on their purposes [13]. To date, there are approximately 40 states with primacy for Class II UIC wells for the injection of more than 2 billion gallons of produced water every day [35].

8.5 DISPOSAL LOCATION SELECTION

The selection of disposal locations is determined largely by criteria based on underground geology, land use, operational conditions, transportation, and environment [35].

Geological character is the dominant factor in the decision of suitable underground disposal sites [36]. Before the selection of injection locations, an area of review (AOR) is performed to identify all active, temporarily abandoned, and plugged oil and gas wells that penetrate the proposed injection area. Private or public water supply wells, water-bearing strata, and drinking water aquifers are also identified in AOR. The ideal injection location is a porous and permeable, non–hydrocarbon-bearing zone that is not considered an aquifer under the UIC program. Limestone, sandstone, and dolomite zones are such kinds of zones with porosity and permeability levels acceptable for brine disposal. In the case where an existing well is used, the well must be completed in a depleted oil and gas reservoir or must penetrate a porous zone that can receive injected wastewater at rates appropriate to meet the disposal requirement (Figure 8.5 [37]). Furthermore, the presence of sufficiently impermeable strata above and below the injection zone is necessary to confine the wastewater in the injection formation and eliminate the possibility of migration to other stratigraphic intervals [35].

For the land use criteria, it is desirable that the prospective disposal well location is owned by the operator; otherwise, agreements need to be achieved between operators and landowners before proceeding with the permitting process. For the

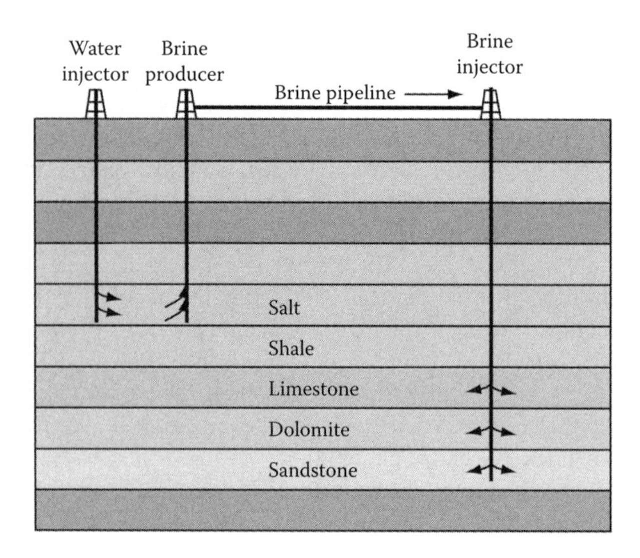

FIGURE 8.5 The locations of on-site water use. (From Smith, L. et al., *Systematic technical innovations initiative brine disposal in the Northeast.* 2005, U.S. Department of Energy.)

operations criteria, proximity of disposal wells to drilling and completion locations is always advantageous, which could reduce transportation distances in between. Environmental reviews of the chosen locations are conducted to ensure no endangered species or rare plants, animals, and natural communities are threatened. Furthermore, groundwater sources, surface waters, wetlands, springs, and surface water intake and discharges must be identified. The disposal site should not be too close to residences, where heavy truck traffic can cause significant wear and tear on roads, and dust and noise may become an issue [35].

Other factors affecting the capacity of disposal are [36] (i) fluid compatibility between formation water and various treatment and disposal fluids; (ii) compatibility of treatment and disposal fluids with the reservoir rock; and (iii) permeability, conductivity, and fracture length of the reservoir rock. Clay swelling and fines migration are among other concerns causing permeability reduction. To address these concerns, extensive tests need to be run. For (i), compatibility tests between reservoir fluid, treatment water, and disposal fluid need to be conducted. For (ii), core tests such as x-ray diffraction, acid solubility, immersion tests, fracture flow capacity and fluid loss tests, and core flow tests (measured permeability vs. pore volume) have to been performed.

8.6 WATER TREATMENT PROCESSES

Adequate treatment of oil and gas wastewater could alleviate the regional water shortage and avoid violations from federal and state regulations [8]. The Marcellus Shale region has unique opportunities to be the "testing ground" of novel treatment technologies because of the unfavorable geological disposal opportunities [23]. Reuse and recycle opportunities with minimal environmental impacts are being actively studied [38].

TABLE 8.3
Selected Treatment Technologies

Categories	Objectives	Treatment Technologies
Primary treatment	Remove suspended solids, oil/grease, iron, unbroken polymers, bacteria	Coagulation/flocculation, hydroclone, gas flotation, filtration (multimedia filtration or cartridge filtration), oxidization
Secondary treatment	Remove scaling formation ions such as Ca, Mg, Ba, Sr, and NORM	Lime softening, ion exchange, nanofiltration
Tertiary treatment	Remove TDS for surface discharge	Ion exchange, reverse osmosis, thermal evaporation
Zero liquid discharge	Convert the liquid waste into solid waste	Evaporation and crystallization

Source: Content adapted with permission from Kuijvenhoven, C. et al., Water management approach for shale operations in North America, in *SPE Unconventional Resources Conference and Exhibition-Asia Pacific*. 2013, Society of Petroleum Engineers: Brisbane, Australia.

Regarding the complex characteristics of oil and gas wastewater mentioned in Sections 8.2 through 8.4, the choice of treatment processes must be preceded by the evaluation of water quality, cost-effectiveness, and regulatory requirements. The constituents in the wastewater can be grouped into four categories: soluble organics, insoluble organics, soluble inorganics, and insoluble inorganics. Insoluble organics, that is, dispersed oil or certain hydrocarbons, can be separated by gravimetric and de-oiling technologies, but soluble organics, usually polar compounds such as formic acid and propionic acid, require more complicated treatment methods such as biological treatment or electrodialysis. Insoluble inorganics (e.g., scales, precipitates, inorganic colloids, etc.) are usually separated by chemical precipitation and filtration, and soluble inorganics such as cations (e.g., Na^+, K^+, Ca^{2+}, Mg^{2+}, $Fe^{2+/3+}$, etc.) and anions $\left(Cl^-, SO_4^{2-}, CO_3^{2-}, HCO_3^-, \text{etc.}\right)$ can be treated by thermal evaporation, membrane filtration, ion exchange, and so on. Generally, a variety of treatment methods are combined following a specific procedure to achieve zero liquid discharge (ZLD). They can be categorized into four main steps: primary, secondary, tertiary, and ZLD [39]. Each step is designed to concentrate or remove a certain group of contaminant with corresponding technologies shown (Table 8.3 [39]).

8.6.1 Primary Treatment

8.6.1.1 Overview

A typical combination of primary treatment processes involves oil–water separation, electrocoagulation (EC) or chemical coagulation (CC), filtration, and oxidation to remove suspended solids, dispersed oil, iron, unbroken polymers, and bacteria in the oil and gas wastewater. The adjustment of pH and the addition of chemicals are usually conducted first to promote coagulation of suspended solids and unbroken polymers and they are then removed by sand or cartridge filtration [38]. Dispersed oil is usually removed by de-oiling technologies such as API oil–water separator, hydroclone, or gas flotation. Iron content can be reduced by aeration and sedimentation, ion exchange, or ozonation. The oxidizers such as chlorine dioxide (ClO_2) could be used to break oil emulsions, destroy FRs, and kill bacteria [40]. The pore size of the filter used in the shale gas industry usually ranges from 0.04 to 3 μm. Another study reported successful filtration of precipitated iron with bacteria by use of 25-μm filters after oxidation, flocculation, and sedimentation [41]. However, the filtration process cannot remove any dissolved salts to lower the TDS level. Therefore, water after primary treatment needs to be diluted with freshwater in order to be directly reused in fracking operations without any treatments.

8.6.1.2 Coagulation

CC and EC are commonly used to remove suspended solids and colloidal particles. For CC, inorganic mixed metal (Fe, Mg, and Al) polynuclear polymer, that is, FMA, has good coagulation, de-oiling, and scaling inhibition properties with high removal efficiency (>92% for suspended solids and >97% for oil) [42]. Spillsorb, calcite, and lime were used to remove heavy metals from produced water with high removal efficiency (>95%) [43]. Oxidant, ferric ions, and flocculants were reported to remove hydrocarbons, arsenic, and mercury [44].

On the other hand, EC removes colloidal (or suspended) solids through direct electric current dissolving iron or aluminum electrodes (anode) for generation of the primary coagulants, that is, ferrous or aluminum hydroxide ($Fe/Al(OH)_3$). As a result, the suspended solids coagulate and fall out of the suspension. While an electric current is transmitted through the water, oxygen, hydrogen, and chlorine gases are generated and these gas bubbles are collected and used for the downstream flotation separation processes [39]. After EC, the pH of the water is adjusted to form a flocculation that is subsequently separated by a weir tank. Then, the water is filtered by multimedia filtration to remove any remaining solids. The sludge after primary treatment from EC needs to be disposed as well. Compared to conventional treatment processes such as RO, EC generates relatively smaller amounts of waste sludge [45].

8.6.1.3 Iron Control

The iron control (IC) strategy introduced here is to chemically complex iron in fluids in the reduced valence state to prevent precipitation or complexation with other mineral or polymer phases. Upon contact with the downhole environment, iron is leached into the flowback water. Released iron can interact with polymer agent added (e.g., FR), and the formed aggregates or precipitates can damage the oil and gas production. Siderite, $FeCO_3$, has been found to be a potentially problematic iron precipitate based on geochemical simulation [46]. Traditional iron-reducing and chelating agents consist of acidic chemicals such as citric acid, acetic acid, and ethylenediaminetetraacetic acid. IC additive works with a blended scale control additive to prevent precipitation of scale such as siderite, preserve effectiveness of other scale inhibitors, and enhance the performance of FRs.

8.6.1.4 Bacteria Treatment

Guar gum used in fracking fluid is easily degraded by bacteria both downhole and at the surface, compromising water reuse and disposal [47]. The bacteria growth in piping systems can reduce the transportation capability and production. Anaerobic bacteria, such as SRB, are also responsible for downhole H_2S souring and equipment corrosion. Since biocides generally have safety and environmental concerns, they must exhibit a short life with known breakdown pathways. A desirable biocide should be able to kill the microorganism instantly and decontaminate a system within a short period, for example, 1 h [46]. Microbial control strategies also need to consider biocide rotation, seasonal loading adjustments, and biocide pulse dosing to enhance microbial control efficacy [48]. The most widely applied antibacterial agents include glutaraldehyde, 2,2-dibromo-3-nitrilopropionamide, and ClO_2.

8.6.1.5 Chlorine Dioxide Treatment

On the basis of the fracturing fluid compatibility and requirement, ClO_2 has been selected as a primary treatment that is generated on location to oxidize bacteria, hydrocarbon chains, hydrogen sulfide, and iron sulfide. The novel properties are attributed to its unique single-electron transfer mechanism, wherein it attacks electron-rich centers in organic molecules. It works quickly and efficiently at low dosages and offers a broad range of bacteria, fungi, and virus destruction. ClO_2 can also oxidize ferrous iron to ferric iron, leading to formation of ferric hydroxide that can be removed by filtration easily. It works over a wide range of pH and therefore is increasingly being used for oil field applications. It is reported that ClO_2 does not react with most NORMs or forms hypochlorous acid or free chlorine [16]. However, other researchers argue that treatment with ClO_2 or hypochlorite will convert the naturally occurring hydrocarbons to chlorocarbons and organobromides. These halogen-containing compounds are very toxic and should be avoided in water treatment [49]. Further investigations are needed to determine if the reactions will occur under downhole conditions or during treatment of produced water.

8.6.2 Secondary Treatment

8.6.2.1 Lime Softening

Chemical precipitation processes such as brine treatment with lime and Na_2SO_4 are often used in secondary treatment to remove scale-forming metal ions and NORM, such as calcium, magnesium, barium, strontium, and so on, and the sludge could be dewatered and disposed through underground injection [40]. Although lime softening removes metals such as barium and NORM, such a process cannot remove halogens (chloride and bromide). In addition, reduction of radioactivity from wastewater and safe disposal of NORM-rich solid waste and residues from treatment of wastewater are critical, complicated, and costly in preventing contamination and accumulation of residual radioactive materials [50]. Hence, many producers do not want to precipitate barium and strontium because of the creation of NORM and the subsequent disposal problem, and therefore they prefer to leave the metals in waters by use of scale inhibitors [41].

Here, we introduce a modified lime softening process that yields aqueous radium and barium concentrate that can be safely disposed by UIC [51]. In the first stage of this process, wastewater is oxidized and contacted with lime to precipitate magnesium, iron, and manganese; the pH is raised from neutral to approximately 10.6–10.8. Then, sodium carbonate is added to precipitate calcium and strontium at which the pH is approximately 11.0–11.3. In the second stage, the wastewater is treated with more sodium carbonate ($NaCO_3$) to precipitate barium and radium as carbonates ($BaCO_3$ and $RaCO_3$). The precipitate from this stage is diverted and treated with concentrated hydrogen chloride (HCl) to form barium and radium chlorides ($BaCl_2$ and $RaCl_2$). Meanwhile, the pH is low enough (pH 2–4) to strip all the carbonates from the aqueous concentrate stream. The concentrate stream is then neutralized and disposed by underground injection. The flowchart of this process is shown in Figure 8.6 [51].

Recently, an advanced oxidation and precipitation process (AOPP) has been reported as an on-the-fly fluid pretreatment technology during fracking operations [52]. This technology provides an on-site, cost-effective (~$1.25/bbl) microbial control and scale that can completely replace the use of biocide or chemical scale inhibitor in the fracking fluid and generate zero liquid waste streams. The major treatment steps in AOPP involve hydrodynamic cavitation, ozone treatment, acoustic cavitation, and electro-oxidation. During hydrodynamic cavitation, extremely high temperature and pressure produces highly reactive hydroxyl radicals that can decompose most organic compounds in water. Ozone is then applied as a highly reactive oxidant to kill bacteria and oxidize heavy metals such as iron to form insoluble salts, which is then precipitated by the passage of electricity in the water. Ozonated water is then

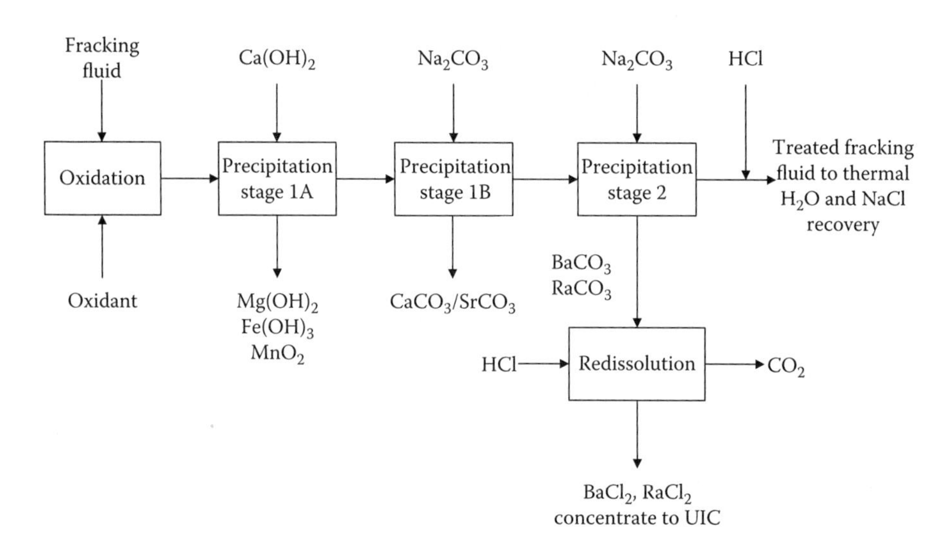

FIGURE 8.6 Schematic of a modified lime-soda process. (Reprinted with permission from J.M. Silva, *Produced Water Pretreatment for Water Recovery and Salt Production [RPSEA Final Report]*. January 2012, New York.)

treated by acoustic cavitation where ultrasound breaks the precipitated salts into nanosized suspended particles that will not cause any scales. The treatment takes only approximately 1 min to complete. The bulky space of a typical AOPP tank was redesigned to fit in a single 53-ft trailer with a treatment capacity of 80 bbl/min [52].

8.6.3 Tertiary Treatment

8.6.3.1 Thermal Evaporation Technologies

Thermal evaporation technologies are a group of technologies aimed at vaporizing and condensing feed water through a series of heat exchanging processes to obtain purified water. Feed flexibility and permeate quality are two distinct advantages of this technology. It works on any type of water, especially water with a very high salinity (TDS at 200,000 mg/L). The process completely removes TSS and reduces almost all of the TDS, producing ultraclean water product (TDS at approximately 2–10 mg/L [53]) and concentrated brine that could be disposed under the UIC program. Also, less extensive pretreatment is involved compared to the membrane processes. Conventional thermal evaporation technologies, such as multistage flash (MSF) and multiple effect distillation (MED), and vapor compression (VC) are mature and well established for seawater desalination at medium to large capacity (19,000–90,000 m^3/day for MSF, 3800–22,700 m^3/day for MED, and 3800 m^3/day for VC [54]). For oil- and gas-produced water treatment, mechanical vapor compression (MVC) has been successfully demonstrated at a small to medium capacity (1715 bbl/day or 204 m^3/day for mobile system and 22,000 bbl/day or 2623 m^3/day for fixed system) [55,56]. Figure 8.7 [23] shows a typical MVC process. Nevertheless,

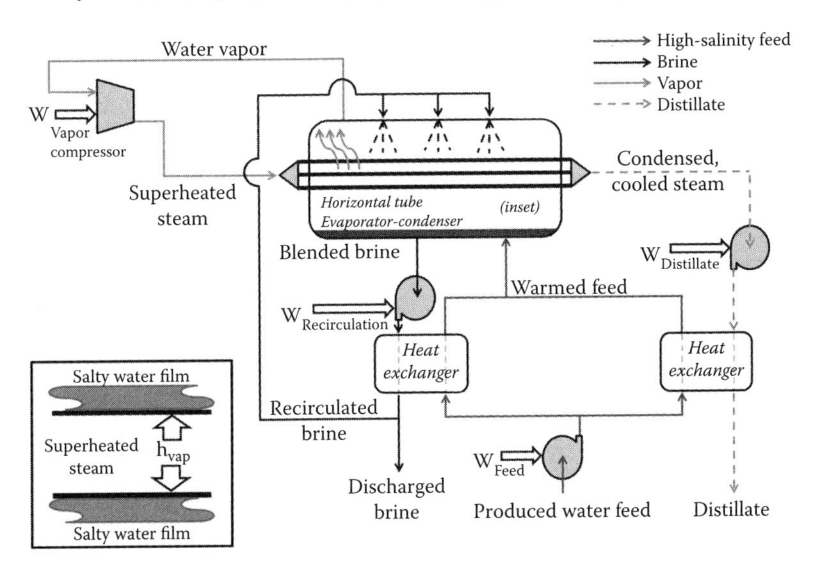

FIGURE 8.7 Schematic of an MVC process. (Reprinted with permission from D.L. Shaffer et al., Desalination and reuse of high-salinity shale gas produced water: Drivers, technologies, and future directions. *Environ. Sci. Technol.*, 2013. 47: 9569–9583. Copyright [2013] American Chemical Society.)

thermal evaporation processes are energy intensive and the treatment costs are usually two to three times higher than membrane separation technologies such as RO [57].

8.6.3.2 Membrane Separation Technologies

Membrane-based technologies are physical separation processes that reject different constituents in the water. Traditional membrane processes can be categorized into microfiltration (MF) (10^{-1} to 1μm), ultrafiltration (UF) (10^{-2} to 10^{-1}μm), nanofiltration (NF) (10^{-3} to 10^{-2}μm), and RO (10^{-4} to 10^{-3}μm) based on pore sizes. Newer developed technologies such as forward osmosis (FO) and membrane distillation (MD) also demonstrate a promising future to treat oil and gas wastewater. In current practices, MF and UF usually serve as pretreatment steps to remove suspended solids, bacteria, and macromolecules. NF and RO are used to remove dissolved multivalent and univalent ions, respectively. RO and NF are gaining popularity recently with advantages such as high removal efficiency, low treatment cost, less space of installation, and easier operation [42,58]. Since membrane materials are usually made of hydrophobic polymers, there is a tendency for oil and grease in the wastewater to foul the membrane surface by hydrophobic interactions. Hence, extensive pretreatment is necessary to remove oil and grease before membrane separation. The modification of membrane surfaces is effective to reduce fouling and scaling potential [59]. We will discuss RO, NF, FO, and MD in more detail in the following sections.

8.6.3.3 Reverse Osmosis and Nanofiltration

RO and NF are pressure-driven processes for saline water treatment. Generally, high hydraulic pressure (e.g., 600–900 psig) is applied on the high salinity feed water side to drive water molecules through a semipermeable membrane that retains salts (Figure 8.8 [60]). The hydraulic pressure must be higher than the osmotic pressure for the water to flow across the membrane against the concentration gradient. NF is normally employed to treat brackish water (TDS 500–30,000 mg/L) while RO is used for seawater with TDS up to 35,000 mg/L. Brine with very high salinity (100,000–200,000 mg/L) may exceed the allowable pressure for RO modules and operational limits on process equipment [23,57]. Therefore, RO is more practical for brackish flowback from wells that have relatively lower salinity than typical produced water.

Commercial RO membranes are typically installed in a spiral wounded configuration with mesh spacers installed in both the feed channel and permeate collection channels of the membrane module (Figure 8.9 [53]). Feed spacers are used to enhance hydrodynamic turbulences in the channel and avoid the concentration polarization, which is a phenomenon in which the feed solution concentrates at the feed–membrane interface, resulting in preferential diffusion of pure water. A permeate spacer is installed to provide mechanical strength at the permeate collection channel.

Although a deionized water product stream of good quality is initially achieved, many operational problems involving membrane fouling arise as a result of the interaction of complex constituents of the produced water with the membrane materials. For instance, oils collected on RO membranes cause them to lose the permeability, and particles and precipitates cause the filter material to break down mechanically.

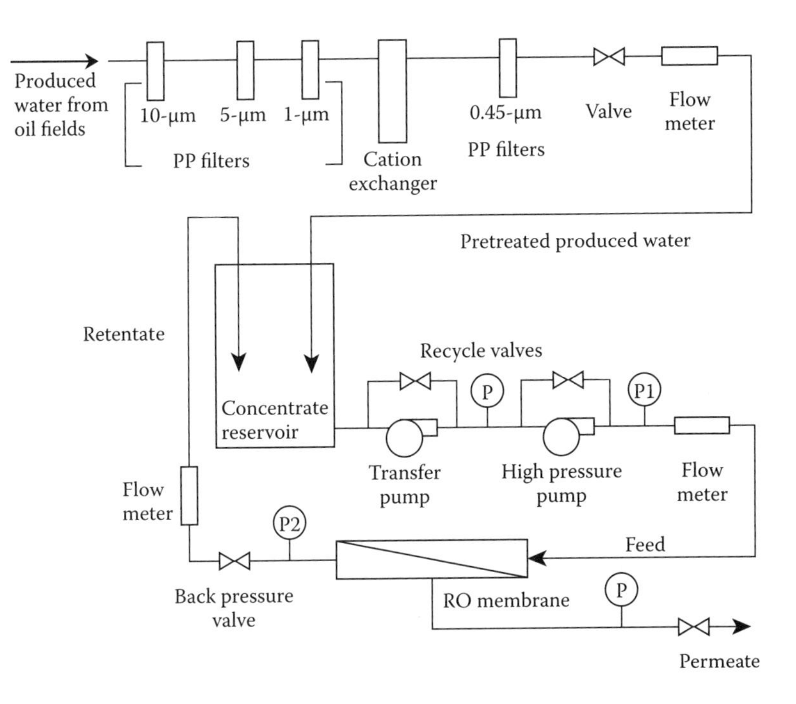

FIGURE 8.8 Schematic of a typical RO process for the brackish oil field–produced water. (Reprinted from *Water Research* 37(3), Murray-Gulde, C. et al., Performance of a hybrid reverse osmosis-constructed wetland treatment system for brackish oil field produced water, 705–713, Copyright [2002], with permission from Elsevier Science Ltd.)

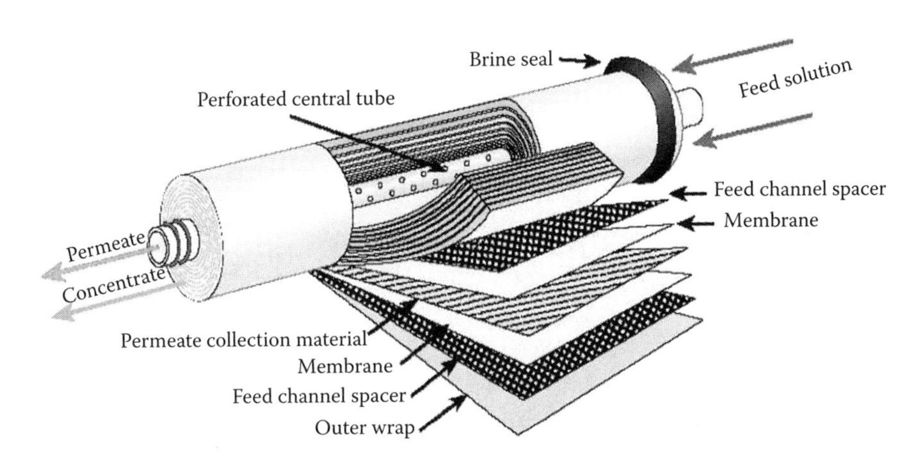

FIGURE 8.9 Schematic of an RO module. (Reprinted with permission from Drewes, J.E., *An integrated framework for treatment and management of produced water: Technical assessment of produced water treatment technologies*. 2009, The Colorado School of Mines: Golden, CO. p. 158.)

Soluble hydrocarbons can promote the growth of microbial films on RO surfaces, decreasing the separation performance. All of these problems are responsible for the lack of deployment success of RO in the oil and gas industry. Nevertheless, rigorous pretreatment and advances in polymeric coatings in the design hold promise in the improvement of performance and cost in wastewater treatment [38].

8.6.3.4 Combined Treatment Scheme

In oil and gas field applications, multiple separation technologies are often combined in a treatment scheme to achieve optimum removal of components in the wastewater. For example, as shown in Figure 8.10, dispersed oil is separated by oil–water separator and gas flotation; MF is applied later to remove dissolved hydrocarbons and then RO membranes serve as the final step to get rid of dissolved solids [61]. However, variations in the contaminant types and concentrations may cause failure of the membrane system if it is not frequently cleaned and recalibrated [62].

8.6.3.5 Forward Osmosis

FO uses the energy from an osmotic gradient across the membrane to "draw" the water through the selectively permeable membrane as shown in Figure 8.11. This is in contrast to RO that uses mechanical energy (hydraulic pressure) to push the water through the membrane. In FO, the solution applied to draw the water is typically a highly concentrated, homogeneous solution (e.g., of salt or sugar), and the resulting produced water is the diluted draw solution [63,64]. The undesirable solids and solutes do not penetrate the FO membrane, while on the effluent side, a pure diluted draw solution is obtained (Figure 8.11 [53]). A major difference between RO and FO is that the water produced by FO is not low TDS freshwater but diluted draw solution that must have a downstream separation to obtain potable water.

Since FO does not create pressure on the membrane, the frequency and intensity of membrane fouling (clogging or contamination) are greatly reduced compared to RO [64]. Another significant benefit of FO is that the chemical energy required to operate this system comes from the salts in the draw solution, the cost of which is already present for many operators. In addition, there is only minimal need for pretreatment and no sizable electrical consumption, and therefore the carbon footprint of the process is small [63].

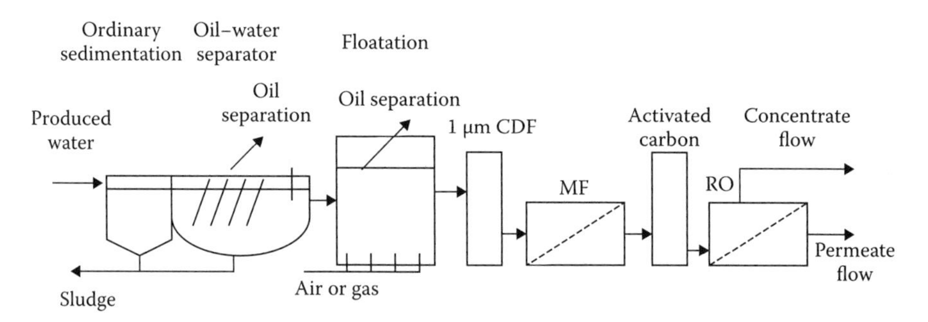

FIGURE 8.10 Combined treatment scheme for the oil field–produced water. (Reprinted from *J. Hazard Mater* 170, F. Ahmadun et al., Review of technologies for oil and gas produced water treatment, 530–551, Copyright [2009], with permission from Elsevier Science Ltd.)

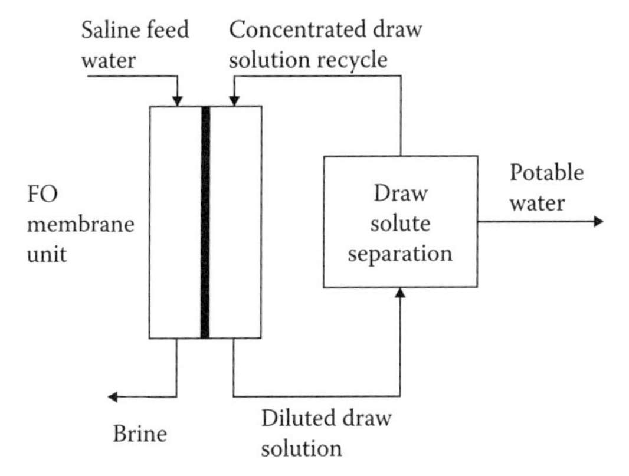

FIGURE 8.11 Schematic of a general FO process. (Reprinted with permission from Drewes, J.E., *An integrated framework for treatment and management of produced water: Technical assessment of produced water treatment technologies*. 2009, The Colorado School of Mines: Golden, CO. p. 158.)

8.6.3.6 Membrane Distillation

MD is a mass transfer process driven by a partial vapor pressure difference attributed to a temperature gradient across the hydrophobic, microporous membrane (Figure 8.12 [65]). Common membrane materials include polytetrafluorethylene, polypropylene, and polyvinylidenedifluoride. A temperature difference of 10°C–20°C between the two sides of the membrane can be sufficient to produce distilled water [66]. There

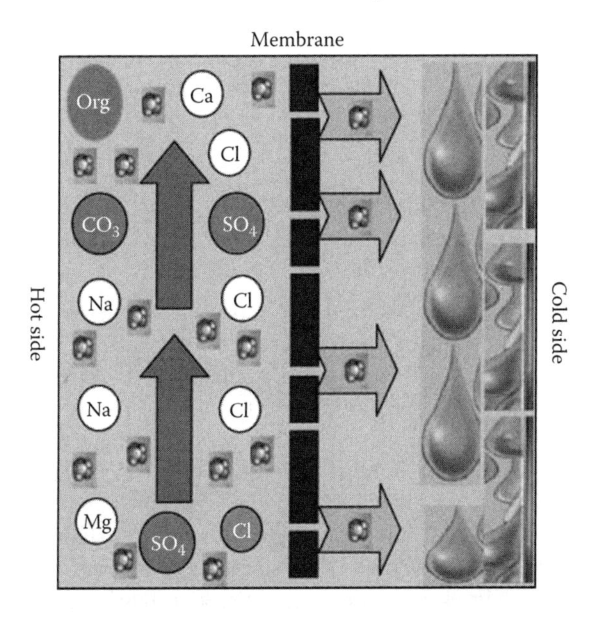

FIGURE 8.12 Schematic of a general MD process. (Reprinted with permission from *Desalination* 314, S. Adham et al., Application of membrane distillation for desalting brines from thermal desalination plants, 101–108, Copyright [2013] Elsevier Science Ltd.)

are four basic configurations: direct contact MD (DCMD), air gap MD (AGMD), vacuum MD (VMD), and sweeping gas MD (SGMD) [67]. Among them, DCMD and AGMD are the most likely to be applied for oil- and gas-produced water owing to the simpler configuration.

As an innovative process, MD can desalinate high saline water such as produced water in oil and gas fields (30,000–100,000 mg/L TDS) using low-grade waste heat that is locally available. It is also extremely flexible for variations in feed water quality and quantity. Excellent TDS rejection (99.9%) on all of the tested water samples was reported with no negative impact on the membrane's flux performance [66]. However, since surfactants can wet the hydrophobic pores of the MD membrane and cause pore flooding that will reduce solute rejection significantly, complete removal of any surfactants in the feed stream is required as the pretreatment.

8.6.4 Zero Liquid Discharge

The aim of ZLD is to eliminate all liquid waste and generate high-purity water and solid salts. The common ZLD system involves the brine concentrator or evaporator, which serves to concentrate the feed, and the crystallizer is then applied to further desiccate the concentrated brine sludge and convert it into solid salts as shown in Figure 8.13 [56]. GE Water & Process Technologies (GEWPT) conducted pilot-scale trials of evaporation and crystallization of Marcellus-produced water, generating solid sodium chloride (NaCl) salt product and distilled water with high water recovery (vol H_2O/vol feed >95% in average) [51]. Current commercial ZLD technologies are relatively energy intensive (100–250 kWh/kgal), and the operation cost is much higher than that of desalination processes [53]. In the short term, the recovery and sale of low-grade salts such as road salts can offset the total cost to a certain degree, while in the long term, improvements in the technologies to reduce energy consumption and increasing salt purity can significantly increase the economic benefits.

8.7 EFFECTS OF TREATED WATER QUALITY

8.7.1 Effects on Fracking Fluids

The treated water quality has a significant impact on the performance of fracking fluids. Leftover contaminants in the treated water can negatively affect the effectiveness of the chemical additives. For instance, colloidal solids can adsorb surfactants and clay stabilizers and thus decrease their effectiveness [41]. Multivalent ions such as Ca^{2+} and Fe^{3+} in water can interact with functional groups of the molecules of anionic FRs, such as polyacrylamide, forming polymer aggregates that are harmful to transportation and production. Hence, multivalent ions need to be removed or controlled by adding complexing agents to improve the performance of FR polymers [68]. In addition, novel FR polymers with salt-tolerant properties have been developed to maintain high levels of performance in produced brine in field trials [22,69]. A high salt-tolerant FR performs effectively at lower dosages than a regular FR in brines, which offers a cost advantage. However, they may have compatibility issues with some biocides by reducing the viscosity [69].

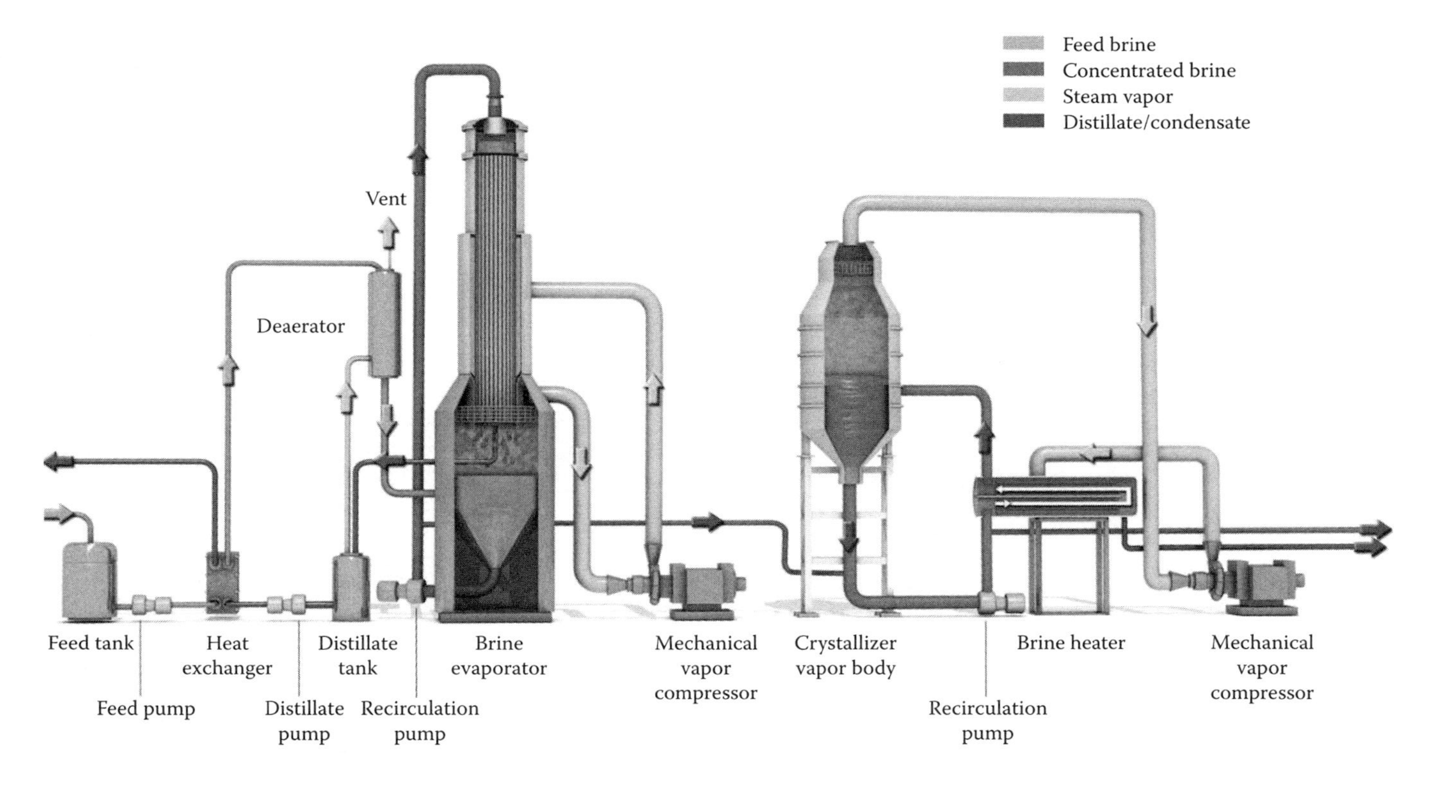

FIGURE 8.13 Schematic of the evaporator and crystallizer. (Reprinted with permission from GE Water & Process Technologies, *Zero Liquid Discharge: Eliminate Liquid Discharge, Recover Valuable Process Water.* 2009; Available from: http://www.gewater.com/products/zero-liquid-discharge-zld-crystalizers.html.)

8.7.2 Effects on Formation

Impurities in the treated water can affect the formation integrity and production as well. For instance, polymers can adsorb and deposit onto the oil-bearing rock formation surface during drilling, completion, or fracking, which results in the plugging of formation pores and impeded production [70,71]. The presence of solids, residual gel, and bacteria can also reduce the permeability of the formation [41]. High pH can cause permeability reduction, plugging of pores, and corresponding formation damage. The permeability reduction in limestone reservoirs can be minimized using brines of NaCl, KCl, and $CaCl_2$ mixtures, compared with brines of NaOH and Na_2SiO_4 at high pH values [72].

REFERENCES

1. Charlez, P.A., *Rock Mechanics, Petroleum Applications. Rock Mechanics Volume 2.* Petroleum applications. Vol. 2. 1997.
2. US Environmental Protection Agency, *Plan to Study the Potential Impacts of Hydraulic Fracturing on Drinking Water Resources.* 2011, Office of Research and Development.
3. Vidic, R.D. et al., Impact of shale gas development on regional water quality. *Science*, 2013. **340**: 6134.
4. Tipton, D.S., Mid-continent water management for stimulation operations, in *SPE Hydraulic Fracturing Technology Conference.* 2014, Society of Petroleum Engineers: Texas.
5. Nicot, J.P. and B.R. Scanlon, Water use for shale-gas production in Texas, U.S. *Environ. Sci. Technol*, 2012. **46**: 3580–3586.
6. Zhu, H. and R. Tomson, *Exploring Water Treatment, Reuse and Alternative Sources in Shale Production.* 2013 [cited May 9, 2015]; Available from: http://www.shaleplaywatermanagement.com/2013/11/exploring-water-treatment-reuse-and-alternative-sources-in-shale-production/.
7. Mauter, M.S. et al., Regional variation in water-related impacts of shale gas development and implications for emerging international plays. *Environ. Sci. Technol.*, 2014. **48**: 8298–8306.
8. Vengosh, A. et al., A critical review of the risks to water resources from unconventional shale gas development and hydraulic fracturing in the United States. *Environ. Sci. Technol.*, 2014. **48**: 8334–8348.
9. Yang, L. and I.E. Grossmann, Optimization models for shale gas water management. *AICHE Journal*, 2014. **60**(10): 3490–3501.
10. Suchy, D.R. and K.D. Newell, *Hydraulic fracturing of oil and gas wells in Kansas.* 2012: Kansas Geological Survey—The University of Kansas.
11. Ground Water Protection Council and ALL Consulting, *Modern shale gas development in the United States: A primer.* 2009, U.S. Department of Energy.
12. Waxman, H.A., E.J. Markey, and D. DeGette, *Chemical used in hydraulic fracturing.* 2011, United States House of Representatives Committe of Energy and Commerce.
13. Clark, C.E. and J.A. Veil, *Produced water volumes and management practices in the United States.* 2009, Environmental Science Division, Argonne National Laboratory.
14. Ma, G., M. Geza, and P. Xu, Review of flowback and produced water management, treatment and beneficial use for major shale gas development basins. *Shale Energy Engineering*, 2014: 53–62.
15. Glazer, Y.R. et al., Potential for using energy from flared gas for on-site hydraulic fracturing wastewater treatment in Texas. *Environ. Sci. Technol. Lett.*, 2014. **1**: 300–304.

16. Seth, K. et al., Maximizing flowback reuse and reducing freshwater demand: Case studies from the challenging Marcellus shale, in *SPE Eastern Regional Meeting*. 2013, Society of Petroleum Engineers: Pittsburgh, Pennsylvania.
17. Information on the quantity, quality, and management of water produced during oil and gas production, in *Energy-Water Nexus*. United States Government Accountability Office.
18. Tomson Technologies Inc., available from: http://tomson.com/.
19. Hayes, T., *Sampling and analysis of water streams associated with the development of Marcellus Shale gas*. 2009, Gas Technology Institute: Des Plaines, IL.
20. Probert, T., Shale gas fracking: Water lessons from the US to Europe. *Water and Wastewater International*, 2012. **27**(2): 22–24.
21. Igunnu, E.T. and G.Z. Chen, Produced water treatment technologies. *International Journal of Low-Carbon Technologies*, 2012. **0**: 1–21.
22. Zhou, J. et al., Water-based environmentally preferred friction reducer in ultrahigh-tds produced water for slickwater fracturing in shale reservoirs, in *European Unconventional Conference and Exhibition*. 2014, Society of Petroleum Engineers: Vienna, Austria.
23. Shaffer, D.L., L.H.A. Chavez, and M. Ben-Sasson, Desalination and reuse of high-salinity shale gas produced water: Drivers, technologies, and future directions. *Environ. Sci. Technol.*, 2013. **47**: 9569–9583.
24. Glazer, Y.R. et al., Potential for using energy from flared gas for on-site hydraulic fracturing wastewater treatment in Texas. *Environ. Sci. Technol. Lett.*, 2014. **1**(7): 300–304.
25. Mauter, M.S. and V.R. Palmer, Expert elicitation of trends in marcellus oil and gas wastewater management. *J. Environ. Eng.*, 2014. **140**(5).
26. Huang, F. et al., Feasibility of using produced water for crosslinked gel-based hydraulic fracturing, in *SPE Production and Operations Symposium*. 2005, Society of Petroleum Engineers: Oklahoma City, OK.
27. Bulat, D. et al., A faster cleanup, produced-water-compatible fracturing fluid: Fluid designs and field case studies, in *SPE International Symposium and Exhibition on Formation Damage Control*. 2008, Society of Petroleum Engineers: Lafayette, Louisiana.
28. Tipton, D.S., Water management for unconventional oil and gas operations while minimizing environmental impact, in *Environmental Challenges and Innovations Conference*. 2012: Houston, Texas.
29. Wastewater treatment requirements, in *Pennsylvania Code, Section 10, Title 25, Chapter 95*. 2010.
30. Slutz, J. et al., Key shale gas water management strategies: An economic assessment tool, in *SPE/APPEA International Conference on Health, Safety, and Environment in Oil and Gas Exploration and Production*. 2012, Society of Petroleum Engineers: Perth, Australia.
31. McCurdy, R. *Underground injection wells for produced water disposal*, in *USEPA technical workshops for the hydraulic fracturing study: Water resources management*. 2011, United States Environmental Protection Agency, Office of Research and Development: Arlington, VA.
32. All Consulting, *Handbook on Coal Bed Methane Produced Water: Management and Beneficial Use Alternatives*. 2003.
33. World Health Organization., *Guidelines for Drinking-Water Quality: Incorporating First Addendum. Vol. 1, Recommendations*. 2006.
34. Arthur, J.D., B. Bohm, and D. Cornue, Environmental considerations of modern shale gas development, in *SPE Annual Technical Conference and Exhibition*. 2009, Society of Petroleum Engineers: New Orleans, Louisiana.

35. Arthur, J.D., S.L. Dutnell, and D.B. Cornue, Siting and permitting of class ii brine disposal wells associated with development of the marcellus shale, in *SPE Eastern Regional Meeting*. 2009, Society of Petroleum Engineers: Charleston, West Virginia.
36. Hunt, J.L. et al., Evaluation and completion procedure for produced brine and waste water disposal wells. *J. Petrol. Sci. Eng.*, 1994. **11**: 51–60.
37. Smith, L. et al., *Systematic technical innovations initiative brine disposal in the Northeast*. 2005, U.S. Department of Energy.
38. Gaudlip, A.W., L.O. Paugh, and T.D. Hayes, Marcellus shale water management challenges in Pennsylvania, in *SPE Shale Gas Production Conference* 2008, Society of Petroleum Engineers: Fort Worth, Texas.
39. Kuijvenhoven, C. et al., Water management approach for shale operations in North America, in *SPE Unconventional Resources Conference and Exhibition-Asia Pacific*. 2013, Society of Petroleum Engineers: Brisbane, Australia.
40. Jenkins, S., Frac water reuse. *Chemical Engineering*, 2012. **112**(2): 14–16.
41. Kaufman, P., G.S. Penny, and J. Paktinat, Critical evaluations of additives used in shale slickwater fracs, in *SPE Shale Gas Production Conference*. 2008, Society of Petroleum Engineers: Irving, Texas.
42. Ahmadun F. et al., Review of technologies for oil and gas produced water treatment. *J. Hazard Mater.*, 2009. **170**: 530–551.
43. Houcine, M., Solution for heavy metals decontamination in produced water/case study in southern Tunisia, in *International Conference on Health, Safety and Environment in Oil and Gas Exploration and Production*. 2002: Kuala Lumpur, Malaysia.
44. Frankiewicz, T.C. and J. Gerlach, *Removal of hydrocarbons, mercury and arsenic from oil-field produced water*. 2000.
45. Bryant, J.E. and J. Haggstrom, An environmental solution to help reduce freshwater demands and minimize chemical use, in *SPE/EAGE European Unconventional Resources Conference and Exhibition*. 2012, Society of Petroleum Engineers: Vienna, Austria.
46. Blauch, M.E. et al., Marcellus shale post-frac flowback waters—Where is all the salt coming from and what are the implications?, in *SPE Eastern Regional Meeting*. 2009, Society of Petroleum Engineers: Charleston, West Virginia.
47. Agrawal, A. et al., Determining microbial activities in fracture fluids compromising water reuse or disposal, in *SPE International Symposium on Oilfield Chemistry*. 2011, Society of Petroleum Engineers: The Woodlands, Texas.
48. Gaspar, J. et al., Microbial dynamics and control in shale gas production. *Environ. Sci. Technol. Lett.*, 2014. **1**(12): 465–473.
49. Maguire-Boyle, S.J. and A.R. Barron, Organic compounds in produced waters from shale gas wells. *Environ. Sci: Processes Impacts*, 2014. **16**(10): 2237–2248.
50. Warner, N.R. et al., Impacts of shale gas wastewater disposal on water quality in western Pennsylvania. *Environ. Sci. Technol.*, 2013. **47**(20): 11849–11857.
51. Silva, J.M. *Produced water pretreatment for water recovery and salt production (RPSEA Final Report)*. January 2012: New York.
52. Ely, J.W. et al., Game changing technology for treating and recycling frac water, in *SPE Annual Technical Conference and Exhibition*. 2011, Society of Petroleum Engineers: Denver, Colorado.
53. Drewes, J.E., *An integrated framework for treatment and management of produced water: Technical assessment of produced water treatment technologies*. 2009, The Colorado School of Mines: Golden, CO. p. 158.
54. Ghaffour, N., T.M. Missimer, and G.L. Amy, Technical review and evaluation of the economics of water desalination: Current and future challenges for better water supply sustainability. *Desalination*, 2013. **309**: 197–207.

55. Heinz, W.F., Is a paradigm shift in produced water treatment technology occurring at sagd facilities?, in *Canadian International Petroleum Conference*. 2007: Calgary, CA.
56. GE Water & Process Technologies. *Zero liquid discharge: Eliminate liquid discharge, recover valuable process water.* 2009, available from: http://www.gewater.com /products/zero-liquid-discharge-zld-crystalizers.html.
57. Greenlee, L.F. et al., Reverse osmosis desalination: Water sources, technology, and today's challenges. *Water Research*, 2009. **43**: 2317–2348.
58. Cheryan, M. and N. Rajagopalan, Membrane processing of oily streams. Wastewater treatment and waste reduction. *J. Membrane Sci.*, 1998. **151**: 13–28.
59. Miller, D.J. et al., Fouling-resistant membranes for the treatment of flowback water from hydraulic shale fracturing: A pilot study. *J. Membrane Sci.*, 2013. **437**: 265–275.
60. Murray-Gulde, C. et al., Performance of a hybrid reverse osmosis-constructed wetland treatment system for brackish oil field produced water. *Water Research*, 2003. **37**(3): 705–713.
61. Xu, P., J.E. Drewes, and D. Heil, Beneficial use of co-produced water through membrane treatment: Technical-economic assessment. *Desalination*, 2008. **225**: 139–155.
62. Horner, P. and B. Halldorson, Shale gas water treatment value chain—A review of technologies including case studies, in *SPE Annual Technical Conference and Exhibition*. 2011, Society of Petroleum Engineers: Denver, Colorado.
63. Hutchings, N.R., E.W. Appleton, and R.A. McGinnis, Making high quality frac water out of oilfield waste, in *SPE Annual Technical Conference and Exhibition*. 2010, Society of Petroleum Engineers: Florence, Italy.
64. Cath, T.Y., A.E. Childress, and M. Elimelech, Forward osmosis: Principles, applications, and recent developments. *J. Membrane Sci.*, 2006. **281**: 70–87.
65. Adham, S. et al., Application of membrane distillation for desalting brines from thermal desalination plants. *Desalination*, 2013. **314**: 101–108.
66. Minier-Matar, J. et al., Treatment of produced water from unconventional resources by membrane distillation, in *International Petroleum Technology Conference*. 2014: Doha, Qatar.
67. Lawson, K.W. and D.R. Lloyd, Review: Membrane distillation. *J. Membrane Sci.*, 1997. **124**: 1–25.
68. Fink, J., *Hydraulic Fracturing Chemicals and Fluids Technology*. 2013, Elsevier: Massachusetts, pp. 1–235.
69. Paktinat, J., B. O'Neil, and M. Tulissi, Case studies: Impact of high salt tolerant friction reducers on fresh water conservation in canadian shale fracturing treatments, in *Canadian Unconventional Resources Conference*. 2011, Society of Petroleum Engineers: Calgary, Alberta, Canada.
70. Maxey, J. and R. van Zanten, Novel method to characterize formation damage caused by polymers, in *SPE International Symposium and Exhibition on Formation Damage Control*. 2012, Society of Petroleum Engineers: Lafayette, Louisiana.
71. Vaidya, R.N. and H.S. Fogler, Formation damage due to colloidally induced fines migration. *Colloids and Surfaces*, 1990. **50**: 215–229.
72. Bagci, S., M.V. Kok, and U. Turksoy, Determination of formation damage in limestone reservoirs and its effect on production. *J. Petrol. Sci. Eng.*, 2000. **28**: 1–12.

9 Membrane Technology for Water Purification and Desalination

Saqib Shirazi and Che-Jen Lin

CONTENTS

9.1 INTRODUCTION

Growing water scarcity is one of the most pressing global challenges. Demand for freshwater will continue to increase significantly (~40% over the next three decades) owing to the increasing population and growing economies around the globe (United Nations 2014). To ensure safe and sustainable water quality and availability for the future, it is critical to develop innovative technological solutions and sound management strategies for the limited freshwater resources. Sustainable water management requires implementation of measures on two fronts. One is to improve the efficiency of the human water consumption cycle through the conservation, harvesting, reuse, and recycling of freshwater and wastewater. The other is to increase freshwater supply beyond what is available from the hydrological cycle, such as desalination of brackish groundwater or seawater. These diverse aspects in providing sustainable water solutions have been described in other chapters of this book.

Pressure-driven membrane processes are considered a versatile technology that uniquely addresses multiple needs in water sustainability. For example, advancement of new-generation, low-pressure-requirement membranes has allowed renewed applications in wastewater treatment that substantially improve the effluent quality for secondary water reuse such as washing and irrigation. Reverse osmosis (RO) is the most rapidly growing and the most energy-efficient technology for seawater desalination (Elimelech and Phillip 2011). Global RO desalination capacity has grown exponentially over the past three decades, producing more than 3.5 billion gallons of high-quality freshwater a day (Lattemann et al. 2010). Recent development in complete recycling of municipal wastewater as a freshwater resource for potable water production, the so-called direct potable (or toilet-to-tap) water reuse, also relies on one or more treatment steps using pressure-driven membrane technology.

Membrane filtration processes reduce the number of unit operations and simplify the treatment train required for water and wastewater treatment. Additional advantages, such as selective separation, continuous and automatic operation, easy scale-up, and low space requirement, make membrane filtration an attractive alternative to address sustainable water needs. This chapter describes the theoretical and operational principles of the pressure-driven membrane filtration process, discusses its applications in providing sustainable water solutions, outlines the major limitations in implementing the technology, and assesses the research needs and technological improvements required for future applications.

9.2 SPECTRUM OF PRESSURE-DRIVEN FILTRATION

Pressure-driven filtration is a process that applies hydraulic pressure as a driving force to accomplish selective separation through flow across the filtration media. The process includes conventional (media) filtration and various types of membrane separation technologies classified as microfiltration, ultrafiltration, nanofiltration, and RO filtration. Figure 9.1 shows a spectrum of pressure-driven filtration processes and their areas of application. Conventional filtration primarily uses sand and anthracite as the media. Membrane filtration uses membranes of various pore sizes. Most membranes are made of organic polymers, such as aromatic polyamide, polysulfonates,

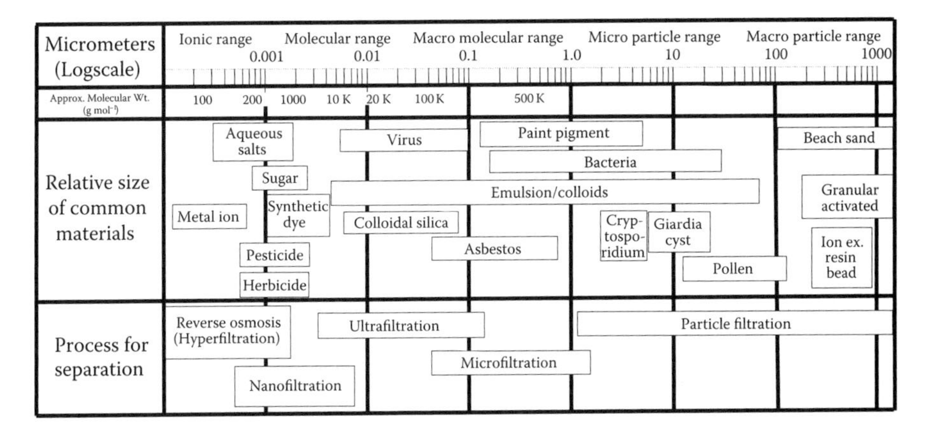

FIGURE 9.1 Spectrum of pressure-driven filtration processes.

polyvinyl alcohol, piperazine amide, polyimide, and polyacetylene (Lee et al. 2011). Inorganic membranes, although less common in the water and wastewater industry, have also been employed for the separation of ions and organic molecules from water (Shirazi et al. 2010).

The implementation of filtration processes can be classified in two configurations: dead-end filtration and cross-flow filtration, as shown in Figure 9.2. Dead-end filtration introduces the feed flow perpendicularly toward the filtration media. It is a simple filtration operation where nearly 100% of the feed water passes through the media as the "filtrate." The filtered matter is either retained within the filtration media or accumulated on the media surface. This results in a relatively rapid increase of filtration resistance that leads to reduced filtration flow and fouling of the filtration media. Under such a circumstance, a backwash of the filter is required to recover the filtration performance. In contrast, cross-flow filtration introduces the feed flow tangentially along the surface of the filter media. This provides a continuous shear flow to alleviate the accumulation of particles and foulants on the membrane surface while the water transports across the filter media. Because of the additional tangential cross-flow, such a configuration results in two streams leaving the filter. The filtrate that transports across the filter media is called "permeate" and the shear flow that carries away the particles and other impurities on the feed side of the

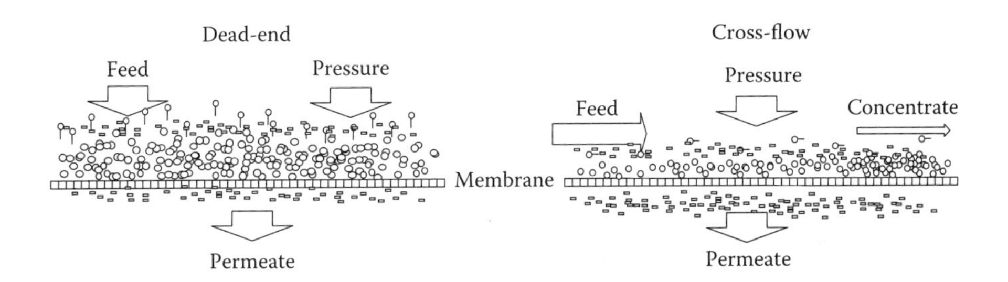

FIGURE 9.2 Dead-end filtration versus cross-flow filtration.

filtration media is called "concentrate" or "retentate." The ratio of the permeate flow to the feed flow is defined as the water recovery of the filtration. Most membrane installations operate in a cross-flow configuration, although recent advancements in membrane manufacturing greatly increase water permeability across membranes and allow the low-resistance ultrafiltration membrane (ultrafilter) and microfiltration membrane (microfilter) to operate in a fashion similar to dead-end filtration. Most media filtration facilities implement the dead-end filtration scheme because of its simplicity and relative ease for filter backwash.

9.2.1 Media Filtration

Media filtration is primarily applied in separating suspended solid particles from water by dead-end filtration. The most common filter media in water treatment are sand and anthracite. The effective grain size (the 10th percentile size of the particle size distribution) is in the range of 0.35–0.5 mm for sand filters and 0.7–0.8 mm for anthracite filters. In comparison to single sand filter media, dual filter media with anthracite over sand permit more penetration of the suspended matter into the filter bed, resulting in more efficient filtration and longer runs between cleaning. The pressure required ranges from 3 to 30 kPa. As shown in Figure 9.1, this filtration regime is effective for removing particles with a size of 1–1000 μm. It is widely used for filtration in conventional drinking water treatment plants for turbidity removal, in water softening plants using lime/soda precipitation, and in polishing of secondary effluents (i.e., clarifier effluents after chemical and biological treatments) in wastewater treatment plant.

9.2.2 Microfiltration

Microfiltration is applied in removing particles with a size of 0.05–10 μm. Microfiltration filter is frequently regarded as the membrane filter that has the largest pore size (0.05–2 μm). It is extensively used for separating small suspended particles, bacteria, and large colloids (Figure 9.1). Dissolved solids and macromolecules can easily pass through a microfilter. Microfiltration typically operates at a water recovery ranging from 85% to 95% and an operating pressure of 10–100 kPa.

9.2.3 Ultrafiltration

Ultrafiltration is a separation process effective for removing colloidal materials and organic and inorganic polymeric molecules (size range, 0.003–0.1 μm) using membranes with a pore size of 2–50 nm. It has a higher removal capacity than microfiltration but operates at higher pressure at 50–250 kPa. Ultrafiltration is also useful for separating high–molecular weight organics from low–molecular weight components. The recovery is similar to that of microfiltration.

9.2.4 Nanofiltration

Nanofiltration is capable of removing multivalent ions and small molecules whose size is in the nanometer range (0.001–0.01 μm), such as sulfate ions, sugar, dye, multivalent salts, and humic substances. It has a much higher operation pressure requirement, typically ranging from 200 to 1000 kPa, and therefore requires more energy than ultrafiltration and microfiltration. The nanofiltration membrane (nanofilter) has a very small pore size (<2 nm). Some tight-end nanofilters are not porous and rely on molecular diffusion, rather than sieving, to achieve separation. The recovery of nanofiltration varies depending on its application, typically 60%–80% and higher than the recovery of the RO process.

9.2.5 RO Filtration

RO filtration uses nonporous semipermeable membranes to achieve material separation based on the difference in solubility and diffusivity. It is widely applied in the production of potable water from brackish water and seawater (the so-called desalination because of its high removal rate for dissolved salts) as well as in industrial molecular separation. RO membranes reject >90% of small molecules and impurities from water (including small organics and dissolved monovalent salts, metal ions, etc.), which results in a permeate (filtrate) of very high water quality. Because of the small pore size of RO membranes, the process requires a very high operating pressure (1000–10,000 kPa) and operates at a relatively low water recovery (40%–80%) except for application in polishing high-quality freshwater.

9.3 FUNDAMENTALS OF PRESSURE-DRIVEN MEMBRANE FILTRATION

9.3.1 Permeation Process

Permeation is the selective transport of materials from one side of the membrane to the other side. The permeation selectivity during membrane filtration depends on the type of membrane, the properties of impurities in water, and the driving forces (hydraulic pressure, concentration gradient, temperature, or electric potential) of the permeation. For pressure-driven membrane processes using a semipermeable membrane (i.e., tight-end nanofilters and RO membranes) for dissolved solid separation, the permeation flux, defined as the amount of water flow permeated through a unit area of membrane, can be estimated as

$$J = K(\Delta P - \Delta\pi), \tag{9.1}$$

where J is the permeation flux (m^3 s^{-1} m^{-2}), K is the specific permeability of a given membrane (m^3 s^{-1} m^{-2} kPa^{-1}), ΔP is the pressure difference across the membrane (kPa), and $\Delta\pi$ is the osmotic pressure difference between the feed and permeate (kPa). The osmotic pressure π can be calculated as

$$\pi = \phi \nu CRT, \tag{9.2}$$

where ϕ is the dimensionless osmotic coefficient, ν is the number of ions formed from one molecule of electrolyte, C is the molar concentration of electrolyte (mol m^{-3}),

R is the universal gas constant (0.0831 Pa m^3 mol^{-1} K^{-1}), and T is the absolute temperature (K).

9.3.2 Water Recovery and Pollutant Rejection in Membrane Filtration

Figure 9.3 shows a schematic of membrane filtration that includes the flow of feed, permeate, and concentrate streams, and the concentration of the feed, permeate, and concentrate streams. Rejection is defined as the pollutant removal efficiency of the membranes:

$$R_i = \frac{C_f - C_p}{C_f}, \tag{9.3}$$

where R_i is the dimensionless pollutant rejection and C_f and C_p are the pollutant concentrations (mg L^{-1}) in the feed and permeate stream, respectively. Feed water recovery is defines as

$$R_f = \frac{Q_p}{Q_f}, \tag{9.4}$$

where R_f is the dimensionless feed water recovery and Q_f and Q_p are the flow rates (m^3 s^{-1}) of the feed and permeate stream, respectively.

Microfilters, ultrafilters, and loose-end nanofilters remove impurities (pollutants) from water based on sieving of particles and molecules. The level of pollutant rejection is nearly 100% for particles and molecules that have a size larger than the membrane pore size. For semipermeable membranes that remove the dissolved solutes of smaller ions and molecules, the rejection of solutes is based on the differential diffusion of water and the solutes across the membrane and highly depends on the operational parameters (Shirazi et al. 2010). As shown in Figure 9.4, a higher rejection of solute can be achieved at a higher operating pressure and a lower recovery.

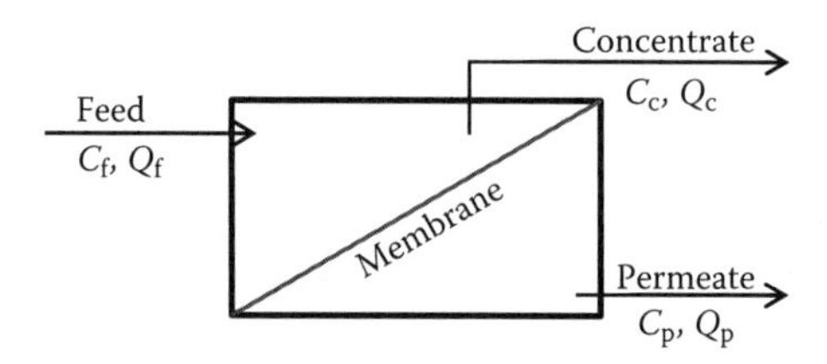

FIGURE 9.3 Schematic of membrane filtration. C denotes the concentration of solutes or pollutant, and Q denotes the volumetric flow rate of the streams.

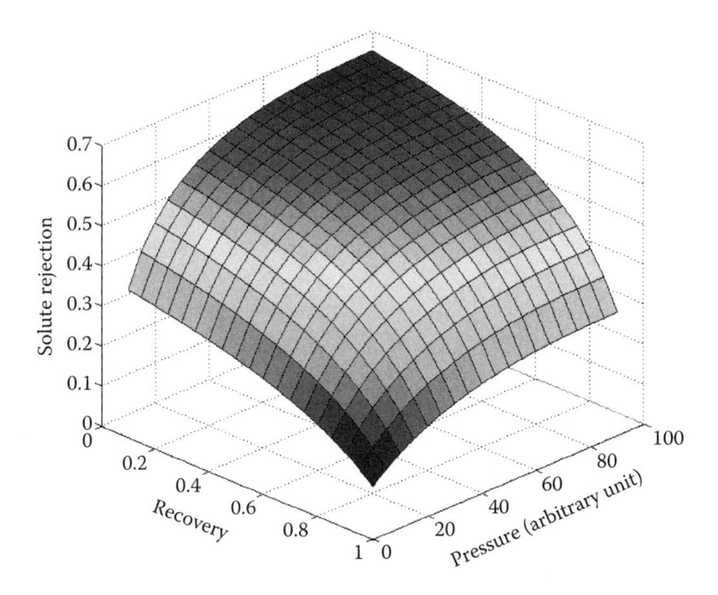

FIGURE 9.4 Relationship between the rejection of pollutants/solutes during membrane filtration and the primary operating parameters (transmembrane pressure and feed water recovery). The semiquantitative relationship is derived based on the film theory of mass transport across a semipermeable membrane described in Shirazi et al. (2010).

9.3.3 Membrane Modules

As seen in Equation 9.1, the production of permeate during membrane filtration is based on the volumetric flow rate per unit surface area of membrane. Therefore, it is critical to package a sufficiently large surface area to provide the desired treated water flow. The housing that packages the membrane is called "membrane module" and typically uses one of the three configurations as shown in Figure 9.5: (a) a spiral wound module, (b) a hollow fiber (tubular) module, and (c) a flat sheet module. Most tight-end nanofilters and RO membranes use the spiral wound module because it allows the insertion of a spacer between the membranes that increases the turbulence on membrane surface and therefore reduces the accumulation of solutes over the membrane. The hollow fiber module is widely utilized for microfilters, ultrafilters, and loose-end nanofilters. Multiple spiral wound and hollow fiber modules can be connected in series or parallel to achieve either high solute rejection or high product water flow (Figure 9.6). The tubular module is a variation of the hollow fiber module but with larger-diameter membranes (AWWA 2005). They have a low packing density rendering higher capital cost, which have been used in food industry (such as cheese production and juice concentration). Recently, flat sheet modules of microfilters and ultrafilters are extensively applied for membrane bioreactors (MBRs) (Sarioglu et al. 2012). In this case, fine-bubble air diffusers are used for aeration to provide dissolved oxygen for the aerobic microorganisms as well as for the scouring of membrane surface to prevent accumulations of suspended particles

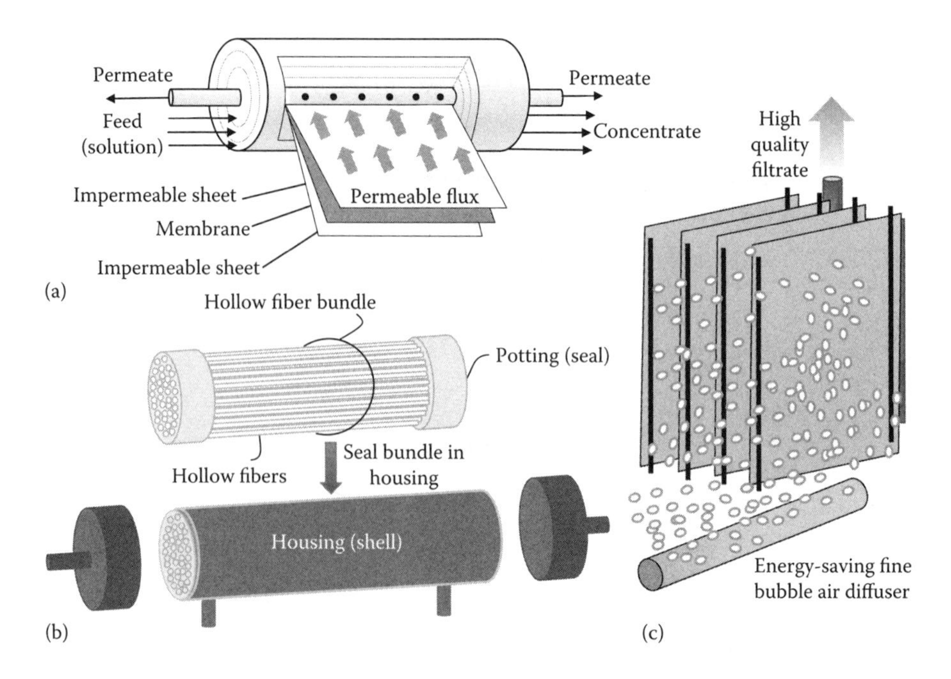

FIGURE 9.5 Common membrane modules: (a) spiral wound, (b) hollow fiber, and (c) flat sheet.

FIGURE 9.6 An RO membrane filtration system using spiral wound modules.

TABLE 9.1
Difference between Constant Flux and Constant Pressure Operation Modes

Parameters	Constant Pressure	Constant Flux
Permeate flux	Varies	Unchanged
Operating pressure	Unchanged	Varies
Energy requirements	Constant	Varies
Pump size	Oversized pumps may be required to meet minimum design flows	No pump oversizing is required

over the membranes. Flat sheet RO membranes are also widely used in vibratory shear–enhanced processes.

9.3.4 Operation of Membrane Systems

Membrane systems can be operated in either constant pressure or constant flux modes. The constant pressure mode keeps the operating pressure constant and allows the permeate flux to be decreased over time as the membrane becomes clogged by contaminants. The advantage of the constant pressure mode is that the energy requirement remains constant, though at the cost of gradual flux decline. Constant flux mode keeps permeate flux constant and gradually increases the operating pressure over time as the membrane becomes dirty. This allows a constant water production at the cost of variable energy requirement. Table 9.1 compares and contrasts the two modes.

9.4 APPLICATIONS OF MEMBRANE FILTRATION IN WATER AND WASTEWATER INDUSTRIES

Pressure-driven membrane processes are extensively used for the removal of suspended solids, dissolved solids, organic matter, silica, and microorganisms from water and wastewater. Low-pressure membranes (microfilters and ultrafilters) are primarily used for the removal of suspended solids and flocs formed from coagulation or activated sludge process, while nanofilters and RO membranes are mainly used for removing dissolved organic and inorganic materials from water and wastewater. Depending on the nature of the contaminants and the target water quality, low-pressure or high-pressure membranes can be applied individually or in combination. Figure 9.7 provides a selection chart for membrane processes. In this section, the applications of low-pressure and high-pressure membrane systems in water and wastewater industries as well as the required pretreatment and posttreatment are discussed.

9.4.1 Applications of Low-Pressure Membranes (Microfiltration and Ultrafiltration)

The first microfiltration membrane was developed in Germany in 1927 (Roesink 1989). Initially, the use of microfiltration was limited in life sciences. The concept

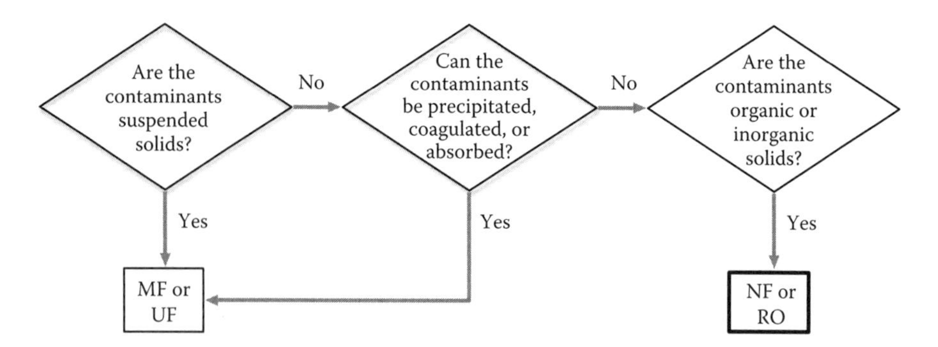

FIGURE 9.7 Selection chart for pressure-driven membrane processes in water and wastewater treatment.

of employing microfiltration and ultrafiltration membranes in the water/wastewater industry began relatively recently (in the 1980s). In the United States, the first large-scale low-pressure membrane filtration plant for water treatment was commissioned in 1994 with a treatment capacity of 3.6 million gallons per day of surface water. Because of the high quality of finished water, low-pressure requirement, and improvement in membrane permeability, microfiltration and ultrafiltration quickly gained popularity and are now widely used in water, wastewater, and water reuse industries. Today, low-pressure membranes are readily commercially available from manufacturers such as Pall Corporation, Zenon Environmental Systems, and Koch Membrane Systems.

Low-pressure membranes are capable of producing extremely high quality filtrate regardless of influent turbidity, membrane configuration, or manufacturers. The finished water of low-pressure membranes typically has turbidity less than 0.1 nepherometric turbidity unit (NTU). The capability of removing pathogenic microbes such as *Giardia*, *Cryptosporidium*, protozoa, bacteria, and viruses also makes low-pressure membranes operationally attractive. Because of the versatility of low-pressure membranes, there are commercial microfiltration products designed for personal use in occasions when potable water is not available (e.g., LifeStraw, http://www.lifestraw.com). Such a personal membrane product provides on-site filtration of natural water without the need of power.

Low-pressure membranes are implemented in two different types of configurations: pressurized membrane systems using hollow fiber configuration (Figure 9.5b) and submerged membrane systems (Figure 9.5c). Most low-pressure membrane modules are mounted vertically. This configuration allows individual modules to be inspected, repaired, and replaced using manual tools. The filtrate flow direction can be either outside-in or inside-out. In an outside-in configuration, feed water surrounds the membrane and filtrate is collected from the inside of the membrane fiber. In an inside-out system, feed water is placed inside the membrane fiber and filtrate is collected on the outside of the membranes. In submerged systems, membrane sheets or fibers are immersed in an open tank and exposed directly to the surrounding feeding water. In this configuration, membranes can be arranged into larger assemblies called cassettes. Submerged systems use vacuum pumps for water suction from the surrounding tank through the membrane and can only be operated using an outside-in configuration.

In drinking water treatment, low-pressure membranes have several advantages over conventional water treatment systems, including smaller footprint, lower sludge production, less chemical requirement, highly selective automated operation, and ease for scaling up (Jacangelo et al. 1998; Saffaj et al. 2004). The single-stage membrane operation provides simultaneous removal of suspended solids, turbidity, taste, odor (caused by particulate matter in water), inorganic oxide particles, and microorganisms.

In the wastewater industry, low-pressure membranes are used in an MBR (Figure 9.8), which combines an aerobic biological treatment process with an integrated membrane system. The advantages of an MBR over a conventional wastewater system include smaller footprint, better product quality, less sludge production, and ease of operation. The first trial of using membranes for liquid–solid separation in wastewater treatment was attempted by Yamamoto et al. (1989). In the early years of MBR development, membranes used to be submerged in a bioreactor, and the filtrate was drawn through the membrane by suction. Currently, commercial MBRs are available in two primary configurations: side-stream and submerged. In side-stream MBRs, the low-pressure membrane module is located outside the aeration tank of the biological treatment. In submerged systems, membrane modules are submerged inside the aeration tank with continuous air scouring. Submerged MBRs are more cost effective for larger-scale applications, and side-stream technologies are more favorable for smaller-scale applications (Hai and Yamamoto 2011).

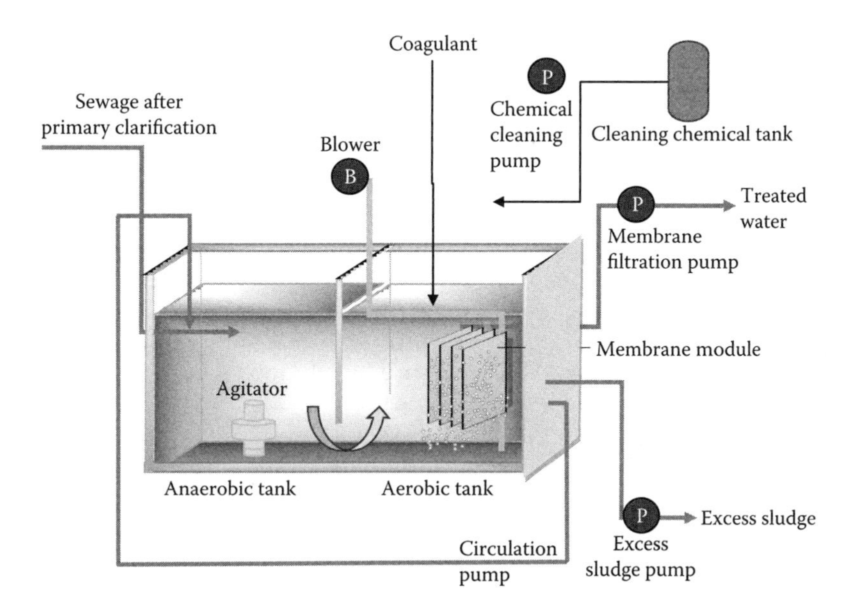

FIGURE 9.8 Schematic of a membrane bioreactor.

9.4.2 Application of High-Pressure Membranes (Nanofiltration and RO Filtration)

In the 1950s, Reid and Bretton first reported the desalination of water through a cellulose acetate membrane (Reid and Bretton 1959). However, the productivity of Reid and Breton's membrane was too low to be applied in a large-scale application. In the 1960s, productivity of the cellulose acetate membranes was greatly improved by reducing the thickness of the membrane (Loeb and Sourirajan 1962). Loeb and Sourirajan's membrane had an asymmetric structure that contained a dense skin at the surface and a porous support layer at the back. The dense skin was responsible for the membrane's selectivity and productivity, while the porous support layer provided mechanical strength to the membrane (Strathmann et al. 2006). In 1970, Cadotte introduced thin-film composite (TFC) membranes in the industry. TFC membranes are made of multiple layers. The top layer is composed of polyamide, followed by a porous layer in the middle, and a supporting layer at the bottom. Multilayer configuration provides TFC membranes higher rejection and better production capacity. Currently, TFC membranes are the most widely used RO and nanofiltration membranes in the water industry. Nanofiltration membranes were first developed in the late 1970s (AWWA 2007; Li et al. 2010). Although nanofilters require much less energy than RO membranes, these membranes are not as effective as RO membranes in rejecting monovalent salts. Therefore, nanofilters are primarily used for the removal of divalent ions (such as calcium and magnesium) from water. High-pressure membrane systems are typically packaged in spiral wound modules that run horizontally (Figures 9.5a and 9.6).

High-pressure membrane systems have a wide variety of application in the drinking water industry (Table 9.2). RO is predominantly used in desalination processes that remove dissolved solids from water. It is the most important technology in regions that lack freshwater resources. Depending on the source of water, desalination can be classified into two major categories: seawater desalination and brackish water desalination. In seawater desalination, RO filtration operates at a recovery of ~50% because of the high TDS (total dissolved solids) concentration (typically >35,000 mg/L) in the feed water. Brackish surface and groundwater contain TDS concentrations between 1000 and 10,000 mg/L (TWDB 2004) and therefore permit a higher water recovery (65%–85%). Because of persistent drought and climate

TABLE 9.2
Applications of High-Pressure Membranes in the Water Industry

Applications	Commonly Used Membrane
Desalination	RO
DBP precursor removal	RO and NF
Hardness removal	NF
Color removal	NF
Inorganic contaminants removal	RO

change, brackish groundwater desalination is gaining importance in the western part of the United States. RO is also considered as one of the best available technologies for removing most inorganic compounds (e.g., nitrates and fluoride) and disinfection by-product (DBP) precursors from water.

In potable water treatment, nanofiltration is mostly used for the removal of DBP precursors that react with chlorine and chloramine to produce DBPs. Some of the common DBPs are trihalomethanes, haloacetic acids, chlorites, and nitrosamines. Natural organic matter and total organic carbon are the primary DBP precursors that can be effectively removed by nanofilters (AWWA 2007). Other common applications of nanofiltration include removal of hardness (presence of divalent ions such as Ca^{2+} and Mg^{2+}) and color. The conventional method for reducing hardness is through lime softening where alkaline chemicals are used to remove hardness from water. Nanofiltration has gained large popularity in removing hardness from water because it does not produce sludge and requires much lower operating pressure than RO.

Typically, nanofiltration can remove >95% of hardness and color in water.

In the wastewater industry, high-pressure membranes have been applied for the removal of dissolved solids, pharmaceutical and personal care products (PPCPs), and endocrine-disrupting compounds (EDCs) from wastewater. PPCPs refer to the products used for personal health (e.g., prescription and over-the-counter drugs) or cosmetic reasons or for enhancing the growth or health of livestock. Endocrine disruptors are substances that interfere with the synthesis, secretion, transport, binding, action, or elimination of natural hormones in the body. PPCPs and EDCs are emerging contaminants that are not readily removed by conventional wastewater treatment processes and have been identified in natural water and in wastewater. RO filtration has been shown to be effective for the removal for a wide variety of PPCPs and EDCs (Snyder et al. 2003). The application is typically implemented before chlorination to prevent the presence of oxidant (e.g., chlorine) from damaging polyamide membranes. Pretreatment that removes turbidity and particulate matter must be accomplished to alleviate membrane fouling.

9.4.3 Application of Membrane Technology in Water Reuse

Pressure-driven membrane systems play an important role in water reuse (Asano et al. 2007). Reuse of wastewater as a freshwater source has received much attention in areas that lack adequate supplies of water. There are two major types of water reuse: indirect reuse and direct reuse (Figure 9.9). In indirect reuse, wastewater effluent is first discharged into an environmental buffer, such as a lake, river, or aquifer, before being retrieved to be used again. Direct reuse refers to the introduction of reclaimed water directly from a water reclamation plant to a distribution system via pipelines, storage tanks, or other necessary infrastructures. Because no environmental barrier is present in direct reuse process, federal, state, and local regulations for direct reuse are very stringent. Both direct and indirect reused water can be applied for potable and nonpotable purposes.

Because of the scarcity of freshwater caused by urbanization, environmental pollution, and changed climate, direct potable reuse (DPR), or the so-called toilet-to-tap water reuse, has been gaining popularity. In DPR, the secondary effluent (e.g., the effluent produced by an activated sludge process or an MBR process) from a

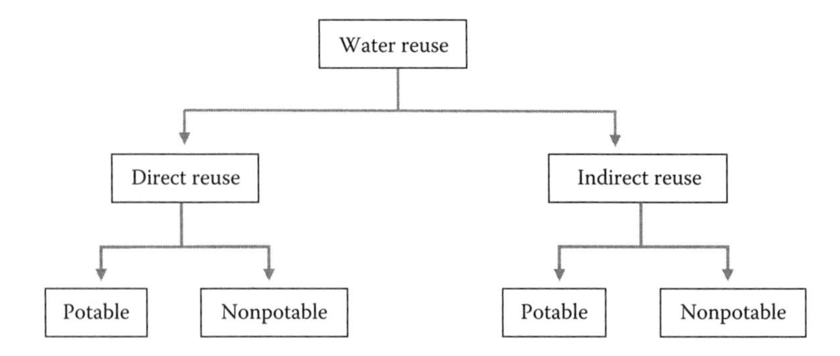

FIGURE 9.9 Classification of water reuse of municipal wastewater plant effluents.

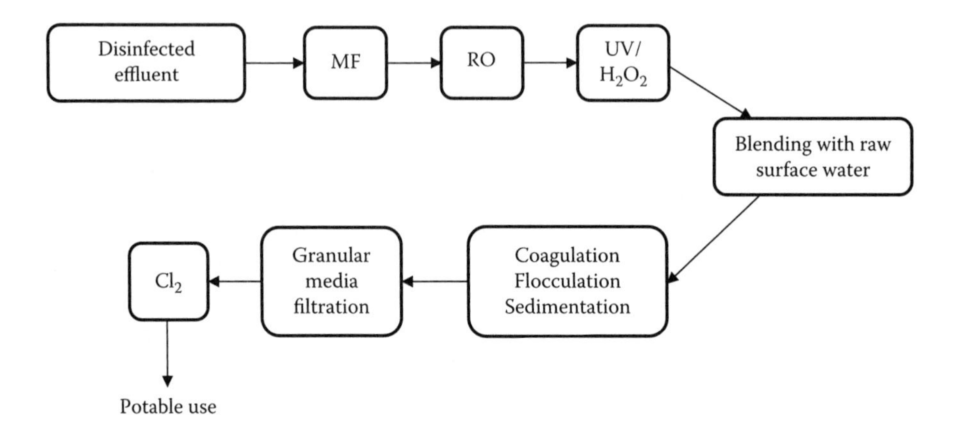

FIGURE 9.10 Processes implemented by the CRMWD's water reuse plant to produce potable water from treated wastewater (toilet-to-tap water reuse).

wastewater treatment plant is further treated with a multistage treatment train incorporating membrane technology to produce potable water. The most widely applied technologies in DPR are microfiltration, RO, activated carbon, and chlorination or UV disinfection. Not every system uses all the mentioned technologies simultaneously. Figure 9.10 shows a schematic of the DPR processes implemented by the Colorado River Municipal Water District's (CRMWD) water reuse plant as an example. The treated water quality meets the National Primary Drinking Water Standards of the United States Environmental Protection Agency (USEPA). Table 9.3 shows the key treatment technologies implemented by several existing DPR water plants.

9.4.4 Pretreatment and Posttreatment Requirements

Prefiltration, pH adjustment, and coagulation are commonly applied as the pretreatment options for low-pressure membrane systems. Prefiltration in the upstream of membrane treatment is accomplished using screens (e.g., rotating disc, drum filters,

TABLE 9.3
Key Treatment Technologies Implemented by Selected DPR Water Plants (Toilet-to-Tap Water Reuse)

Location	Start-Up Date	Capacity (m^3/Day)	Technologies Used in DPR
Windhoek, Namibia	2002	21,000	• Oxidation and pre-ozonation • Powdered activated carbon dosing • Coagulation and flocculation • Dissolved air flotation (DAF) • Dual media filtration • Biological activated carbon filtration (BAC) • Granular activated carbon filtration (GAC) • Ultrafiltration • Ozone disinfection and stabilization
Big Spring, Texas (CRMWD Water Reuse Plant)	2013	7500	• Rapid mixing • Flocculation • Sedimentation • Media filtration • MF • RO • UV/H_2O_2 • Disinfection with chlorine
Village of Cloudcroft, New Mexico	2011	379	• MBR • RO • GAC • UF • UV disinfection
Wichita Falls, Texas	2014	37,854	• MF • RO • UV disinfection

Note: Pressure-driven membrane technology plays an important role in these treatment facilities.

and bag filters) with nominal openings ranging between 50 and 500 μm. pH in the low-pressure membranes is adjusted to keep the pH of water in operating range. It is more important if the membrane materials are cellulosic because cellulosic materials cannot tolerate pH below 5 and over 8. In some cases, coagulation is performed before the membrane stage to improve rejection performance. Coagulation also reduces membrane fouling and increases filtrate production. Other pretreatment methods such as oxidation and biofiltration can also be employed to reduce the concentration of organic or biological compounds in raw water. Low-pressure membrane systems require little posttreatment, except water disinfection, as required by drinking water regulations in the United States.

For high-pressure membrane systems, common pretreatments include prefiltration and chemical addition for fouling control. For surface water containing suspended and organic materials, a conventional treatment (e.g., coagulation, flocculation, sedimentation, and sand filtration) or low-pressure membrane filtration is frequently used. In almost all cases, cartridge filters are used as a prefiltration stage in RO or nanofiltration systems. Cartridge filters serve as a barrier to prevent particulate matters entering the membrane module. The most common cartridge filters used in the industry have a 5-μm opening. During the operation, differential pressure across cartridge filters is monitored regularly to determine when cartridge filters should be replaced. Typically, cartridge filters are changed when the differential pressure reaches 103 kPa or higher.

Sulfuric or hydrochloric acid is added in raw water to control precipitation of carbonate and bicarbonate salts on membrane surface. Addition of acid converts bicarbonates into carbon dioxide, which passes through the membrane and is removed in the posttreatment process. In addition to acid, proprietary anti-scalants are also used to protect membranes from precipitation of sparingly soluble salts and silica. Section 9.5.1.1 discusses in details on chemical addition for the pre-treatment of high-pressure membrane systems.

Permeates of RO membranes and nanofilters contain a very low concentration of minerals and are considered abrasive. If not further treated, they can cause dissolution of minerals and corrosion of the components in a water distribution system. To maintain appropriate alkalinity and increase the stability of treated water, degasification and chemical addition are often employed as posttreatment options. Degasification reduces the concentration of carbonic acid formed during acid addition in the pretreatment step. It also removes hydrogen sulfide, radon, and other volatile chemicals present in raw water. Degasification can be achieved by packed tower aeration, tray aeration, or hollow fiber membrane aeration. Chemical addition using caustic soda (NaOH), sodium bicarbonate, soda ash (Na_2CO_3), and lime ($Ca(OH)_2$) restores the alkalinity of the finished water. In brackish water desalination, raw water can also be blended with permeate to increase the stability of finished water, followed by disinfection using chlorination, UV, or ozonation.

9.5 LIMITATIONS AND CHALLENGES

Several limitations prohibit the implementation of membrane processes in sustainable water applications. These limitations include membrane fouling, concentrate disposal, and treatment cost associated with energy requirements of membrane filtration systems. This section discusses these limiting factors and the ongoing developments addressing these limitations.

9.5.1 Membrane Fouling and Mitigation

Membrane fouling is the process of accumulation of particulate, suspended, or dissolved materials on the membrane surface that causes a decrease in membrane productivity (AWWA 2007). The adverse effects of membrane fouling include flux decline, increased feed pressure owing to the increased differential pressure across the membrane, membrane degradation, reduced product water quality owing to the accumulation of foulants (materials that cause membrane fouling), and increased energy consumption. The characteristics and mitigation of membrane fouling depend

TABLE 9.4
Different Types of Membrane Fouling

Type of Fouling	Major Foulants	Membranes Affected
Organic and colloidal	Polysaccharides, proteins, aminosugars, humic and fulvic acids, and silica	Both low-pressure (MF and UF) and high-pressure (NF and RO) membranes
Biological	Bacteria, algae, protozoa, and fungi	Both low-pressure and high-pressure membranes
Inorganic fouling	Calcium carbonate, calcium fluoride, calcium phosphate, calcium sulfate, barium sulfate, and strontium sulfate	Primarily high-pressure membranes

TABLE 9.5
Summary of Methods for Membrane Fouling Mitigation

Type of Fouling	Methods	Applications	Limitations
Organic and colloidal	Pretreatment using conventional treatment, microfiltration, and ultrafiltration	Primarily on high-pressure membranes	Costly
Inorganic	Addition of acids and anti-scalants (hydrochloric and sulfuric acid, polysulfates, polyphosphates)	Primarily on high-pressure membranes	Reduces pH Excessive use of anti-scalants may cause fouling
	Ion exchange, lime softening, and green sand filtration		Costly, sludge production from lime softening
Biological	Addition of disinfectants (chlorine, chloramines, sodium hypochlorite, hydrogen peroxides)	Low-pressure and high-pressure membranes	Not applicable to the membranes that are sensitive to oxidants

on foulant characteristics, feed water chemistry (pH, ionic strength, foulant concentration), membrane properties (surface charge, hydrophobicity, roughness) and hydrodynamic conditions in membrane modules (Li and Elimelech 2004). Membrane fouling can be categorized into three different categories: organic and colloidal, inorganic, and biological fouling. More than one type of fouling can occur simultaneously on the membrane surface (Amjad 1992). Table 9.4 shows a summary of membrane fouling types and their characteristics. Table 9.5 outlines the methods of fouling mitigation.

9.5.1.1 Inorganic Fouling

Inorganic fouling (mineral scale) is caused by the precipitation of sparingly soluble salts on the membrane surface. Major salts that precipitate on the membrane surface include calcium carbonate, calcium fluoride, calcium phosphate, calcium sulfate, barium sulfate, and strontium sulfate (Xie et al. 2004). Because low-pressure membranes (microfilters and ultrafilters) rarely reject dissolved inorganic substances from source water, these membranes are not affected by inorganic fouling. The presence of inorganic salts in the feed water is primarily responsible for inorganic scaling. For nonporous, semipermeable membrane (tight-end nanofilters and RO membrane), the inorganic salts tend to accumulate within the mass transfer boundary layers near the membrane because of the selective permeation of water across the membrane, known as "concentration polarization" under a cross-flow filtration scheme. This leads to a much increased salt concentration on the membrane surface exceeding the aqueous solubility of the sparingly soluble salts and causes inorganic fouling (Shirazi et al. 2010). Depending on the operating conditions, the index of concentration polarization, defined as the ratio of the salt concentration on the membrane surface to the salt concentration in the feed water, can increase drastically at high water recovery.

Chemical addition is the most common approach in reducing inorganic fouling on high-pressure membranes. In this method, chemicals such as acids and anti-scalants are added in feed water to mitigate mineral scales. Addition of acids lowers the pH of raw water and reduces the precipitation of carbonate and bicarbonate salts at the cost of decreasing feed water pH. Anti-scalants such as polysulfates and sodium hexametaphosphate are frequently applied to inhibit scaling on high-pressure membrane systems. These anti-scalants work by inhibiting the rate of formation of crystalline precipitates under supersaturated conditions; chelation of metal ions, permitting them to stay in solution at higher concentrations; or dispersion, which retains colloids in suspension until discharged in the concentrate stream. However, excessive use of anti-scalants may cause chemical fouling on membrane surface (AWWA 2007). Other methods such as lime softening, ion exchange, and greensand filtration are also employed to mitigate inorganic fouling. Lime softening and ion exchange are primarily used for controlling inorganic fouling caused by calcium and magnesium, while greensand filtration is used for controlling inorganic fouling caused by iron. One limitation of these techniques is that they are more costly than chemical addition. Lime softening also generates sludge that needs to be disposed of properly.

9.5.1.2 Organic and Colloidal Fouling

This type of fouling is caused by the accumulation of organic and colloidal substances on the membrane surface caused by either particle sieving or concentration polarization. Major organic materials that cause membrane fouling include polysaccharides, proteins, aminosugars, humic and fulvic acids, and silica (Ang and Elimelech 2009). The physical and chemical characteristics that affect organic/colloidal fouling include charge on membrane surface, affinity of organic matter to the membrane, and molecular weight of organic matter. Negatively charged organic molecules (such as humic and fulvic acids) are repulsed by negatively charged membranes (e.g., polysulfone, cellulose acetate, and TFC membranes carrying negative charge). A greater charge density on the membrane surface is associated with greater membrane hydrophilicity. Hydrophobic interactions cause the accumulation of organic matter on membranes, which, in turn, leads to adsorptive fouling (Mallevialle et al. 1996).

Organic and colloidal fouling is typically mitigated by reducing the level of particulate matter in feed water. Two parameters, turbidity and silt density index (SDI), are applied to determine the fouling potential of the feed water by organic materials. Turbidity is the cloudiness of a fluid caused by colloidal suspended particles, expressed in terms of NTU. SDI is a parameter that measures the rate at which a 0.45-μm filter is plugged when subjected to a constant water pressure of 206.8 kPa. Feed water should be pretreated if SDI and turbidity levels exceed the recommended level for a given system. Pretreatment using coagulation–flocculation (conventional) or microfiltration/ultrafiltration has been applied to reduce turbidity and SDI levels in feed water for nanofiltration and RO systems. Each of the pretreatment alternatives bears additional capital and operational costs. Conventional pretreatment also produces sludge that needs to be properly disposed of.

9.5.1.3 Biological Fouling

Biological fouling is caused by the attachment of microbial cells on the membrane surface, which subsequently form biofilm (Flemming and Schaule 1988). The biofilm is composed of different types of microorganisms including bacteria, algae, protozoa, and fungi (Nguyen et al. 2012). During biological fouling, microbial cells excrete extracellular polymeric substances that help the microbial cells anchor to the membrane surface. The attachment of microorganisms to the membrane surface is affected by factors including membrane material, surface roughness of the membrane, hydrophobicity, and surface charge of the membrane (Park et al. 2005). Biological fouling is typically mitigated by adding chemical disinfectants (e.g., chlorine, chloramines, sodium hypochlorite, and hydrogen peroxide) to the feed water, which controls biological growth. One limitation of this technique is that it may also damage polyamide membranes that are sensitive to oxidants.

9.5.1.4 Membrane Cleaning after Fouling

Cleaning of low-pressure membranes is typically achieved through backwash and periodic chemical cleaning. Low-pressure membranes can be backwashed using filtrate, chlorinated, or unchlorinated water with or without air scour depending on the membrane system. Although the frequency of backwashing varies depending on source water quality, a backwash every 15 to 60 min for a period of 3 to 180 s is commonly practiced. Spent backwash water may be recycled to increase system recovery or discharged into a sanitary sewer or receiving stream. Periodic chemical cleaning in low-pressure membrane systems is conducted much less frequently than backwashing. During chemical cleaning, caustic solutions are used for removing organic fouling. For systems with oxidant-tolerant membrane materials, chlorine or other oxidants may be used to remove biological fouling. The chemical cleaning cycles can last from 30 min to several hours including soak and rinse time, depending on the extent and nature of the fouling.

In high-pressure membrane systems, only periodic cleaning is performed. It is advisable to perform an autopsy of a spent membrane to identify the type (organic, colloidal, inorganic, or biological) and extent of fouling before the chemical cleaning. For inorganic fouling, acids are used as the cleaning agent. The most common acidic cleaning agents include 2% citric acid and a small amount of commercial cleaning chemical (e.g., 0.1% Triton X-100). For organic and biological fouling, a solution with high pH (alkaline) is used as a cleaning agent. Trisodium phosphate, sodium triphosphate, and EDTA (~2% solution) are the most common alkaline cleaning agents. The cleaning solutions are introduced from the feed side of membrane modules and continue to circulate until completion of the cleaning cycle.

9.5.2 Energy and Sustainability Issues

9.5.2.1 Energy Footprint of Membrane Systems

Membrane filtration is an energy-intensive process where energy consumption is one of the largest contributors toward the total costs of water production (AWWA 2007). Higher energy consumption increases power cost for a membrane facility, which, in turn, raises unit water production cost. For low-pressure membrane systems, ultrafiltration requires approximately 0.5 kWh (kilowatt-hour) of energy to produce 1000 gallons of water, while MF requires approximately 0.1 kWh per 1000 gallons water produced (AWWA 2012). For these low-pressure systems, more frequent backwash, longer backwash duration, and lower membrane permeability (K in Equation 9.1) result in higher energy consumption.

Technological advancement has reduced the energy requirement for high-pressure membrane systems by 75% in the past 40 years (Elimelech and Phillip 2011). Presently, RO desalination of brackish water consumes approximately 5 kWh per 1000 gallons water produced, while RO desalination of seawater consumes approximately 13 kWh per 1000 gallons water produced. In addition to the operational parameters of the membrane systems, other factors such as process design, equipment selection, and regulation of feed water temperature also affect the energy footprint. Modern process flow designs allow an energy recovery device to harvest energy from concentrate stream and deliver it back to the feed stream. Since most high-pressure membrane systems must be operated with somewhat variable outputs, the use of variable frequency devices also improves energy efficiency. The increase in temperature of the RO feed water reduces energy cost since the water production increases approximately 2.5% for each degree Celsius of temperature increase.

Recently, the use of renewable energy as an alternative to power grid for running brackish and seawater RO desalination is gaining popularity. Membrane systems driven by renewable energies avoid dependency on fossil fuel and subsequent greenhouse gas emissions (Richards et al. 2014). Therefore, renewable energy-driven membrane systems are considered more environment friendly than grid-powered membrane systems. Using renewable energy sources is particularly favorable in remote regions where grid power is either not available or not cost effective. Solar and wind are the predominant alternatives for RO desalination. Solar photovoltaic powered RO is considered one of the most promising forms of renewable energy–powered small-scale desalination for remote sunny areas (Bilton et al. 2011; Qiblawey et al. 2011). The world's largest solar-powered desalination plant is being built in Al-Khafji, Saudi Arabia. Studies are being carried out to make solar-powered desalination technically and economically more feasible (Buonomenna and Bae 2015). Wind turbines can also be used to supply electricity to RO plants in coastal areas where wind energy resources are abundant. The major limitation of wind power is the inherent discontinuous availability of the resource. For continuous operation of an RO plant using wind energy, an energy storage or a backup system is required.

The successful deployment of renewable energy sources will be especially important for developing countries that are experiencing water scarcity and do not have access (geographically or economically) to sufficient conventional energy resources to implement desalination systems.

9.5.2.2 Concentrate Management

As discussed in Section 9.2, low-pressure membranes are primarily implemented in the form of dead-end filtration mode that does not produce a concentrate stream. Therefore, concentrate management strategy is primarily applicable to high-pressure membranes. Concentrate stream contains only those chemicals that are present in feed water. However, the chemical concentration in concentrate stream is much higher than that of feed water. Depending on the feed water recovery (50% to 80%), the concentrate stream can be two to five times more concentrated than the feed water. Because of the presence of high concentration of chemicals, the concentrate needs to be disposed of in an environmentally friendly manner. Disposal of concentrate generated from high-pressure desalination systems is particularly a concern because of the comparatively lower recovery that gives a large concentrate stream. Table 9.6 provides a summary for the available concentrate management options.

TABLE 9.6
Options for Concentrate Disposal

Method	Technology	Advantages	Limitations	Cost
Surface water discharge	Direct discharge of concentrate to surface water body	Simple	Potential impact to environment Stringent regulatory requirements	Low
Sanitary sewer discharge	Direct discharge of concentrate into sanitary sewer system	Simple	Potential impact to treatment process and microbial growth in treatment plants Stringent regulatory requirements	Low
Evaporation pond	Water evaporation for volume reduction	Does not discharge liquid outside plant boundary	Requires dry climate Requires large land area	High
Land application	Using concentrate stream for irrigation	Help conserve water	Potential to impact environment Requires large land area	Medium
Deep well injection	Store concentrate stream in confined underground space	Permanent storage	Stringent regulatory requirements Site selection	Medium to high
Zero liquid discharge (ZLD)	Treat concentrate to produce dry solids	Does not discharge liquid outside plant boundary	High energy requirements	High

Surface water discharge is the simplest and least expensive method among all available concentrate disposal alternatives (USBR 2003). In most cases, discharge of surface water consists of a simple pipe extending from the treatment plant to a location that has minimal environmental impacts. The discharge may increase TDS and other constituent levels of the receiving surface water. Therefore, the receiving water bodies are marine or brackish in nature to provide adequate dilution of the concentrate stream. This method is particularly popular in seawater RO desalination where the concentrate stream is discharged into the nearby sea or ocean. Because of the potential for contaminating the environment and the risk to marine organisms, the regulatory requirements for this option are very stringent.

Solar evaporation uses solar energy to allow the water content of concentrate to evaporate into the atmosphere. An evaporation pond with an impervious liner at the bottom of the pond is constructed to hold the concentrate. This is a viable alternative in relatively warm, dry climates with high evaporation rates, level terrain, and low land costs. Because it has a potential to contaminate underlying groundwater, regulations for this method are very stringent. Multiple monitoring wells are often required, which increases costs of evaporation ponds. The cost of a solar evaporation pond depends on the cost of land excavation, height of the dike, and liner thickness.

Land application uses the concentrate stream directly for irrigation. This option could be attractive where water conservation is of great importance. The cost depends on the loading rate, storage period, and cost of land. A major drawback of this option is that the high TDS concentration may be detrimental to vegetation and potentially contaminate the underlying groundwater. Therefore, concentrate stream is typically diluted before application.

Underground injection stores the concentrate stream in a confined aquifer that cannot be used for drinking water or irrigation. The storage must not be located near a fault zone or in an area that has seismic activity. Because of the threat for contaminating underground raw water sources, USEPA has established a set of requirements for the construction, operation, and maintenance of injection wells. Additionally, injection wells need to be tested periodically for integrity and monitored continuously for possible contamination. The cost of this method depends primarily on the depth and size of the injection well. Regulatory requirements are also stringent because this method has the potential to raise the salinity of receiving aquifer if the native water in the receiving aquifer has less salinity than the concentrate stream. Figure 9.11 shows a schematic of a concentrate injection well.

Zero liquid discharge (ZLD) desalination recovers the entire concentrate so that no liquid leaves the plant boundary (USBR 2009). In ZLD facilities, concentrate stream is treated to produce solid products (such as dry salts) for land disposal. Additional processes could also be used to treat concentrate and produce salts. These processes include thermal evaporation, crystallization, electrodialysis, and electrodialysis reversal. ZLD is an energy-intensive, costly process. Therefore, this process is rarely used in municipal water and wastewater industry unless other alternatives are not possible.

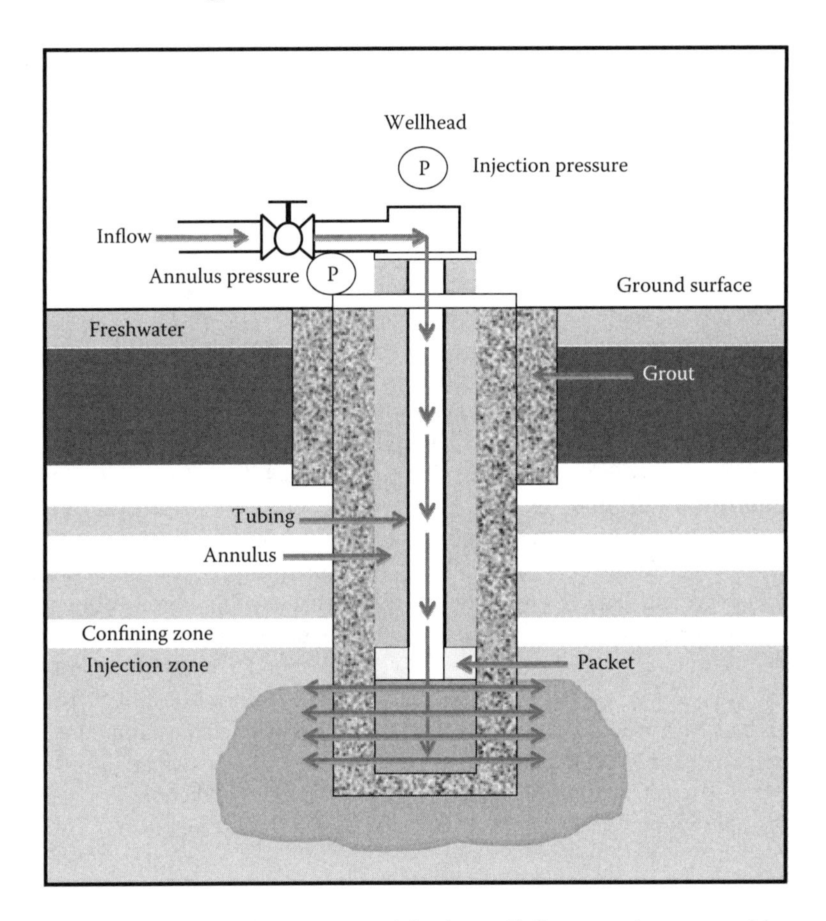

FIGURE 9.11 Schematic of a concentrate injection well (figure not drawn to scale).

9.5.3 Cost Competitiveness

The cost of membrane processes depends on a number of parameters including feed water source, feed water quality, target finished water quality, plant size, process design, pre- and posttreatment processes, concentrate disposal method, regulatory issues, plant location, proximity of the installation to raw water source, and finished water distribution systems.

Several cost estimation tools are available to estimate the cost of water production using a membrane filtration process. One such tool is the US Bureau of Reclamation's planning level estimating procedures for seawater and surface brackish water desalination facilities. Estimating procedures include nomographs to calculate the impact of selected variables, such as the cost of power, on the cost of desalination projects (USBR 2003). Another of Reclamation's products is WTCost, a database and computer program with cost algorithms for different types of desalination pretreatment and treatment technologies.

Membrane system costs are most commonly described as the unit water production cost, defined as the sum of the amortized capital cost and the annual operating/maintenance cost per volume of water produced. Capital cost includes both direct and indirect cost. Direct capital cost includes the spending for land acquisition, membrane equipment installation, building and structures, site development, electric utilities, finished water storage, pumping and piping, and concentrate/residual disposal facilities. Indirect capital costs cover legal/administrative fees and interest. The fixed cost for operation and maintenance includes the spending for labor and equipment/membrane replacement. The spending for power and chemicals (e.g., for membrane cleaning) is considered variable operating cost (Bergman 2012).

The cost of producing freshwater using low-pressure membranes is very competitive to that of conventional systems (typically \$1–\$2 per 1000 gallons of water produced in the United States). However, high-pressure membrane filtration is more costly compared to conventional processes. Cost surveys show that the cost for water production from RO seawater desalination ranges from \$5 to \$10 per 1000 gallons produced and that a larger treatment apacity generally reduces the unit treatment cost (Water Reuse Association 2012). The treatment cost of RO desalination for brackish groundwater is substantially lower because of the lower salinity in the feed water that results in lower pressure requirement and higher water recovery. For example, the potable water plants using RO desalination for brackish water (TDS = 2500–5000 mg/L) in Texas (United States) report a treatment cost of \$1 to \$3 per 1000 gallons (Arroyo and Shirazi 2012). Although significantly more costly compared to the treatment cost of conventional processes, RO desalination is a highly cost-competitive alternative for seawater desalination, and therefore, it has been growing rapidly in regions where freshwater is scarce (Lee et al. 2011).

9.6 CONCLUSIONS

As the world population and economy grow in the future, and as climate forces unfavorable changes to the global hydrologic cycle (Nielsen-Gammon 2015; Yang and Metchis 2015), the pressure on the world's limited water resource will continue to intensify, particularly when the public and industries view the supply of clean and safe freshwater as one of the primary risks in the coming decades (World Economic Forum 2015). Addressing the availability and quality of freshwater is a complex social, economic, and technological issue that must be addressed in multiple fronts. Conventionally, managing water availability and quality has not been determined by the economic elements of supply and demand as for other commodities (Salzman 2013). Rather, it has been focused on the supply and treatment side, that is, construction of storage, conveyance, and treatment facilities. Such a water management scheme is NOT sustainable. As water stress continues to grow in many regions of the world, increasing efforts to obtain additional freshwater resources will certainly be attempted. Desalination is an attractive option to be considered, along with other alternatives such as water conservation and reuse programs, for addressing future water supply shortages. In this perspective, pressure-driven membrane technology expands water availability through desalination of seawater and brackish water and also serves as a centerpiece of potable and nonpotable reuse of wastewater in a world that has depleting freshwater resources.

Given the niche that pressure-driven membrane technology brings forward in the sustainable water management of the future, there are challenges in its potential to effectively address the water demand. In the past few decades, the technology of membrane filtration processes has been improved drastically and the implementation of pressure-driven membrane systems is not constrained by its technological capabilities (i.e., performance of pollutant removal). Instead, it is predominantly limited by financial and environmental factors. The cost-intensive nature of membrane treatment is probably the most significant roadblock. This is reflected in two areas: one is the high energy requirement; the other comes from the need for the pretreatment and maintenance of membrane systems (such as fouling mitigation). Recent developments in the increase of unit capacity, improvement in process design and membrane materials, and the use of hybrid systems have contributed to significant reduction in capital cost and energy consumption (Ghaffour et al. 2015). However, the cost of membrane treatment, particularly for seawater desalination, must be further reduced to facilitate implementation in regions that are financially restricted. Fundamentally, the development of high-permeability, fouling-resistant, high-selectivity and high-rejection, oxidant-resistant membranes is still critically needed. System designs that eliminate the need for pretreatment and that improve system performance will substantially enhance the market penetration of membrane technology. Methods that reduce energy footprints and apply renewable energy sources are also necessary. Furthermore, assessment of the social and environmental impact of desalination operation is needed. For seawater desalination, the impact of seawater intake and concentrate discharge back to the marine environment needs further assessment. Other concentrate management options for high-pressure membrane systems (deep well injection, evaporation, and land disposal) also need to be investigated to understand the fate and transport of the contaminants in concentrates.

REFERENCES

Amjad Z. 1992. *Reverse Osmosis, Membrane Technology, Water Chemistry and Industrial Application*; Van Nostrand Reinhold: New York.

Ang W.S. and Elimelech M. 2009. Optimization of Chemical Cleaning of Organic-Fouled Reverse Osmosis Membranes. A report published by the United States Bureau of Reclamation (USBR).

Arroyo J. and Shirazi S. 2012. Cost of brackish groundwater desalination in Texas. A white paper published at TWDB.

Asano T., Burton F.L., Leverenz H.L., Tsuchihashi R. and Tchobanoglous G. 2007. *Water Reuse: Issues, Technologies, and Applications*, 109 pp.

AWWA. 2005. Microfiltration and ultrafiltration membranes for drinking water. *Manual of Water Supply Practices*, M53, pp. 4–5.

AWWA. 2007. Reverse osmosis and nanofiltration. *Manual of Water Supply Practices*, M46, pp. 8–9.

AWWA. 2012. Evaluation of dynamic energy consumption of advanced water and wastewater treatment technologies. The study was jointly sponsored by AWWA Research Foundation, and California Energy Commission. A Report prepared by YuJung Chang, David J. Reardon, Pierre Kwan, Glen Boyd, Jonathan Brant, Kerwin L. Rakness David Furukawa. pp. 18–19.

Bergman R. 2012. Cost of membrane treatment: Current costs and trends, presented in the Membrane Technologies Conference, Glendale, AZ.

Bilton A.M., Wiesman R., Arif A.F.M., Zubair S.M. and Dubowsky S. 2011. On the feasibility of community-scale photovoltaic-powered reverse osmosis desalination systems for remote locations. *Renewable Energy*, 36, 3246–3256.

Buonomenna M.G. and Bae J. 2015. Membrane processes and renewable energies. *Renewable and Sustainable Energy Reviews*, 43, 1343–1398.

Elimelech M. and Phillip W.A. 2011. The future of seawater desalination: Energy, technology, and the environment. *Science*, 333 (6043), 712–717.

Flemming H.-C. and Schaule G. 1988. Biofouling on membranes—A microbiological approach. *Desalination*, 70, 95–119.

Ghaffour N., Bundschuh J., Mahmoudi H. and Goosen M.F.A. 2015. Renewable energy-driven desalination technologies: A comprehensive review on challenges and potential applications of integrated systems. *Desalination*, 356 (15), 94–114.

Hai F.I. and Yamamoto K. 2011. Membrane biological reactors. In P. Wilderer (Ed.), *Treatise on Water Science* (pp. 571–613). UK: Elsevier.

Jacangelo J.G., Chellam S. and Trussell R.R. 1998. The membrane treatment. *Civil Engineering*, 68 (9), 42–45.

Lattemann S., Kennedy M.D., Schippers J.C. and Amy G. 2010. Global desalination situation. In Escobar I.C. and Schäfer A.I. (Eds.), *Sustainability Science and Engineering*, Vol. 2, 7–39.

Lee K.P., Arnot T.C. and Mattia D. 2011. A review of reverse osmosis membrane materials for desalination—Development to date and future potential. *Journal of Membrane Science*, 370 (1–2), 1–22.

Li Q. and Elimelech M. 2004. Organic fouling and chemical cleaning of nanofiltration membranes: Measurements and mechanisms. *Environmental Science and Technology*, 38, 4683–4693.

Li C., Levy E. and Wang P. 2010. Nanofiltration—An attractive alternative to RO. *Water Technology*.

Loeb S. and Sourirajan S. 1962. Sea water demineralization by means of an osmotic membrane. *Advances in Chemistry Series*, 38, 117.

Mallevialle J., Odendaal P.E. and Wiesner M.R. 1996. Permeation behavior of clean membranes. In *Water Treatment Membrane Processes*. McGraw-Hill.

Nguyen T., Roddick F.A. and Linhua F. 2012. Biofouling of water treatment membranes: A review of the underlying causes, monitoring techniques and control measures. *Membranes*, 2, 804–840.

Nielsen-Gammon J. 2015. Climate change and future water supply, Chapter 11. In Daniel H. Chen (Ed.), *Sustainable Water Management (Volume I)*, Taylor & Francis/CRC Press, Boca Raton, FL (to be published in 2015).

Park N., Kwon B., Kim I.S. and Cho J. 2005. Biofouling potential of various NF membranes with respect to bacteria and their soluble microbial products (SMP): Characterization, flux decline, and transport parameters. *Journal of Membrane Science*, 258, 43–54.

Qiblawey H., Banat F. and Al-Nasser Q. 2011. Performance of reverse osmosis pilot plant powered by photovoltaic in Jordan. *Renewable Energy*, 36, 3452–3460.

Reid C. and Breton E. 1959. Water and ion flow across cellulosic membranes. *Journal of Applied Polymer Science*, 1, 133.

Richards B.S., Park G.L., Pietzsch T. and Schäfer A. 2014. Renewable energy powered membrane technology: Safe operating window of a brackish water desalination system. *Journal of Membrane Science*, 468, 400–409.

Roesink H.D.W. 1989. Microfiltration Membrane Development and Module Design. Printed by ALFA Enschede, The Netherlands.

Saffaj N., Loukili H., Alami-Younssi A., Albizane A., Bouhria M., Persin M. and Larbt A. 2004. Filtration of solution containing heavy metals and dyes by means of ultrafiltration membranes deposited on support made of Moroccan clay. *Desalination*, 168, 301–306.

Salzman J. 2013. *Drinking Water—A History*, Overlook Duckworth, Peter Mayer Publishers, Inc., New York.

Sarioglu M., Insel G. and Orhon D. 2012. Dynamic in-series resistance modeling and analysis of a submerged membrane bioreactor using a novel filtration model. *Desalination*, 285, 285–294.

Shirazi S., Lin C.-J. and Chen D. 2010. Inorganic fouling of pressure-driven membrane processes—A critical review. *Desalination*, 250 (1), 236–248.

Snyder S.A., Westerhoff P., Yoon Y. and Sedlak D.L. 2003. Pharmaceuticals, personal care products, and endocrine disruptors in water: Implications for the water industry. *Environmental Engineering Science*, 20 (5), 449–469.

Strathmann H., Giorno L. and Drioli E. 2006. An Introduction to Membrane Science and Technology. Institute on Membrane Technology, CNR-ITM, at University of Calabria, Via P. Bucci 17/C, 87036 Rende (CS), Italy.

TWDB. 2004. Report 360, *Aquifers of the Edwards Plateau*. Robert E. Mace, Edwards S. Angle, and William F. Mullican (Eds.), Chapter 15, p. 293.

United Nations. 2014. World Water Development Report 2014—Water and Energy, Vol. 1: Water and Energy.

US Bureau of Reclamation. 2003. *Desalting Handbook for Planners*. 3rd Edition. 187–233.

US Bureau of Reclamation. 2009. Treatment of concentrate. A report published by the Desalination and water purification research and development program, Report No. 155.

Water Reuse Association. 2012. White Paper on Seawater Desalination Cost, https://www.watereuse.org/sites/default/files/u8/WateReuse_Desal_Cost_White_Paper.pdf, accessed March 2015.

World Economic Forum. 2015. *Global Risks Report, 10th Edition*, Geneva, Switzerland.

Xie R.J., Gomez M.J., Xing Y.J. and Klose P.S. 2004. Fouling assessment in a municipal water reclamation reverse osmosis system as related to concentration factor. *Journal of Environmental Engineering Science*, 3, 61–72.

Yamamoto K., Hiasa H., Talat M. and Matsuo T. 1989. Direct solid liquid separation using hollow fiber membranes in activated sludge aeration tank. *Water Science and Technology*, 21, 43–54.

Yang J. and Metchis K. 2015. Water supply sustainability issues in time of climate and land use changes, Chapter 12. In Daniel H. Chen (Ed.), *Sustainable Water Management (Volume I)*, Taylor & Francis/CRC Press, Boca Raton, FL (to be published in 2015).

10 Biotechnology for Water Sustainability

Ingo Wolf and Yen Wah Tong

CONTENTS

10.1 INTRODUCTION

Water is essential for all living organisms and by far the most abundant compound on our planet. Unfortunately, only at first glance does water seem like an unlimited resource. Its unique chemical properties make it an almost universal solvent, allowing accumulation of chemical compounds beyond tolerable concentrations. Moreover, it creates an environment for microorganisms and viruses that may exert severe health issues. Clearly, this generates serious concerns about hazard risks associated with poor water quality for human health, which are currently acknowledged by policy makers through extended guidelines and regulations. In the future, however, the situation will come to a head as only urban regions will have to cope with world population growth [1], leading to an increasing number of extreme urban environments called megacities. Hence, within a very short time, established and proven urban water treatment facilities will be challenged to the limit by denser waste streams, new emerging contaminants, and generally increasing costs for wastewater monitoring and treatment. Biotechnology takes advantage of the creativity shown in biological systems to solve engineering problems and to fulfill industrial tasks. It represents a promising possibility to quickly broaden the availability of powerful applications for water monitoring and treatment necessary to guarantee safe and also sustainable water under those circumstances. This chapter will first introduce biosensor technologies as a means to detect and evaluate potentially toxic compounds in water bodies. It will then highlight two remarkable biotechnological methods to eradicate contaminants for sustainable drinking water and a healthy environment.

10.2 BIOSENSOR TECHNOLOGY

In many developed countries, people regard provision of safe water as a matter of course. An increasing number of frameworks and directives have therefore been brought into operation that identify and group substances together, which are of major concern for the aquatic environment. Such high-risk substances are controlled on a regular basis by a multitude of sophisticated analytical methods, of which the various types of mass spectrometry are of particular importance [2]. However, the supply of safe water is continuously exacerbated through emerging contaminants that originate as products of our everyday lifestyle (e.g., consumption of pharmaceuticals and use of personal health products) and under certain conditions such as the reuse of water in extreme environments like megacities [3]. Even though progress has been made in recent years, mass spectrometric techniques still encounter difficulties when it comes to identification of nontarget compounds and unknown degradation products because either analytical standards or appropriate reference databases are missing [2].

The use of biotechnology in the form of biosensor technology could help overcome this problem by taking advantage of inherent functional properties of biological elements. The following sections aim to introduce the major characteristics of typical biosensors and to give a compact overview on available recognition element technologies.

10.2.1 Biosensor Design

A biosensor is an analytical device that is able to generate a response signal proportional to the concentration of a specific analyte or group of analytes. Its schematic follows the rather simple design of general sensor systems. It consists of only three parts, namely, the sensing or recognition element, the transducer element, and the processing element.

In a biosensor, as its name implies, the recognition element is a biological component that targets a specific molecule or family of molecules. Two types that rely on either affinity- or catalytic-based interaction with the target can be distinguished, even though modern biosensors can combine both principles [4]. Technological and scientific progress allows researchers to select the biological recognition element from nucleic acids, proteins (receptors or enzymes), whole cells, or even tissues. Even though choice is diverse, the final criteria in the development of an optimal biosensor are usually the same (Table 10.1).

The recognition element itself is immobilized on the surface of the physical transducer element that converts and delivers a signal to the processing element, which, in

TABLE 10.1
Characteristics of an Optimal Biosensor

Handling	Functionality	Cost
Easy to use	Adaptability to many types of analytes	High reusability
Speed of operation	High affinity to an analyte or group of analytes	High stability during storage
Continuous operation capability	High specificity to an analyte or group of analytes	Low production cost
Portability (sometimes)	Real-time or near real-time monitoring	Low operational cost
Not prone to interferences	Low level of detection	Miniaturization
High stability during measurement	Reproducibility of results	Low power requirements

Source: Ronkainen, N.J., H.B. Halsall, and W.R. Heineman, *Chem Soc Rev*, 2010. **39**(5): 1747–1763; Grieshaber, D. et al., *Sensors*, 2008. **8**(3): 1400–1458.

turn, functions in the amplification and visualization of the signal [5]. On the basis of different physical principles, several types of transducer elements, which allow one to choose the optimal transduction method for a certain recognition element, have been developed (Table 10.2).

The following section will give an introduction into the most common types of biological recognition elements and their immobilization on the transducer surface.

10.2.2 Biosensor Recognition Elements

10.2.2.1 Nucleic Acid–Based Sensors

In nucleic acid–based biosensors, aptamers, which are RNA or single-stranded DNA molecules of usually less than 50 nucleotides, are applied as a biorecognition element. They represent the newest member in biosensor technology. The name originates from the Latin word *aptus* ("to fit") and the Greek word *meros*, the part [6] that describes the mode of action that, despite the small size of aptamers, enables binding of a great variety of different molecules from low–molecular weight compounds such as amino acids and peptides to proteins and whole cells. In unbound form, aptamers exhibit an essentially amorphous structure. Upon contact to the target molecule, a defined three-dimensional aptamer–target molecule complex is formed in which either a small molecule is assimilated into the aptamer structure or the aptamer tightly attaches to the larger target molecule [6]. In contrast to many enzyme and immuno-based biosensors, a time-consuming, costly, and often error-prone sample pretreatment is usually unnecessary [7].

Using aptamers as a biological recognition element has become a very attractive alternative to immunosensors not only because affinity is highly competitive to that of antibodies (picomolar to nanomolar K_d range) [8] but mainly because of several advantages offered by the sophisticated aptamer selection process. Even though

TABLE 10.2
Different Types of Biosensor Transducer Elements

Mode of Transduction	Physical Principle
Potentiometric	Measurement of the potential difference between working and reference electrodes that are connected electrically by the sample analyte.
Amperometric	A timely constant potential is used to create a current that is directly proportional to the electrochemical oxidation reduction reactions of the analyte at the working and reference electrodes.
Conductometric	Changes in the number and charge of molecules (ions) in a solution are determined by measuring its ability to conduct electrical current between two electrodes of known surface area and distance.
Impedimetric	Applied in the form of electrochemical impedance spectroscopy. A small amplitude sinusoidal ac excitation signal perturbs a system at equilibrium to obtain information about the resistive and capacitive properties of materials.
Mechanical (converse piezoelectric effect)	Piezoelectric materials (e.g., quartz crystals) oscillate when an electrical field is applied. A mass change at the material surface causes a measurable alteration in oscillation frequency.
Optical	Measurement is based on the effect of the interaction between a biorecognition element and its target on various optical parameters, for example, refractive index (SPR) or fluorescence.

Source: Ronkainen, N.J., H.B. Halsall, and W.R. Heineman, *Chem Soc Rev*, 2010. **39**(5): 1747–1763; Grieshaber, D. et al., *Sensors*, 2008. **8**(3): 1400–1458; Lagarde, F. and N. Jaffrezic-Renault, *New Trends in Biosensors for Water Monitoring*. 2011.

research on synthetic antibody production is making progress, the common process of raising antibodies is time-consuming, is laborious, involves costly purification steps, and is dependent on the use of animals [9–11].

The aptamer selection and production process, in contrast, is solely performed in vitro. Systematic evolution of ligands by exponential enrichment (SELEX) is a powerful method to select for aptamers that target against almost any molecule with low cross-reactivity [12,13]. It was designed to select and enrich highly affine and extremely specific molecular recognition processes between aptamers and target. Among others, those that have been shown to involve intermolecular contacts such as formation of hydrogen bonds as well as structural electrostatic interactions thereby allow differentiation of molecules by just a single group (e.g., $-CH_3$) and even chirality [6,14].

In SELEX, the molecule of interest is first exposed to a library of random oligonucleotide sequences. It is advantageous in that specific physiological conditions can be applied at this step to reflect the environmental conditions in which the biosensor is expected to perform later. By that, only those three-dimensional aptamer–target molecule complexes that are enriched can optimally form under such specific conditions. Washing steps are then applied to remove any oligonucleotides that show low affinity or have not bound the target. Next, aptamer–target complexes are disintegrated and the nucleic acid molecules are amplified by PCR (polymerase chain reaction). Double-stranded PCR products are subsequently separated into single strands and used for the next round of selection in which more stringent conditions are applied. Generally, less than 15 rounds of SELEX enrich aptamers that exhibit the desired binding properties [5,12]. Final steps comprise molecular cloning followed by capillary sequencing to reveal the exact nucleotide sequences. Once this information is obtained, it can be used for synthetic production of aptamers in high quantity and quality [12]. Even though the SELEX method very efficiently produces high affine and specific aptamers, it shares at least one major drawback that antibody production is also suffering from. For SELEX to be performed, necessary amounts of target molecules need to be available in adequate quality. Accordingly, target purification procedures or synthetic production must be available.

Many different approaches have been developed to immobilize aptamers to transducer surfaces. Direct, noncovalent attachment by exploiting inherent features of the DNA molecule has been used. For example, the negative charges of the phosphate backbone can be used for physical immobilization by electrostatic attachment. A more advanced and frequently applied direct attachment method uses aptamers in which the 3′ or 5′ carbon of the sugar moiety is labeled by a thiol group. After labeling, aptamers can form into a monolayer by interaction of its thiol group with the gold or silver electrode surface. Another possibility is to use intermediate molecules that have a high affinity to each other such as the well-known biotin/streptavidin interaction. Its main advantages are its ability to couple streptavidin to nonmetal electrodes and the high density of biotinylated aptamers on the transducer surface attributed to a 4:1 biotin/streptavidin binding ratio [15]. However, covalent attachment is generally preferred because of its higher stability but it is also more complex to achieve. Most commonly, chemically modification of transducer surfaces aims to provide various functional groups (e.g., hydroxyl or carboxyl) that can be used to form covalent bonds with amino-labeled aptamers [15]. Moreover, ester bond formation between phosphate or hydroxyl groups of aptamers and hydroxyl or carboxyl groups of modified surface layers, respectively, represents an alternative to accomplish a stable surface immobilization [7].

Aptamers are compatible with different types of transduction modes. As for other affinity-based biosensors, electrochemical transduction is common and mechanical and optical systems have also been investigated [16]. Optical and electrochemical transduction often differs from mechanical approaches in that they rely on the specific conformational change of aptamers upon binding to their target. For example, fluorescence-based optical systems use fluorophore–quencher systems. Unbound aptamers either form intramolecular double strands in a way that terminally fused fluorophore and quencher molecules are in proximity to each other or form an intermolecular double strand with a complementary DNA single strand that is labeled with the quencher [16]. Either way, upon binding of the target molecule, the aptamer changes its three-dimensional structure, thereby spatially separating fluorophore and quencher. Subsequently, light emitted from the fluorophore can be detected in a target concentration-dependent manner. Recently, colorimetric detection using aptamer-conjugated gold nanoparticles has come into focus because of, in principle, simple application [17]. For mechanical transduction, quartz crystal microbalance

(QCM) should be mentioned, in which the aptamers are conjugated to the surface of an oscillating crystal. Formation of aptamer–target molecule complexes increases the crystals' surface mass, thereby leading to a measurable change in oscillation frequency [18].

Aptamer-based biosensors are superior to immunosensors in terms of regeneration since they are less prone to denaturation [12]. Another reason could be that depending on the transduction method, denaturation of the single-stranded nucleic acid should not affect reestablishment of an ordered aptamer–target molecule complex once physiological conditions have been normalized. Aptamers are known to be unstructured in solution until interaction with the target molecule occurs [8]. This represents a major advantage to immunosensors where laborious regeneration procedures can lead to reduced reproducibility and higher costs. Nucleic acids are known to be very susceptible to hydrolytic degradation especially when catalyzed by nucleases, enzymes that can cleave phosphodiester bonds between nucleotides. Modification of 2′ carbon in RNA ribose moieties with functional groups or the use of DNA-based aptamers shows higher stability against hydrolytic degradation. Current research is focused on optimizing stability under certain environmental conditions by modification with functional groups [19].

Taken together, aptamer-based biosensors are still a young technology. However, optimization of the SELEX procedure and introduction of automated steps have gradually reduced selection time from weeks to several days [5,12]. In combination with great progress in the field of synthetic biology in recent years in terms of large-scale, cost-efficient, and precise synthesis of even longer nucleic acid sequences and its compatibility with different transducer types, this has built a strong foundation for the development and application of new aptamer-based biosensors.

10.2.2.2 Immuno-Based Sensors

Immuno-based biosensors take advantage of the major players of the adaptive immune system in vertebrates called antibodies or immunoglobulins. A tremendous amount of knowledge related to biosynthesis, structure, and function of antibodies has been gathered within the last few decades, and researchers have quickly recognized the massive inherent potential that subsequently led to the development of new analytical methods (e.g., Western blot analysis) as well as to enhanced established techniques (e.g., fluorescent microscopy) [20]. In either case, the main property conferred by the antibody molecule relies on its binding to a unique part (epitope) of a target molecule (antigen) with high affinity and specificity [21]. Structurally, antibodies are made of four polypeptide chains, two heavy (50 kDa) and two light (25 kDa) chains, which, connected via disulfide bridges, form a distinctive Y-shaped structure. The C-termini of the heavy chains are overall constant in sequence and form the base of the Y-shaped structure. It has important functions in complement binding during an immune response; however, regarding the use of antibodies as a biological recognition element in biosensors, the arms of the Y-shaped structure are much more important. This has caused some approaches to partially or completely omit the antibody C-termini but is not commonly used yet [22]. In each arm, one heavy chain is combined with one light chain. Since the N-termini of those polypeptide chains exhibit extreme sequence variability within the first 150 amino acids, an antigen-binding site can be formed to recognize and bind any structural motif. This ability is actually the central advantage antibodies have over other, in particular, biocatalytic recognition elements in biosensor technology. Antibodies are synthesized and released in large amounts by B-type immune cells upon injection of a foreign target macromolecule (antigen) into an animal (mouse, rat, rabbit, goat, chicken, donkey, horse, etc.) followed by recognition and polyclonal response of the adaptive immune system. In a polyclonal response, various antibodies targeted against the same antigen molecule but directed to different epitopes on it are produced. A polyclonal antibody mixture can be purified and used to make a biosensor, but the more epitopes are being targeted, the higher the chance that an epitope also exists in other molecules. This could result in an increased background signal and reduced level of detection owing to response of the biosensor to nontarget molecules. Therefore, great care has to be taken during development to guarantee high specificity for the target analyte only. Nevertheless, once established, those types of biosensors can be produced in an easy and cost-effective manner.

Monoclonal antibody technology can help improve the specificity of biosensors. As for polyclonal antibody production, the antigen is first injected into an animal to induce an adaptive immune response. In a second step, antibody-producing B-cells are isolated from the animal's spleen. Köhler and Milstein [23] invented a technique to make those short-lived B-cells immortal by electrofusing them with mouse myeloma cells that have lost antibody production and secretion functionality. In principle, the product is a collection of immortal cells (hybridomas) that can be used to generate indefinite amounts of a single type of antibody instead of a polyclonal mixture. In the last step, single hybridoma cells are regrown (clones), screened, and selected for optimal selectivity and affinity against the antigen. With increasing efficiency of cell lines in antibody production, the major cost is progressively imposed on purification steps [10].

After extraction of pure antibodies, the efficient immobilization of the molecules on the transducer surface is of utmost importance. The antigen-binding Y-shaped arms should thereby be oriented to the target analyte and the immobilization should not interfere with the antibodies' binding activity and specificity. Strategies for immobilization can be distinguished into physical and chemical methods. Physical methods are based on adsorption of the antibody to the transducer surface. Formation of covalent bonds is not involved, decreasing the risk for changes in the native conformation and accompanied loss of sensitivity. Instead, binding is achieved by low-energy bonds such as van der Waals forces or ionic bonds. To better control orientation and density of the recognition element on the transducer surface, chemical modifications of the recognition element and the transducer surface are carried out [24]. Apart from using electrostatic coupling via charged amino acids like arginine or metal binding amino acids like histidine and cysteine, a variety of well-established biochemical methods are in use. The most commonly used methods are coupling of the antibody or antibody fragment via a genetically attached 6× His-tag to the Ni-NTA–covered transducer surface and the very strong biotin–streptavidin linkage. Moreover, natural antibody binding proteins such as protein G and protein A are frequently used [22,24]. Even though these modifications are efficient to prevent random orientation of the recognition element on the transducer surface, there is a high susceptibility to leakage of the recognition element because of the weak nature of the physical forces involved, which is subsequently accompanied by loss of sensitivity, reproducibility, and shorter shelf life [5,18]. Chemical modifications that involve formation of covalent bonds are therefore applied for increased stability. Natural or genetically engineered lysine side chains have been frequently used to link the amine group to functionalized polymers [25]. Since lysine residues are quite common in polypeptide chains, making a specific orientation difficult to achieve, other procedures such as UV cross-linking via special aromatic amino acids or carbohydrate moieties of antibodies and chemical cross-linking after oriented adsorption using intermediate proteins (protein A or G) are used [26–28].

Today, electrochemical transducers (e.g., electrochemical impedance spectrometry) are most frequently used, but other precise and reliable methods based on either mechanical (e.g., QCM) or optical (surface plasmon resonance [SPR]) principles are also available to monitor the formation of antibody–antigen complexes without the need for additional labels as commonly used in immunoassays [18,22].

The major disadvantage of using antibodies as recognition elements is caused by their immanent function as part of the immune system. Once specifically bound to a foreign target molecule, they are not supposed to detach until elimination via the complement system occurs [21]. This, however, significantly hampers the reusability of the biosensor and increases its cost. Usually, rather harsh dissociation procedures are applied during regeneration to avoid insufficient removal of antigen but may lead to significant release of antibody molecules from the transducer surface, which again reduces biosensor performance especially in terms of reproducibility [5]. Moreover, detection using immunosensors is restricted to known target molecules, which have been used to raise a specific, high-affinity antibody. Generally, this demands both sufficient amounts of antigen in very pure form and the triggering of a strong immune response. However, very often, the molecule of interest cannot be synthetically produced, is toxic, or does not cause an efficient immune response, for example, because of its low molecular weight or high similarity to an endogenous occurring molecule. The amount of the resulting antibody titers then becomes very low (if not totally nonexistent), which subsequently increases production costs. Current research focuses on synthetic antibody production and improved antibody selection procedures that have the potential to overcome those restrictions, making immunosensors an even more important type of biosensor in the future [9].

10.2.2.3 Enzyme-Based Sensors

Enzymes are a special class of proteins that exhibit many characteristics that render them extremely suitable as biological recognition elements in biosensors. The general principle of an enzyme-based biosensor has first been proposed by Clark in 1962 and is currently the most successful recognition element in biosensor technology [5,29]. After successful folding, enzymes constitute functional biochemical catalysts that exhibit very high specificity and activity [20]. Biosensor technology exploits this inherent catalytic property by covering the surface of an electrode with a layer of immobilized enzymes. The detection mechanism itself is based on the catalytic reaction of the enzyme with its substrate, which leads to a change in the number of charged molecules and is ultimately detected by the underlying electrode [5,18,30].

Specificity can be attributed to the unique three-dimensional arrangement of amino acid side chains of the protein and the formation of an active site, which allows formation of interactions with a certain substrate (enzyme–substrate complex) [20]. Often, before conventional analytical techniques can be used with complex environmental samples, costly, time-consuming pretreatments such as solid-phase extraction (SPE) need to be performed to prevent interferences during measurement or to increase the concentration of key analytes [31]. In contrast, formation of an enzyme–substrate complex is usually very robust, making time-consuming pretreatments of complex environmental samples unnecessary [5].

Moreover, the enzyme–substrate complex considerably accelerates the chemical transformation of a substrate by lowering the necessary activation energy leading to high biocatalytic activity. It is very important to note that after the reaction, the enzyme itself is left unchanged [20]. Therefore, in contrast to biosensors with affinity-based biological recognition elements, the application of often difficult and labor-intensive regeneration procedures can be avoided. This opens up the possibility for easy biosensor reuse and reduces operating costs [5].

Enzyme-based biosensor technology has now reached its third generation. First, second, and third generations basically differ in the mode of electron shuttling to the electrode.

First-generation sensors apply enzymes that use O_2-dependent redox reactions to shuttle electrons to an amperometric electrode. Generally, those sensors suffer under low solubility of O_2 in aqueous media, which restricts the formation of measurable currents and eventually the limit of detection. Changing concentrations of O_2 also raise concerns about consistency of measurement when using different samples [32]. Therefore, to provide a constant and reliable detection, high concentration of O_2 in solution needs to be ensured.

The second generation of enzyme-based biosensors was considerably improved through extended application of other oxidoreductases and the uncoupling of the dependency for oxygen. This was achieved by the introduction of just one new component to the system, a redox-active mediator. Artificial redox mediators are small molecules that can easily assist in the rapid shuttling of electrons between the active site of the enzyme and the electrode surface [33]. Here, improved quality of detection results from the diversity of available mediators that exhibit different oxidation potentials. Use of artificial redox mediators with a low oxidation potential allowed reduction of the working electrode potential. Thereby, unwanted oxidation–reduction reactions of other molecules, which would be subjected to interference under high potentials at the electrode, could be significantly reduced [33]. Chaubey and Malhotra [33] summarized what the ideal mediator molecule should be: ready to conduct rapid and reversible redox reactions with the enzyme, nonreactive with oxygen, stable in oxidized and reduced forms, and pH independent. Several molecules have been found and tested to function as such kind of redox mediators in biosensors like conducting salts, quinones, ferrocene, tetrathiafulvalene, and ferrocyanide [33]. Despite all amendments that redox mediators have added to enzyme-based biosensors, one problem still exists, they lack specificity. Thus, they may interfere in the high specific detection of the enzyme by committing unspecific redox reactions at a downstream step [32].

The development of the third generation of biosensors is therefore focused on waiving redox mediators by direct electron transfer between the enzyme and the electrode. However, this task is far from being trivial. It has been found that the speed of electron transfer exponentially decreases over distance [30,34]. This needs to be considered because the active site of enzymes can generally be found deeply buried within the three-dimensional structure [20] and the pathway for electrons to reach the electrode surface might thus be restrained. In order to prevent slow electron transfer followed by weak signal detection, a very close arrangement between enzyme and electrode is compulsory, which is usually approached by advanced immobilization of the enzyme on the electrode surface or by surface functionalization technologies such as protein-film voltammetry or electroactive nanotubes [32]. However, only a rather small number of enzymes have been successfully involved in direct electron transfer reactions in biosensors, namely, cellobiose dehydrogenase [35], cytochrome c, hemoglobin, and horseradish peroxidase [32].

This already indicates one of the major drawbacks enzyme-based biosensor technology is facing. Currently, the restriction to naturally occurring enzymes identified and extracted from a variety of living organism is of capital importance. Used as detecting elements, those enzymes are only capable of reacting with the corresponding biological substrate molecules. Therefore, if nonbiological analytes have to be detected, affinity-based biosensors are currently the better choice. However, great effort has been made in biotechnological research to develop and produce tailor-made enzymes. Since there is a tremendous lack of natural enzymes able to perform direct electron transfer, this might be especially helpful for extending the application and accompanied cost of third-generation sensors [32]. Apart from extending the number of detectable analytes, this approach is very promising to overcome other general drawbacks that can occur, for example, difficulties in manufacturing, unstable enzymatic activity owing to susceptibility to activators and inhibitors, and other interferences, as well as reduced enzyme stability followed by loss of function. The latter is often observed when enzymes are used under assay conditions that cannot fully provide necessary requirements such as pH, ionic strength, or a lipid environment [20].

Although inactivation or inhibition of the recognition element activity is usually not desired, it can be exploited for the detection of unknown inhibitors or toxic compounds in aquatic systems. Many sensors that use inhibition of immobilized enzymes (e.g., glucose oxidase, alkaline phosphatase, or acetylcholinesterase) for the detection of pesticides (such as carbofuran) or heavy metals (such as mercury and cadmium) have been developed [36–38]. However, since many factors can negatively affect enzyme activity when environmental samples are being investigated, assessment and data obtained always demand critical control and evaluation [39].

10.2.2.4 Whole Cell–Based Sensors

Nucleic acid–, immuno-, or enzyme-based biosensors introduced in Sections 10.2.2.1 through 10.2.2.3 respectively usually lacks the ability to recognize and respond to pollutants or toxic substances of unknown nature [5,18,30]. In contrast, whole-cell biosensors aim to respond to a wide range of environmental effects and can therefore be especially suitable for ecotoxicological evaluation of treated and reused water in megacities where emerging contaminants pose a possible threat. Further advantages provided by whole-cell biosensors are assessment of the degree of toxicity, possible automated real-time monitoring, and the fact that the biorecognition element is usually inside the cell. The latter allows assessment for cellular penetration of pollutants and makes the biorecognition element more durable since the inner cellular environment provides important cofactors and can better protect it from outer influences (e.g., ionic strength, pH). Moreover, difficult, expensive, time-consuming extraction and purification procedures can also be avoided [10,18]. In contrast, the existence of other cellular components (e.g., enzymes) runs the risk of having negative influences on specificity or affinity.

Cell-type selection is usually made according to the origin of the sample (freshwater, marine water, wastewater, etc.) and certain cell-/organism-specific characteristics (e.g., prokaryotic vs. eukaryotic, generation time, metabolic features, amenability to genetic modifications, immobilization tolerance, etc.). Most frequently, microorganisms such as bacteria (*Escherichia coli*, *Pseudomonas putida*), cyanobacteria (*Synechocystis* sp. PCC6803), and yeasts (*Saccharomyces cerevisiae*) have been used because they are easy to handle, are compatible with different immobilization procedures, and allow stable genetic modifications. Eukaryotic cells should be preferred when toxicological effects on higher organisms such as humans are being investigated. However, available cell lines (e.g., fibroblasts) tend to be less tolerant to immobilization techniques or certain transduction methods, leading to cellular stress responses or cell death [40]. The proper transduction method therefore needs to be carefully decided to avoid misinterpretation. Two principal analytical techniques have been employed based on optical transduction electrochemical transduction.

For optical transductions, usually genetically modified cells are engineered and a change in fluorescence, luminescence, or color is monitored [41]. Such approaches have been more extensively used for immobilization and electrochemical transduction–sensitive higher cell types because immobilization is not mandatory here [41]. A reporter gene construct is first introduced into the cell. Expression of the reporter gene is subsequently controlled by a short DNA sequence called promoter [20] and can either be continuous or induced by intracellular transcription factors. In the first case, reduction of the reporter signal upon exposure to a target analyte indicates a general metabolic effect. The use of a promoter that is controlled by certain transcriptional regulators, on the other hand, allows one to have a closer look on the implications a sample might exert on the cell. In fluorescent detection, the reporter is a fluorescent protein such as green fluorescent protein. Luminescent detection is most commonly based on the measurement of the catalytic activity of the bacterial luciferase or the firefly luciferase that produces light from oxidation of long-chain fatty aldehyde, reduced riboflavin and oxygen, or reduced luciferin and ATP-Mg^{2+}, respectively [42]. Generally, luminescence approaches are superior to fluorescence ones in terms of response time and sensitivity, whereas colorimetric detection on the other hand is known for its simplicity [42]. It uses an enzymatic conversion of a chromogen into a dye or pigment that can be subsequently detected. In either case, the response to effectors in the sample is in a concentration-dependent manner.

As for other enzyme-based biosensors, amperometry is the most commonly used technique [18,40]. Upon exposure of a sample to the whole-cell biosensor, amperometry is used to monitor changes in respiration/photosynthesis or metabolic-related redox processes occurring at the cell surface or within the cytoplasm of a cell. Whole-cell biosensors have also successfully been developed to shorten the rather long-term determination (5 days) of the biological oxygen demand, which is an important water quality parameter [40,43].

10.3 BIOSORPTION TECHNOLOGY

Analytical techniques and methodologies that identify both natural and anthropogenic hazardous contaminants that may display risks for environment and human health have significantly been improved in recent years (see Section 10.2). However, awareness of existing hazards can only be the first step that should be followed to prevent actual contact and eventually to eliminate the hazard. Various conventional methods using either chemical or physical approaches are available to tackle such contaminations. The most commonly used methods are those based on reverse osmosis, ion exchange, and electrodialysis, and combined application is under investigation [44,45]. However, related operational costs in conventional water treatment are still high [46]. In this regard, biotechnology in the form of biosorption is a promising approach for cost-effective removal of contaminants from aqueous solutions followed by either recovery or elimination [47].

Biosorption constitutes a naturally occurring physicochemical process in which a certain substance, the sorbate, can passively bind to the surface of dead biomass, the biosorbent, dependent on their chemical and structural characteristics [45,47,48]. It is completely independent of metabolic processes and can therefore be distinguished from the process of bioaccumulation [49]. In principle, any dead biomass shows some degree of biosorptive potential. Consequently, a multitude of tested biosorbent and sorbate combinations can be found in literature. Much work has been carried out on the treatment of heavy metal–contaminated industrial or municipal wastewater, but there is still an increasing interest in determining the potential for removal of organic compounds such as dyes, phenolic compounds, and pesticides as well as the underlying mechanism [50].

10.3.1 Mechanism Behind Biosorption

In terms of application, biosorption constitutes a rather simple, passive, and reversible process in which the reaction establishes an equilibrium after a sufficient time, which is usually within a few minutes [47,49]. However, determination of the underlying mechanisms that cause biosorption is complicated by the diverse mixture of chemical compounds in natural biomass [51]. It has been known for a while that the equilibrium is strongly dependent on physicochemical factors such as temperature, pH, type and concentration of sorbate and sorbent, and existence of other ions [49,50]. Initiated by those observations, several fundamental interacting forces were suggested to constitute and contribute to the binding mechanism in biosorption such as physisorption (van der Waals forces), chemisorption (complex formation), microprecipitation, and ion exchange [45,51]. Today, most experts agree on ion exchange as the major governing mechanism especially for metal binding; however, occurrence of multiple binding mechanisms is expected when complex biomass is involved [45,49,51]. It is important to note that for ion exchange to occur, the sorbate needs to be available in ionic form. However, especially in terms of many metals (e.g., Zn, Pb, Cu, Cd, Ni), this is not necessarily true since they are able to form metal hydroxides, sulfides, or carbonates as a function of pH and other forms of complexes. Ion exchange and biosorption in general are enabled by a variety of functional chemical groups located and exposed mainly on cell wall surfaces. Among others, those functional groups can be hydroxyl, carboxyl, sulfate, phosphate, amino, amine, thiol, imidazole, or phosphodiester groups [45,49,51]. The binding capacity of a sorbent for a certain sorbate is generally determined by the quantity, accessibility, and type of those functional groups and also the chemical state of the sorbate/sorbent under given conditions [47]. For example, metal cation biosorption can occur via interaction with negatively charged surface groups that can donate electrons (Lewis bases) [45,49]. The availability of such sites is often affected by protonation or deprotonation events owing to pH changes or to the number of competing ions [45,47].

In recent years, biosorption research expanded to organic target molecules but less is known about the underlying mechanisms. Organic molecules are very diverse in chemical properties that may affect biosorption [45]. Research on biosorption of dyes, phenolic compounds, and pesticides suggests though that similar functional groups as for biosorption of heavy metals may be important. Among others, hydrophobic interactions as well as carboxyl, amino, and phosphate groups of cell surface polysaccharides, amino acids, and lipids may be involved in interaction between organic sorbate and biomass [50].

Desorption describes the release of the bound sorbate from the sorbent, which optimally completely regenerates the sorbent for repeated use. However, the complexity of biosorbents is high, thereby leading to reduced regeneration efficiency and reusability compared to synthetic conventional sorbents [45], which is probably the major drawback of this technology. Desorption can be achieved by a variety of chemicals such as acids (HCl, HNO_3) [52], bases (e.g., NaOH) [52], surfactants (e.g., Tween), solvents (e.g., methanol, ethanol) [50], and chelating agents (e.g., ethylenediaminetetraacetic acid) and is considered optimal when maximum sorbate release is achieved in a minimum elution volume.

10.3.2 Characteristics of Biosorbents

Biosorption occurs at the surface of dead biomass, the biosorbent, depending on the availability and chemistry of functional groups, the binding sites [45]. Organisms such as Gram-positive and Gram-negative bacteria, algae, fungi, and plants have a high amount of those biosorptive sites located in their encompassing cell walls [47]. However, each organism has special requirements for its environment; therefore, a manifold of different surface structures with greatly varying chemistries can be found in nature, which may be more or less suited for binding of a certain sorbate. As an example, brown algae such as *Sargassum* that are rich in metal-binding alginate or bacteria, which have negatively charged cell wall components under low pH conditions, show good heavy metal binding performance, whereas the performance of many fungi is rather limited [47]. Thus, many biomass and sorbate combinations have been tested over the years. Cost efficiency was proposed as a major property of biosorbents when it comes to comparison with existing conventional technologies such as ion exchange chromatography, and research has focused on obtaining low-cost sorbents accordingly. A low-cost biosorbent should be highly available and easy to obtain, that is, without the need for expensive processing [53]. Obviously, this applies for biomass, which occurs in large quantity in nature (e.g., seaweed), although another interesting source is the utilization of biomass that is obtained as waste product from industrial processes (e.g., yeast cell biomass from fermentation processes). In particular, biomass derived from textile, agricultural, and food industries were investigated (e.g., cotton fibers, pineapple peel, peanut or rice husk, sugarcane bagasse, etc.) [54–57]. A further option is to gain biomass from organisms that have short generation times and can be artificially cultured in an inexpensive manner such as cyanobacteria, bacteria, some algae, and fungi [47].

To characterize biosorbents in terms of their biosorptive potential, equilibrium contact experiments can be performed. These are batch experiments used to determine sorbent metal uptake by measuring the residual metal concentration after establishment of a reaction equilibrium under constant experimental conditions [47]. Moreover, for a detailed analysis of type and distribution of available binding sites, advanced analytical methods based on spectroscopy can be used, for example, nuclear magnetic resonance, Raman, x-ray diffraction, infrared, and electron microscopy [49].

Because of the many sorbate–sorbent combinations that have been already tested and the diminishing differences in binding abilities and affinities between closely related organisms, many researchers now focus more on premodification or pretreatment instead of combination evaluation to increase binding performance. Chemical or physical treatment can be used to change the sorbent chemistry in order to increase its binding capacity and affinity for a certain sorbate [54,55]. Premodifications, in contrast, aim to increase the number of available binding sites or to introduce new types into the sorbent. This is achieved by genetic engineering and can therefore only be applied to organisms that are amenable to genetic modifications, for example, fungi, cyanobacteria, and bacteria [58].

10.3.3 Future of Biosorbent Application for Wastewater Treatment

After 30 years, biosorption research with focus on its potential for water treatment is now facing a crucial crossroad. On the one hand, the application of biosorption for water purification seems to be a very promising technology, which is distinguished mainly by its potential of remarkably reduced capital and operational costs [46]. On the other hand, even though this scientific field has received much attention, leading to an ever-increasing number of publications in recent years, knowledge transfer into industrial products has been deflating, thus raising concerns about the general need for research in this field [45,49]. One important reason might be the lack of representative studies on the economic margin for the application of biosorption and especially the proof of its practicality and scalability for real water treatment situations

[49]. Thus, many possible customers are afraid of the risk and rather rely on conventional well-established processes that have been shown to be reliable and at the moment have more specific binding properties and better regeneration performance, in particular ion exchange, regardless of the higher costs [45].

Biosorbents with remarkably different specificity and affinity compared to already tested biosorbents are not expected to be found [45,49]. Gadd [45] recently suggested to instead focus research on modification or pretreatment of known promising biosorbents as well as on optimization of reaction conditions and reactors. Future steps should also aim on making biosorption more competitive to conventional approaches by specifically addressing research regarding its drawbacks, specificity, and reusability. A promising technology is cell surface engineering. Here, genetic modifications are used to express and display heterologous proteins on the cell surface of recombinant strains [58]. According to the function of the expressed protein, they can be used to increase specificity for certain toxic heavy metals such as mercury or cadmium and help improve biomass immobilization.

Even though this approach may have a good prospect, it is still in its infancy [58]. Therefore, some researchers see a favorable application of hybrid technologies, for example, by combining biosorption either with bioreduction or bioprecipitation (biotechnological processes) or with electrochemical/chemical processes (nonbiotechnological) [59]. However, if those developments cannot promote increased industrial application of biosorption in wastewater treatment, there might still be a prospective future for it in purification of high-value molecules (pharmaceuticals, etc.) by application of manufactured specific antibodies or aptamers instead of less specific cell components such as cell walls [49,60].

10.4 BIOLEACHING TECHNOLOGY

Bioleaching is a biotechnological technique that uses various microorganisms for solubilization and recovery of heavy metals from insoluble ores (minerals that contain high concentrations of certain metals) by means of oxidation and complexation. It can be distinguished from the process of biooxidation where bioleaching is merely used to remove metal sulfides from gold- or silver-containing ores without concurrent recovery [61]. In contrast to biosorption, bioleaching is already well established and widely used on an industrial scale [62]. Most commonly, bioleaching is used to extract copper, cobalt, nickel, and zinc, which exist in the form of insoluble metal sulfides and are important raw materials for products in our modern lifestyle (electrical wires, batteries, alloys, etc.). The decreasing availability of high-grade ores and stricter environmental standards and regulations have led to increased interest in this technology.

10.4.1 Organisms in Bioleaching

Bioleaching microorganisms are chemolithotrophs, which means they can oxidize inorganic reduced compounds (particularly reduced inorganic sulfur compounds [RISCs] or Fe^{2+} ions) to harvest energy [63]. This process of energy generation therefore appears to be very restricted in available substrate types. Moreover, the character of habitats that are rich in inorganic resources is often extreme in terms of pH, ionic strength, and temperature [61,63]. Nevertheless, a variety of different organisms have successfully adapted to such conditions, although most bioleaching organisms belong to one of two domains: bacteria or archaea. What they generally have in common is that they thrive best under extremely acidic environmental conditions ($pH < 4$) and accordingly are referred to as acidophiles [64]. However, single groups of leaching microorganism can significantly differ in oxygen dependency (obligate aerobes, facultative anaerobes, and obligate anaerobes), in structural features (Gram-positive and Gram-negative), and in their metabolic performance (autotrophic, heterotrophic, or mixotrophic growth; reduced sulfur oxidizing or iron(II) oxidizing; etc.) [63]. Moreover, differences exist in terms of optimal growth temperature, metal resistance, and carbon assimilation pathways [65]. In recent years, genetic determination, such as genomic sequencing and metagenomic approaches, has led to a better understanding of the phylogenetic relationship among bioleaching organisms as well as their ecophysiology [65,66].

The first microorganisms isolated and shown to have bioleaching capability were mesophiles that grow optimally at moderate temperatures between 20°C and 40°C [67].

The best studied in this group is *Acidithiobacillus ferrooxidans*, which is a Gram-negative proteobacterium. Research using it as a model organism has contributed greatly to our current understanding of the fundamental principles and mechanisms underlying bioleaching processes [68]. Species of the genera *Acidithiobacillus*, *Acidiphillium*, and *Leptospirillum* are among the most prominent microorganisms in various natural and anthropogenic bioleaching sites. In the case of *A. ferrooxidans*, the reason could be the ability to perform chemotaxis and that its metabolism apart from RISCs and Fe^{2+} ions is able to oxidize a multitude of other substrates including molecular hydrogen and formic acid [68,69].

Moderate and extreme thermophiles that can optimally thrive at temperatures between 40°C and 60°C and >60°C [67], respectively, have now come into focus because they may provide advantages for industrial bioleaching applications. First, some data suggest that bioleaching at high temperatures could influence sulfide oxidation and metal recovery [70]. More importantly for biotechnological applications, sulfide oxidation is an exothermic reaction that increases the temperature in waste heaps or stirred-tank systems up to 75°C [70]. Since such temperatures are at least inhibiting but usually lethal for mesophiles, expensive cooling measures have to be applied, thus reducing the overall economy of the bioleaching process. Here, hope lies on archaea species such as *Acidianus sulfidivorans* (T_{opt}, 74°C) as an interesting alternative to most bacterial species because they can thrive at very high temperatures and are less affected by high concentrations of dissolved metals [62,71].

10.4.2 Mechanism of Bioleaching

Bioleaching is a biologically catalyzed process in which metal cations are released from an insoluble metal sulfide containing ore by means of oxidation and complexation [61,66,72]. Several classes of oxidizing enzymes exist in biological systems, and for a long time, it was not clear if organisms involved in bioleaching express specialized enzymes that can catalyze the direct metal sulfide oxidation reaction (direct mechanism). However, to date, relevant data for the existence of such enzymes are not available, suggesting that the main catalytic function is achieved using an indirect mechanism by recycling the oxidant instead (the agent that causes metal sulfide dissolution) [61,73,74].

It is important to distinguish that the indirect leaching mechanism cannot be equated with obligatory spatial separation between cells and mineral. In fact, a tight contact can usually be observed in which extracellular polymeric substances (EPS) produced and secreted by the cells seem to have a strong effect both on attachment to the mineral surface and on the bioleaching process [73,75]. The EPS production rate and chemical composition were further shown to be adapted to the mineral's physicochemical properties upon cell to mineral surface contact [76]. Considerable work has been done to elucidate and to understand the underlying mechanisms and kinetics of bioleaching, which is especially important for optimization of industrial applications. There is increasing consent on two mechanisms for sulfide mineral oxidation.

The thiosulfate mechanism is named after the first soluble sulfur intermediate of the pathway [77]. The reaction occurs on the surface of metal sulfide minerals, which are attributed to the properties of their chemical bonds resistant to acid dissolution [61,73]. The most common metal sulfide minerals are pyrite (FeS_2), molybdenite (MoS_2), laurite (RuS_2), and tungstenite (WS_2). The mechanism can be summarized by the following equations [77]:

$$FeS_2 + 6\,Fe^{3+} + 3\,H_2O \rightarrow S_2O_3^{2-} + 7\,Fe^{2+} + 6\,H^+ \quad (10.1)$$

$$S_2O_3^{2-} + 8\,Fe^{3+} + 5\,H_2O \rightarrow 2\,SO_4^{2-} + 8\,Fe^{2+} + 10\,H^+ \quad (10.2)$$

In a first step, ferric ions oxidize metal sulfides (MS) like pyrite, thereby releasing the metal as a cation (M^{2+}) and sulfur as a soluble primary species like thiosulfate (Equation 10.1). Microorganisms such as *A. ferrooxidans* are able to catalyze the reoxidation of ferrous in which they harvest energy and in addition recycle the ferric form for a new cycle of metal sulfide oxidation. Via several intermediates, the primary sulfur compounds are then either abiotically oxidized under acidic conditions or biotically oxidized by the microorganisms to harvest energy. Although part of the primary sulfur compounds can be converted to and remain as elemental sulfur, the principal sulfur product of the thiosulfate mechanism is sulfuric acid (Equation 10.2).

The polysulfide mechanism applies to metal sulfide minerals in which the chemical bond between metal and sulfur moieties can be broken by proton attack. In contrast to the thiosulfate pathway, this means that ferric ions and protons can dissolve those metal sulfides [77]. Among others, sphalerite (ZnS), hauerite (MnS_2), orpiment (As_2S_3), and so on belong to this group [73]. The mechanism can be summarized by the following equations [77]:

$$MS + Fe^{3+} + H^+ \rightarrow M^{2+} + 0.5\ H_2S_n + Fe^{2+}\ (n \geq 2) \quad (10.3)$$

$$0.5\ H_2S_n + Fe^{3+} \rightarrow 0.125\ S_8 + Fe^{2+} + H^+ \quad (10.4)$$

$$0.125\ S_8 + 1.5\ O_2 + H_2O \rightarrow SO_4^{2-} + 2\ H^+ \quad (10.5)$$

Products are the corresponding metal cations (M^{2+}), ferrous ions, and polysulfide species (Equation 10.3). Again, iron-oxidizing microorganisms harvest energy by reoxidation of ferrous to ferric ions. In contrast, a major sulfur product of the polysulfide mechanism is elemental sulfur, which cannot be further oxidized abiotically (Equation 10.4). Only when sulfur-oxidizing microorganisms are present can elemental sulfur be further oxidized to sulfuric acid, thereby making protons available for a new attack on metal sulfide bonds.

On the basis of the two mechanisms, the principal catalytic functions of microorganisms involved in bioleaching processes can be summarized to control and maintain a high redox potential and low pH, to provide ferric iron (Fe^{3+}), and to regenerate it from ferrous iron (Fe^{2+}) and lastly the oxidation of sulfur species [61,77].

10.4.3 Bioleaching in Water Treatment

Apart from the immense rise of bioleaching applications in mineral industries [62], great efforts are also made on its application for environmental protection and remediation.

Acid mine drainage (AMD) is a common environmental problem that can be observed at many current and former mining sites, particularly in places such as abandoned mineshafts/tunnels as well as in mine dumps composed of fractions of mining material in which the metal concentration was originally too low for economic extraction using conventional processes. AMD occurs when material abundant in sulfide minerals is exposed to oxygen and water, thereby creating an environment that offers all requirements to promote growth of leaching microorganisms and the bioleaching reactions introduced in Sections 10.4.1 and 10.4.2 respectively. Eventually, this results in groundwater and surface water acidification and contamination by dissolved toxic metals such as arsenic, causing serious detrimental effects on connected ecosystems. Measures to prevent AMD are usually focused on controlling oxygen availability, for example, by refilling or flooding of abandoned mines and covering of waste dumps, and application of mitigating materials (e.g., acid-neutralizing limestone) or antimicrobials (e.g., sodium dodecyl sulfate) [78]. However, since some of those measures may not show long-lasting effects or by themselves are of environmental concern, comprehensive waste management should be preferred. In the future, the decreasing availability of high-grade ores will further shift mining activities from conventional physical and chemical processes to bioleaching approaches that have lower capital and energy costs, show superior recovery, and thereby extract most metal sulfides before deposition [79]. Today, many ecosystems close to abandoned mining sites are already suffering from AMD. Here, the root problem should be addressed to start environmental cleanup. Indeed, a current trend can be seen in companies like BacTech that are trying to combine bioleaching for remediation of highly toxic mine tailings and the processing of valuable metals especially gold.

Heavy metals originated from AMD, acid rock drainage, atmospheric deposition, or discharge from industrial and urban sources may eventually end up in rivers and harbors. Once introduced, the heavy metals can adsorb to suspended organic matter and settle down to form contaminated aquatic sediments. Uptake of contaminated organic particles by organisms and acidic conditions within the digestive system could cause release followed by bioaccumulation of metals in animal organs [80]. Furthermore, delayed release of adsorbed metals to the overlaying water column as a result of environmental changes pose a potential hazard to even distant ecosystems [81]. Simple stripping and disposal of contaminated sediment even worsens the problem because abiotic and biotic oxidative leaching processes benefit from exposure to a high-oxygen environment. Therefore, current research focuses on elucidating the optimal conditions for controlled application of bioleaching processes before disposal to remediate heavy metal–contaminated sediments [81].

The current trend of population movement from rural to urban areas not only challenges water treatment facilities to provide safe water but also causes an immense increase in arising wastewater treatment sludge [82]. Direct drainage of sewerage into rivers and the direct use of sewage sludge as fertilizer in agriculture are a major source for toxic metal contamination in soil and water bodies. It was estimated that approximately half of the sludges cannot be used as environmentally safe fertilizer without preceding treatment [83]. An integral part to biologically leach metals from sludges is the application of sulfur-oxidizing bacteria, mainly *Acidithiobacillus thiooxidans* and *A. ferrooxidans* [82,83]. In contrast to bioleaching of metals from minerals, the leaching process is initiated by supplementation of elemental sulfur as energy source. Sulfur-oxidizing bacteria harvest energy by oxidizing elemental sulfur, thereby producing sulfuric acid as a final oxidation product, which significantly reduces the pH [74] [61]. Establishment of an acidic environment is the main factor for metal mobilization [82]. The efficiency of bioleaching process from sludge and its economic application is very much dependent on a set of parameters such as the type of microorganism used; the tolerance of the microorganisms to toxic sludge substances such as dyes [82]; the type, availability, and price of the supplement; the type, form, and concentration of metals in the sludge; the sludge solids concentration; and the process pH and temperature. Recently, iron-based bioleaching experiments addressed to exploit the iron-oxidizing bacterium *A. ferrooxidans* and using $FeSO_4$ as a supplement to mobilize heavy metals were shown to be promising [84]. Using this approach, the authors hope to reduce the risk of secondary pollution and to eliminate the formation of residual sulfur, both known to be disadvantages of the sulfur-based bioleaching process [84].

In conclusion, bioleaching is an established biotechnology that has successfully found its way into commercial application in the mineral industry. Moreover, it is a promising technology that processes sludge from wastewater treatment into safe fertilizer for agriculture.

REFERENCES

1. WWAP, U., *The United Nations World Water Development Report 4: Managing Water under Uncertainty and Risk (Vol. 1), Knowledge Base (Vol. 2) and Facing the Challenges (Vol. 3)*. 2013, United Nations Educational, Scientific and Cultural Organization: Paris, France.
2. Aguera, A., M.J. Martinez Bueno, and A.R. Fernandez-Alba, New trends in the analytical determination of emerging contaminants and their transformation products in environmental waters. *Environ Sci Pollut Res Int*, 2013. **20**(6): 3496–3515.
3. Lapworth, D.J. et al., Emerging organic contaminants in groundwater: A review of sources, fate and occurrence. *Environ Pollut*, 2012. **163**: 287–303.
4. Liu, Y. et al., Immune-biosensor for aflatoxin B1 based bio-electrocatalytic reaction on micro-comb electrode. *Biochemical Engineering Journal*, 2006. **32**(3): 211–217.
5. Ronkainen, N.J., H.B. Halsall, and W.R. Heineman, Electrochemical biosensors. *Chem Soc Rev*, 2010. **39**(5): 1747–1763.
6. Ferapontova, E. and K.V. Gothelf, Recent advances in electrochemical aptamer-based sensors. *Current Organic Chemistry*, 2011. **15**(4): 498–505.
7. Van Dorst, B. et al., Recent advances in recognition elements of food and environmental biosensors: A review. *Biosens Bioelectron*, 2010. **26**(4): 1178–1194.
8. Collett, J.R. et al., Functional RNA microarrays for high-throughput screening of antiprotein aptamers. *Anal Biochem*, 2005. **338**(1): 113–123.
9. Miersch, S. and S.S. Sidhu, Synthetic antibodies: Concepts, potential and practical considerations. *Methods*, 2012. **57**(4): 486–498.
10. Low, D., R. O'Leary, and N.S. Pujar, Future of antibody purification. *J Chromatogr B Analyt Technol Biomed Life Sci*, 2007. **848**(1): 48–63.
11. Birch, J.R. and A.J. Racher, Antibody production. *Adv Drug Deliv Rev*, 2006. **58**(5–6): 671–685.
12. Tombelli, S., M. Minunni, and M. Mascini, Analytical applications of aptamers. *Biosens Bioelectron*, 2005. **20**(12): 2424–2434.
13. Hansen, J.A. et al., Quantum-dot/aptamer-based ultrasensitive multi-analyte electrochemical biosensor. *J Am Chem Soc*, 2006. **128**(7): 2228–2229.
14. Geiger, A., RNA aptamers that bind L-arginine with sub-micromolar dissociation constants and high enantioselectivity. *Nucleic Acids Research*, 1996. **24**(6): 1029–1036.
15. Zhou, L. et al., Application of biosensor surface immobilization methods for aptamer. *Chinese Journal of Analytical Chemistry*, 2011. **39**(3): 432–438.
16. Song, S. et al., Aptamer-based biosensors. *TrAC Trends in Analytical Chemistry*, 2008. **27**(2): 108–117.
17. Zhang, J. et al., Aptamer-conjugated gold nanoparticles for bioanalysis. *Nanomedicine (Lond)*, 2013. **8**(6): 983–993.
18. Lagarde, F. and N. Jaffrezic-Renault, *New Trends in Biosensors for Water Monitoring*. 2011.
19. Wang, R.E. et al., Improving the stability of aptamers by chemical modification. *Current Medicinal Chemistry*, 2011. **18**(27): 4126–4138.
20. Alberts, B. et al., *Molecular Biology of the Cell*. 5th ed. 2007, New York: Taylor & Francis. 1392.
21. Murphy, K., *Janeway's Immunobiology*. 8th ed. 2011, Garland Science.
22. Zeng, X., Z. Shen, and R. Mernaugh, Recombinant antibodies and their use in biosensors. *Anal Bioanal Chem*, 2012. **402**(10): 3027–3038.
23. Köhler, G. and C. Milstein, Continuous cultures of fused cells secreting antibody of predefined specificity. *Nature*, 1975. **256**(5517): 495–497.
24. Trilling, A.K., J. Beekwilder, and H. Zuilhof, Antibody orientation on biosensor surfaces: A minireview. *Analyst*, 2013. **138**(6): 1619–1627.

25. Teles, F.R.R. and L.P. Fonseca, Applications of polymers for biomolecule immobilization in electrochemical biosensors. *Materials Science and Engineering: C*, 2008. **28**(8): 1530–1543.
26. Yuan, Y. et al., Site-directed immobilization of antibodies onto blood contacting grafts for enhanced endothelial cell adhesion and proliferation. *Soft Matter*, 2011. **7**(16): 7207.
27. Alves, N.J., T. Kiziltepe, and B. Bilgicer, Oriented surface immobilization of antibodies at the conserved nucleotide binding site for enhanced antigen detection. *Langmuir*, 2012. **28**(25): 9640–9648.
28. Bergström, G. and C.-F. Mandenius, Orientation and capturing of antibody affinity ligands: Applications to surface plasmon resonance biochips. *Sensors and Actuators B: Chemical*, 2011. **158**(1): 265–270.
29. Clark, L.C. and C. Lyons, Electrode systems for continuous monitoring in cardiovascular surgery. *Annals of the New York Academy of Sciences*, 2006. **102**(1): 29–45.
30. Grieshaber, D. et al., Electrochemical biosensors—Sensor principles and architectures. *Sensors*, 2008. **8**(3): 1400–1458.
31. Gómez, M.J. et al., *Analysis of Organochlorine Endocrine-Disrupter Pesticides in Food Commodities*. 2011: 75–125.
32. Zhang, W. and G. Li, Third-generation biosensors based on the direct electron transfer of proteins. *Analytical Sciences*, 2004. **20**(4): 603–609.
33. Chaubey, A. and B.D. Malhotra, Mediated biosensors. *Biosensors and Bioelectronics*, 2002. **17**(6–7): 441–456.
34. Kuznetsov, B.A. et al., On applicability of laccase as label in the mediated and mediatorless electroimmunoassay: Effect of distance on the direct electron transfer between laccase and electrode. *Biosensors and Bioelectronics*, 2001. **16**(1–2): 73–84.
35. Zafar, M.N. et al., Characteristics of third-generation glucose biosensors based on Corynascus thermophilus cellobiose dehydrogenase immobilized on commercially available screen-printed electrodes working under physiological conditions. *Anal Biochem*, 2012. **425**(1): 36–42.
36. Guedri, H. and C. Durrieu, A self-assembled monolayers based conductometric algal whole cell biosensor for water monitoring. *Microchimica Acta*, 2008. **163**(3–4): 179–184.
37. Samphao, A. et al., Alkaline phosphatase inhibition-based amperometric biosensor for the detection of carbofuran. *International Journal of Electrochemical Science*, 2013. **8**: 3254–3264.
38. Samphao, A. et al., Indirect determination of mercury by inhibition of glucose oxidase immobilized on a carbon paste electrode. *International Journal of Electrochemical Science*, 2012. **7**: 1001–1010.
39. Luque de Castro, M.D. and M.C. Herrera, Enzyme inhibition-based biosensors and biosensing systems: Questionable analytical devices. *Biosensors and Bioelectronics*, 2003. **18**(2–3): 279–294.
40. Bentley, A. et al., Whole cell biosensors—Electrochemical and optical approaches to ecotoxicity testing. *Toxicology in Vitro*, 2001. **15**(4–5): 469–475.
41. Su, L. et al., Microbial biosensors: A review. *Biosens Bioelectron*, 2011. **26**(5): 1788–1799.
42. Close, D.M., S. Ripp, and G.S. Sayler, Reporter proteins in whole-cell optical bioreporter detection systems, biosensor integrations, and biosensing applications. *Sensors (Basel)*, 2009. **9**(11): 9147–9174.
43. Kara, S., B. Keskinler, and E. Erhan, A novel microbial BOD biosensor developed by the immobilization of *P. syringae* in micro-cellular polymers. *Journal of Chemical Technology & Biotechnology*, 2009. **84**(4): 511–518.
44. Mahmoud, A. and A.F. Hoadley, An evaluation of a hybrid ion exchange electrodialysis process in the recovery of heavy metals from simulated dilute industrial wastewater. *Water Res*, 2012. **46**(10): 3364–3376.

45. Gadd, G.M., Biosorption: Critical review of scientific rationale, environmental importance and significance for pollution treatment. *Journal of Chemical Technology & Biotechnology*, 2009. **84**(1): 13–28.
46. Ali, I. and V.K. Gupta, Advances in water treatment by adsorption technology. *Nat Protoc*, 2006. **1**(6): 2661–2667.
47. Vieira, R.H. and B. Volesky, Biosorption: A solution to pollution? *Int Microbiol*, 2000. **3**(1): 17–24.
48. Volesky, B. and Z.R. Holan, Biosorption of heavy metals. *Biotechnol Prog*, 1995. **11**(3): 235–250.
49. Chojnacka, K., Biosorption and bioaccumulation—The prospects for practical applications. *Environ Int*, 2010. **36**(3): 299–307.
50. Aksu, Z., Application of biosorption for the removal of organic pollutants: A review. *Process Biochemistry*, 2005. **40**(3–4): 997–1026.
51. Naja, G. and B. Volesky, *The Mechanism of Metal Cation and Anion Biosorption*, in *Microbial Biosorption of Metals*, P. Kotrba, M. Mackova, and T. Macek, Editors. 2011, Springer Netherlands. pp. 19–58.
52. Gong, R. et al., Lead biosorption and desorption by intact and pretreated spirulina maxima biomass. *Chemosphere*, 2005. **58**(1): 125–130.
53. Bailey, S.E. et al., A review of potentially low-cost adsorbents for heavy metals. *Water Research*, 1999. **33**: 2469–2479.
54. Paulino, Á.G. et al., Chemically modified natural cotton fiber: A low-cost biosorbent for the removal of the Cu(II), Zn(II), Cd(II), and Pb(II) from natural water. *Desalination and Water Treatment*, 2013: 1–11.
55. Xu, M. et al., Utilization of rice husks modified by organomultiphosphonic acids as low-cost biosorbents for enhanced adsorption of heavy metal ions. *Bioresour Technol*, 2013. **149**: 420–424.
56. Krishni, R.R., K.Y. Foo, and B.H. Hameed, Food cannery effluent, pineapple peel as an effective low-cost biosorbent for removing cationic dye from aqueous solutions. *Desalination and Water Treatment*, 2013: 1–8.
57. Sadaf, S. and H.N. Bhatti, Evaluation of peanut husk as a novel, low cost biosorbent for the removal of Indosol Orange RSN dye from aqueous solutions: Batch and fixed bed studies. *Clean Technologies and Environmental Policy*, 2013.
58. Li, P.S. and H.C. Tao, Cell surface engineering of microorganisms towards adsorption of heavy metals. *Crit Rev Microbiol*, 2013.
59. Tsezos, M., Biosorption of metals. The experience accumulated and the outlook for technology development. *Hydrometallurgy*, 2001. **59**(2–3): 241–243.
60. Volesky, B., Biosorption and me. *Water Res*, 2007. **41**(18): 4017–4029.
61. Vera, M., A. Schippers, and W. Sand, Progress in bioleaching: Fundamentals and mechanisms of bacterial metal sulfide oxidation—Part A. *Appl Microbiol Biotechnol*, 2013. **97**(17): 7529–7541.
62. Brierley, C.L. and J.A. Brierley, Progress in bioleaching: Part B: Applications of microbial processes by the minerals industries. *Appl Microbiol Biotechnol*, 2013. **97**(17): 7543–7552.
63. Rawlings, D.E., Characteristics and adaptability of iron- and sulfur-oxidizing microorganisms used for the recovery of metals from minerals and their concentrates. *Microb Cell Fact*, 2005. **4**(1): 13.
64. Saro, F.J.L. et al., *The Dynamic Genomes of Acidophiles*. 2013. **27**: 81–97.
65. Valdés, J. et al., Comparative genomics begins to unravel the ecophysiology of bioleaching. *Hydrometallurgy*, 2010. **104**(3–4): 471–476.
66. Cardenas, J.P. et al., Lessons from the genomes of extremely acidophilic bacteria and archaea with special emphasis on bioleaching microorganisms. *Appl Microbiol Biotechnol*, 2010. **88**(3): 605–620.

67. Johnson, D.B., Biodiversity and ecology of acidophilic microorganisms. *FEMS Microbiology Ecology*, 1998. **27**(4): 307–317.
68. Valdes, J. et al., *Acidithiobacillus ferrooxidans* metabolism: from genome sequence to industrial applications. *BMC Genomics*, 2008. **9**: 597.
69. Rohwerder, T. et al., Bioleaching review part A: Progress in bioleaching: fundamentals and mechanisms of bacterial metal sulfide oxidation. *Appl Microbiol Biotechnol*, 2003. **63**(3): 239–248.
70. Brierley, J.A., Response of microbial systems to thermal stress in biooxidation-heap pretreatment of refractory gold ores. *Hydrometallurgy*, 2003. **71**(1–2): 13–19.
71. Plumb, J.J. et al., *Acidianus sulfidivorans* sp. nov., an extremely acidophilic, thermophilic archaeon isolated from a solfatara on Lihir Island, Papua New Guinea, and emendation of the genus description. *Int J Syst Evol Microbiol*, 2007. **57**(Pt 7): 1418–1423.
72. Rawlings, D.E. and D.B. Johnson, The microbiology of biomining: Development and optimization of mineral-oxidizing microbial consortia. *Microbiology*, 2007. **153**(Pt 2): 315–324.
73. Tributsch, H., Direct versus indirect bioleaching. *Hydrometallurgy*, 2001. **59**(2–3): 177–185.
74. Sand, W. et al., (Bio)chemistry of bacterial leaching—Direct vs. indirect bioleaching. *Hydrometallurgy*, 2001. **59**(2–3): 159–175.
75. Kinzler, K. et al., Bioleaching—A result of interfacial processes caused by extracellular polymeric substances (EPS). *Hydrometallurgy*, 2003. **71**(1–2): 83–88.
76. Gehrke, T. et al., Importance of extracellular polymeric substances from *Thiobacillus ferrooxidans* for bioleaching. *Applied and Environmental Microbiology*, 1998. **64**(7): 2743–2747.
77. Schippers, A. and W. Sand, Bacterial leaching of metal sulfides proceeds by two indirect mechansism via thiosulfate or via polysulfides and sulfur. *Applied and Environmental Microbiology*, 1999. **65**(1): 319–321.
78. Sand, W. et al., Long-term evaluation of acid rock drainage mitigation measures in large lysimeters. *Journal of Geochemical Exploration*, 2007. **92**(2–3): 205–211.
79. Schippers, A. et al., The biogeochemistry and microbiology of sulfidic mine waste and bioleaching dumps and heaps, and novel Fe(II)-oxidizing bacteria. *Hydrometallurgy*, 2010. **104**(3–4): 342–350.
80. Malik, N. et al., Bioaccumulation of heavy metals in fish tissues of a freshwater lake of Bhopal. *Environ Monit Assess*, 2010. **160**(1–4): 267–276.
81. Akinci, G. and D.E. Guven, Bioleaching of heavy metals contaminated sediment by pure and mixed cultures of *Acidithiobacillus* spp. *Desalination*, 2011. **268**(1–3): 221–226.
82. Wen, Y.-M. et al., Bioleaching of heavy metals from sewage sludge by *Acidithiobacillus thiooxidans*—A comparative study. *Journal of Soils and Sediments*, 2012. **12**(6): 900–908.
83. Lombardi, A.T., O. Garcia, and W.A.N. Menezes, The effects of bacterial leaching on metal partitioning in sewage sludge. *World Journal of Microbiology and Biotechnology*, 2006. **22**(10): 1013–1019.
84. Wen, Y.-M. et al., Bioleaching of heavy metals from sewage sludge using indigenous iron-oxidizing microorganisms. *Journal of Soils and Sediments*, 2013. **13**(1): 166–175.

11 Biodegradation/ Bioremediation for Soil and Water

Siddharth Jain and Yen Wah Tong

CONTENTS

11.1 INTRODUCTION

The large-scale buildup, dispensation, and handling of chemicals have led to serious surface and subsurface soil contamination with a wide variety of harmful and toxic hydrocarbons. Many of the chemicals that have been synthesized in great volume, including polychlorinated biphenyls (PCBs), trichloroethylene, and others, differ substantially in chemical structure from natural organic compounds and are designated as xenobiotics because of their relative recalcitrance to biodegradation. Other compounds, for example, the polycyclic aromatic hydrocarbons (PAHs), are also toxic and are typically intractable to biodegradation. Intensification of energy-related and other industrial processes with associated production of wastes and by-products, rich in PAHs, has led to soil contamination in most of the industrial sites. The resultant accumulations of the various organic chemicals in the environment, particularly in soil, are of significant concern because of their toxicity, carcinogenicity, and potential to bioaccumulate in living systems. A wide variety of nitrogen-containing industrial chemicals are produced for use in petroleum products, dyes, polymers, pesticides, explosives, and pharmaceuticals. Major chemical groups involved include different nitro-aromatics, nitrate-esters, and nitrogen-containing hetero-cycles. Most of these chemicals are toxic and adversely affect human health and are classified as hazardous by the United States Environmental Protection Agency (EPA).

As it is documented that microbes are able to degrade toxic xenobiotic compounds, which were earlier believed to be resistant to the natural biological processes occurring in the soil, it has increased the interest of researchers in bioremediation of polluted soil and water. Although microbial activity in soil accounts for most of the degradation of organic contaminants, chemical and physical mechanisms can also provide significant transformation pathways for these compounds. Bioremediation is generally considered a safe and less expensive method for the removal of hazardous contaminants and production of nontoxic by-products.

Literature have reported many experimental successes with the more difficult to degrade contaminants and at the same time have been shown many notable failures. However, it has been suggested that, although microorganisms have the primary catalytic role in bioremediation, the knowledge of the alterations occurring in microbial communities remains limited and the microbial community is still treated as a "black box." In the best terms, bioremediation remains a developing field that is done in the natural environment without detailed characterisation of the organisms involved (Verstraete, 2002).

Biotechnology has the potential to play an immense role in the development of treatment processes for contaminated soil. Optimization of the environmental conditions in bioremediation processes is a central goal in order that the microbial, physiological, and biochemical activities are directed toward biodegradation of the targeted contaminants. Environmental factors influencing microbial growth and bioactivity include moisture content, temperature, pH, soil type, contaminant concentrations, and oxygen for aerobic degradation. Deviations of these parameters away from optimal conditions will reduce rates of microbial growth and transformation of target substrates and perhaps cause premature cessation and failure of the bioremediation process. Biodegradation potential is also limited by the toxicity of the pollutants to the degrading microbes. Some species have developed cellular defenses, enabling them to tolerate high concentrations of toxic contaminants.

Understanding the biochemical and physiological aspects of bioremediation processes will provide the requisite knowledge and tools to optimize these processes, to control key parameters, and to make the processes more reliable. Since the majority of bioremediation processes rely on the activities of complex microbial communities, there is need to learn about the interactive and interdependent roles played by individual species in these communities. There is a need to develop strategies for improving the bioavailability of the many hydrophobic contaminants that have an extremely low water solubility and tend to be adsorbed by soil particles and persist there. We need to continue to clarify the complex aerobic and anaerobic metabolic pathways that microbes have evolved to degrade organic contaminants. There is a need to continue to characterize many of the key enzymatic reactions that participate in contaminant transformation and to relate contrasting reaction rates, substrate specificities, and enzyme mechanisms to differences in protein structures. Such new knowledge can provide us with the requisite information to test, design, and engineer biocatalysts with improved substrate specificities, reaction rates, or other desired catabolic properties and ultimately to engineer improved catabolic pathways for bioremediation. We must recognize that some chemical species are inherently intractable to enzyme transformation and we should be open to the possibility of combining chemical or physical strategies with biological systems to achieve overall effective remediation. We must also continue to devise better methods for monitoring and assessing the progress and effectiveness of microbial biodegradation processes at both the research and process implementation level. Clearly, the availability of advanced molecular techniques provides a new impetus and enhances our abilities to address many of these issues.

11.2 BIODEGRADATION

Biodegradation involves the breakdown of organic compounds either through biotransformation into less complex metabolites or through mineralization into inorganic minerals, H_2O, CO_2 (aerobic), or CH_4 (anaerobic). Both bacteria and fungi have been extensively studied for their ability to degrade a range of environmental pollutants including recalcitrant PAHs, halogenated hydrocarbons, and nitroaromatic compounds. The biochemical pathways/enzymes required for the initial transformation stages are often specific for particular target environmental contaminants, converting them to metabolites that can be assimilated into more ubiquitous central bacterial pathways. An overview of some of the biodegradation systems used by microorganisms in the catabolism of key organic contaminants in soil is shown in Table 11.1.

The extent and rate of biodegradation depend on many factors such as pH, temperature, oxygen, microbial population, degree of acclimation, accessibility of nutrients, chemical structure of the compound, cellular transport properties, and chemical partitioning in growth medium. Some recalcitrant chemicals contain novel structural elements that seldom occur in nature and which may be incompletely transformed as microbes lack the degradative pathway for the complete degradation of these xenobiotics. While microbes may not have the metabolic pathways for mineralization of certain newly introduced synthetic chemicals, there is evidence that microorganisms have the capacity to evolve such catabolic systems over time. In bioremediation processes, it is generally an objective to exploit microbial technology to accelerate the rate of pollutant removal.

Many contaminants in soil exist in anaerobic environments. A couple of decades ago, by observing the anaerobic dechlorination of PCBs over time in Hudson River sediments, it became clear that microbes could transform contaminants under

TABLE 11.1
Most Commonly Organic Constituents Found in Groundwater

Grade	Chemicals	Chemical Formula
1	Dichloromethane	CH_2Cl_2
2	Trichloroethene	C_2Cl_3H
3	Tetrachloroethene	C_2Cl_4
4	*trans*-1,2-Dichloroethene	$C_2H_2Cl_2$
5	Chloroform	$CHCl_3$
6	1,1-Dichloroethane	$C_2Cl_2H_2$
7	1,1-Dicholoroethene	$C_2Cl_2II_2$
8	1,1,1-Trichloroethane	$C_2Cl_3H_3$
9	Toluene	C_7H_8
10	1,2-Dicholoroethane	$C_2CL_2H_4$
11	Benzene	C_6H_6
12	Ethylbenzene	C_6H_{10}
13	Phenol	C_6H_5OH
14	Chlorobenzene	C_6H_5Cl
15	Vinyl chloride	C_2ClH_3
16	Carbon tetrachloride	CCl_4
17	Bis(20ethylhexyl)phthalate	$C_{24}H_{38}O_4$
18	Naphthalene	$C_{10}H_8$
19	1,1,2-Trichloroethane	$C_2Cl_3H_3$
20	Chloroethane	C_2ClH_5

anaerobic conditions. By the late 1980s, there was conclusive evidence that hydrocarbons could be degraded in the absence of oxygen. These anaerobic degradation systems required terminal electron acceptors such as iron, manganese oxide, or nitrate to replace that function of oxygen in aerobic systems. We have now entered a period of intensive research and discovery focused on the catalytic mechanisms that facilitate the anaerobic catabolism of pollutants.

The intensity of biodegradation is affected by various factors, such as nutrients, oxygen, pH value, composition, concentration and bioavailability of the contaminants, chemical and physical characteristics, and the pollution history of the contaminated environment. Bioremediation attempts to accelerate the naturally occurring biodegradation of contaminants through the optimization of limiting conditions (Alexander 1999; Allard and Neilson 1997; Norris 1994).

11.3 CONTAMINANTS

It is reported in the literature that more than 200 substances have been found in US groundwater. These substances can be naturally occurring or can be from anthropogenic sources that include industrial and agricultural organic chemicals, metals, and radioactive material. Although site contamination is generally reported as groundwater contamination, soil contamination is just as prevalent. Groundwater contamination typically occurs via the initial discharge of chemicals onto the ground surface. Moreover, many sites are contaminated with more than one hazardous chemical or toxic compound. Therefore, any remediation method or combination of methods selected should be capable of cleaning up all contaminants in the groundwater.

11.3.1 Contaminant Types and Their Biodegradation Potentials

Table 11.1 shows the 20 most abundant organic chemicals reported in groundwater at solid and hazardous waste disposal sites. Out of these 20 chemicals, 7 are chlorinated aliphatic hydrocarbons. Some of the others include benzene, toluene, and ethylbenzene, which are components of fuel products and also serve as feedstock for production of other chemicals. Other examples of common organic chemicals that have contaminated soil and groundwater include multiple ring aromatic hydrocarbons, nitroaromatics, and aromatic and nonaromatic hydrocarbons that contain nitrogen, phosphorus, sulfur, chlorine, and bromine substituent groups.

Depending on site-specific characteristics, bioremediation may be used for a variety of contaminants, such as pesticides, fertilizers, metals, and radioactive materials. However, not all compounds in one chemical class are biodegradable. For instance, petroleum-based compounds such as benzene, toluene, ethylbenzene, and xylenes are biodegradable under certain environmental conditions, whereas other petroleum-based compounds such as methyl tertiary butyl ether are not readily biodegradable.

11.3.2 Effect of Chemical Structures on Biodegradation

Many compounds are considered defiant to biodegradation and generally remain in the environment. One reason for this is related to chemical hydrophobicity. In general, the more hydrophobic a chemical is, the less it is biodegradable. This becomes evident with the increased molecular weight in a series of similarly structured chemicals, such as PAHs. For instance, biodegradation rates generally decrease going from benzene to naphthalene as a result of their increasing hydrophobicity and water solubility. A further reason is that not all chemical structures are modifiable to biodegradation. For example, the addition of a halogen or nitro group to a readily degradable compound is thought to decrease the compound's susceptibility to biodegradation. The position of the additional substituent is also important. However, integration of oxygen into the compound in hydroxyl and carboxyl substituent form has been shown to increase biodegradability. Branching of hydrocarbons also results in lesser biodegradation potential. This can be evidenced by the straight-chain octadecane (18 carbons), which has a much greater potential to biodegrade than branched phytane (18 carbons).

Competitive and noncompetitive inhibition can also affect bioremediation. A molecule that resembles the contaminant substrate molecule might bind to the active site of the enzyme. This would decrease overall enzyme activity, which would cause completive inhibition. Noncompetitive inhibition occurs when a molecule binds to a nonactive site, which results in changing the enzyme's shape. If the change in the shape is sufficient, the enzyme may be inactivated.

11.4 ENVIRONMENTAL FACTORS FOR BIODEGRADATIONS

The applicability and success of in situ bioremediation processes are primarily determined by the geology and hydrology of the contaminated site. For instance, both play critical roles in determining contaminant allocation. Hydraulic conductivity is also a restrictive factor for applying the bioremediation process. Contaminated locations with high porosity and less hydraulic conductivities are found to be poor candidates for the bioremediation process as delivery of nutrients and an electron acceptor to the contaminated location becomes complex. Therefore, there are a numbers of environmental factors that must be considered in evaluating the application of bioremediation.

11.4.1 Subsurface Heterogeneity and Abiotic Factors

Soil properties can vary from one region to another. They also can vary within the same region. The characteristics that define a soil type are the same abiotic factors that influence biodegradation such as cation exchange capacity, clay type, gradation, liquid limit, organic matter content, particle size, pH, porosity, and soil texture. Each of these factors influences the occurrence, rate, and product of biodegradation. Microorganisms have a range of tolerances to these factors, which affect their growth and activity. Consideration of these factors is necessary, because if they are outside the tolerance limits of the active microorganism, no biodegradation will occur.

Since groundwater and vapors follow the path of least resistance, regions in the subsurface with high permeability will become preferential flow paths. Regions with low permeability, such as clays and silts, will remain contaminated. These subsurface heterogeneities play important roles in the contaminated transport and the delivery of nutrients and electron acceptors in engineered systems as water is the primary delivery mechanism. Contaminated sites with a high degree of geological complexity are often poor candidates for in situ bioremediation.

11.4.2 Sorption and Bioavailability

Chemical transport, reactivity, and toxicity can be strongly influenced by the compound's interaction with the soil surfaces that exist in the environment. Sorption is defined as the uptake of a solute by a sorbent. Natural sorbents include soils, sediments, and microorganisms. Sorption includes the processes of adsorption of the solute onto surface or interior voids and the partitioning of the solute into an organic medium, usually organic coatings found on soils, sediments, and clays. Because these organic coatings tend to accumulate on charged surfaces, soils such as clays typically contain a disproportional amount of these coatings.

Chemicals that strongly sorb to soil may not transport in the unsaturated zone easily. That is why they are more likely to accrue in the surface soils and sediments. PCBs and PAHs are examples of such contaminants.

Because sorption strongly influences the bioavailability of an organic chemical, sorption intensity influences whether the chemical is readily accessible to microorganisms for biodegradation. In addition, the chemicals that have partitioned into a separate phase or present themselves as a separate phase, such as oil, non–aqueous-phase liquids, or dense non–aqueous-phase liquids, may also not be bioavailable.

11.4.3 Moisture Content

The microorganisms carrying out metabolic transformation require sufficient moisture for their growth and activity. Therefore, the drying of surface soils can harshly restrict biodegradation. For example, decrease in the moisture content will decrease the rate of degradation. The optimum moisture level will depend on the properties of the soil and contaminant. It will also depend on whether degradation is targeted under aerobic or anaerobic conditions, because an abundance of water may cause anaerobic conditions if there is an active enough microbial community.

11.5 BIOREMEDIATION

Bioremediation is a technique to manage waste that involves the use of organisms to remove or neutralize pollutants from a contaminated site. Bioremediation technologies can be generally classified as in situ or ex situ. The in situ bioremediation technique involves treating the contaminated material at the site; on the other hand, the ex situ technique involves the removal of the contaminated material to be treated somewhere else. Examples of bioremediation techniques are bioventing, phytoremediation, bioleaching, bioreactor, land farming, composting, rhizofiltration, bioaugmentation, and biostimulation.

Bioremediation may occur by its own, which is called natural attenuation or intrinsic bioremediation, or it may efficiently occur through the addition of fertilizers, oxygen, and so on that generally help in encouraging the growth of pollutant-eating microbes within the medium. Recent advancements have also been successful owing to the addition of matched microbe strains to the medium to enhance the resident microbe population's ability to break down contaminants. Microorganisms that are used to perform the function of bioremediation are known as bioremediators.

11.5.1 Biostimulation

Biostimulation is the modification of the environment to stimulate existing bacteria that are capable of bioremediation. This can be done by the addition of several forms of rate-limiting nutrients and electron acceptors, such as phosphorus, nitrogen, oxygen, or carbon additives. These additives are usually added to the subsurface through injection wells; injection well technology for biostimulation purposes is an emerging technology. Removal of the contaminated material is also one of the options but an expensive one. Biostimulation can be enhanced by bioaugmentation. This overall process is referred to as bioremediation and is an EPA-approved method.

The basic and primary advantage of biostimulation is that bioremediation will be undertaken by already present native microorganisms that are well suited to the subsurface environment and are well distributed within the subsurface. The disadvantage of biostimulation is that the delivery of additives allows the additives to be readily available to subsurface microorganisms and is based on the local geology of the subsurface. Tight, impermeable subsurface lithology makes it difficult to spread additives throughout the affected area. Fractures in the subsurface create preferential pathways in the subsurface that additives preferentially follow, thus preventing smooth distribution of additives.

Recently, a number of products that allow the use of bioremediation using biostimulative methods have been introduced. They may bind local bacteria using biostimulation by creating a hospitable environment for hydrocarbon-devouring microorganisms or may introduce foreign bacteria into the environment as a direct application to the hydrocarbon. While the jury is out as to whether either is particularly more effective than the other, prima facie consideration suggests that the introduction of foreign bacteria to any environment stands a chance of mutating organisms already present and affecting the biome.

Investigations that determine subsurface characteristics, such as hydraulic conductivity of the subsurface, natural groundwater velocity during ambient conditions, and lithology of the subsurface, are considered important in developing a successful biostimulation system. In addition, a pilot-scale study of the potential biostimulation system should be undertaken before full-scale design and implementation.

However, some biostimulative agents may be used in chaotic surfaces such as open water and sand so long as they are oleophilic, meaning that they bond exclusively to hydrocarbons and basically sink in the water column, bonding to oil, where they then float to the water's surface, exposing the hydrocarbon to more abundant sunlight and oxygen where greater microorganic aerobic activity can be encouraged. Some consumer-targeted biostimulants possess this quality and others do not.

11.5.2 Bioaugmentation

Bioaugmentation is defined as the process of addition of essential nutrients required to speed up the rate of degradation of contaminants. Usually, the steps involve studying the indigenous varieties present in the location to determine the possibility of biostimulation. If the indigenous variety does not have the metabolic capability to perform the remediation process, then exogenous varieties with sophisticated pathways are introduced.

Bioaugmentation is generally used in municipal wastewater treatment plants to restart activated sludge bioreactors. Activated sludge systems are generally based on microorganisms; for example, protozoa, bacteria, nematodes, fungi, and rotifers are able to degrade biodegradable organic matter.

11.5.3 Intrinsic Bioremediation

Intrinsic bioremediation is defined as the conversion of environmental pollutants into risk-free forms through the inherent capabilities of naturally occurring microbial population. However, there is increasing interest on intrinsic bioremediation for control of the contamination. The intrinsic capacity of microorganisms to metabolize the contaminants should be tested at the laboratory scale before use for intrinsic bioremediation. Through site supervision, the progress of intrinsic bioremediation should be recorded from time to time. The favorable conditions of sites for intrinsic bioremediation include groundwater flow throughout the year, supply of electron acceptors and nutrients for microbial growth, and absence of toxic elements. Other environmental factors like pH, concentration, temperature, and nutrient availability determine whether or not biotransformation takes place.

11.5.4 Land Farming

Land farming is defined as a bioremediation treatment process that is performed in the upper zone of soil or in biotreatment cells. Contaminated soils, sediments, and sludge are integrated into the soil surface and periodically turned over to aerate the same. This technique has been successfully used for a number of years for the management of oily sludge and other petroleum wastes. In situ systems have been used to take care of near surface soil contamination for hydrocarbons and pesticides. The equipment employed in land farming is typical of that used for agricultural purposes. These land farming activities cultivate and improve microbial degradation of harmful elements. As a rule of thumb, the higher the molecular weight, the slower the degradation rate, and the more chlorinated or nitrated the compound, the more difficult to degrade it.

11.5.5 Compost

Compost is defined as the organic matter that has been decomposed and recycled as a fertilizer for soil improvement. Compost is a key ingredient in organic farming. At the most basic level, the process of composting simply requires making a heap of wetted organic matter (leaves, "green" food waste) and waiting for the materials to break down into humus after a period of weeks or months. Modern, methodical composting is a multistep, closely monitored process with measured inputs of water, air, and carbon-rich and nitrogen-rich compounds. The decomposition process is assisted by shredding of yard waste, adding water, and ensuring appropriate aeration by regular mixing. Worms and fungi species further break down the matter. Aerobic bacteria manage the activity by converting the inputs into heat, CO_2, and ammonium. The ammonium is further converted into nitrites and nitrates through the nitrification process.

Compost can be rich in nutrients. It is mostly used in gardens, landscaping, and agriculture. The compost itself is beneficial to the soil in several ways, namely, for soil conditioning, as a fertilizer, and as a natural insect killer for soil. In ecosystems, compost is useful for erosion control, for land and stream recovery, for wetland construction, and as landfill cover. Organic ingredients intended for composting can also be used for biogas generation through anaerobic digestion. In some parts of the world, anaerobic digestion is fast overtaking composting as the primary means of treating waste organic matter.

11.5.6 Bioventing

Bioventing is defined as an in situ bioremediation technique that uses microorganisms to biodegrade organic constituents generally adsorbed in the groundwater. Bioventing enhances the activity of aboriginal bacteria and simulates the natural in situ biodegradation of hydrocarbons by inducing air/oxygen flow into the unsaturated zone or by adding nutrients. During bioventing, oxygen may be supplied through direct air injection into the remaining contamination in soil. Bioventing not only assists in the degradation of adsorbed fuel residuals but also assists in the degradation of volatile organic compounds.

11.5.7 Rhizofiltration

Rhizofiltration is a type of phytoremediation that refers to the approach of using hydroponically cultivated plant roots to remediate contaminated water through absorption, concentration, and precipitation of pollutants. It also filters through water and dirt. The contaminated water is either collected from a waste site or brought to the plants, or the plants are planted in the contaminated land, where the roots then take up the water and the contaminants dissolved in it. Many plant species naturally ingest heavy metals and excess nutrients for a variety of reasons such as sequestration, drought resistance, disposal by leaf abscission, interference with other plants, and protection against pathogens and herbivores. Some of these species can accumulate extraordinary amounts of these contaminants. Identification of such plant species has led environmental researchers to realize the potential for using these plants for remediation of contaminated soil and wastewater.

This process is very similar to phytoextraction in which contaminants are removed by trapping them into harvestable plant biomass. Both phytoextraction and rhizofiltration follow the same basic path for remediation. First, plants are put in contact with the contamination; then, they absorb contaminants through their root systems and store them in root biomass or transport them up into the stems or leaves. The plants continue to absorb contaminants until they are harvested. The plants are then replaced to continue the growth/harvest cycle until satisfactory levels of impurity are achieved. Both processes are also aimed more toward concentrating and precipitating heavy metals than organic pollutants. The major difference between rhizofiltration and phytoextraction is that rhizofiltration is used to treat aquatic environments while phytoextraction deals with soil remediation.

11.6 FACTORS AFFECTING BIOREMEDIATION

11.6.1 Energy Sources

One of the primary variables that influence the activity of bacteria is the ability and availability of reduced organic materials to serve as energy sources. Whether a contaminant will serve as an effective energy source for an aerobic heterotrophic organism is a function of the average oxidation state of the carbon in the material. In general, higher oxidation states correspond to lower energy yields, thus providing less energetic incentive for microorganism degradation. The outcome of each degradation process depends on microbial (e.g., biomass concentration, population diversity, and enzyme activities), substrate (e.g., physicochemical characteristics, molecular structure, and concentration), and a range of environmental factors such as pH, temperature, moisture content, E_h, availability of electron acceptors and carbon, and energy sources. These parameters

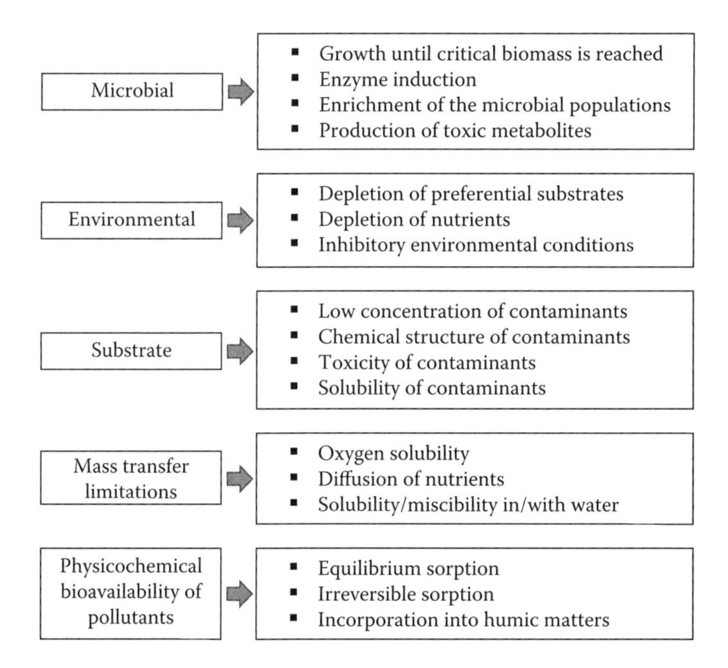

FIGURE 11.1 Factors affecting bioremediation.

affect the acclimation period of the microbes to the substrate. The molecular structure and contaminant concentration have been shown to strongly affect the possibility of bioremediation and the type of microbial alteration occurring and whether the compound will serve as a primary, secondary, or co-metabolic substrate (Figure 11.1).

11.6.2 Bioavailability

The rate at which microbial cells can convert contaminants during bioremediation depends on the rate of contaminant uptake and metabolism and the rate of transfer to the cell (mass transfer). Increased microbial conversion capacities do not lead to higher bioconversion rates when mass transfer is a limiting factor. This is the case in most contaminated soils and sediments (e.g., the contaminating explosives in soil did not undergo the biodegradation process even after 50 years). Treatments involving rigorous mixing of the soil and breaking up of the larger soil particles drastically stimulated biodegradation. The bioavailability of a contaminant is controlled by a number of physicochemical processes such as sorption and desorption, diffusion, and dissolution. A reduced bioavailability of contaminants in soil is caused by the slow mass transfer to the degrading microbes, and contaminants become unavailable when the rate of mass transfer is zero. The decrease of bioavailability in the course of time is often referred to as aging.

In nature, the ability of organisms to transfer contaminants to both simpler and more complex molecules is very dissimilar. In light of our current limited ability to measure and control biochemical pathways in complex environments, favorable or unfavorable biochemical conversions are estimated in terms of whether individual or groups of parent compounds are removed, whether increased toxicity is a result of the bioremediation process, and sometimes whether the elements in the parent compound are converted to quantifiable metabolites. These biochemical activities can be controlled in an in situ operation when one can manage and optimize the conditions to achieve a desirable result.

11.7 ADVANTAGES AND DISADVANTAGES OF BIOREMEDIATION

For bioremediation to be successful, one must have the right microbes in the right place with the right environmental factors for degradation to occur. The right microbes are bacteria or fungi, which have the physiological and metabolic abilities to degrade the pollutants. Bioremediation offers several advantages over conventional techniques such as land filling or incineration:

- Bioremediation can be done on site.
- Bioremediation is less expensive.
- Site disruption is negligible.
- Bioremediation eliminates waste permanently.
- Bioremediation eliminates long-term liability.
- Bioremediation has greater public acceptance.
- Bioremediation can also be coupled with other physical or chemical treatment methods.

Bioremediation has also its limitations:

- Some chemicals are not suitable for biodegradation (e.g., heavy metals, chlorinated compounds, etc.).
- In some cases, microbial metabolism of contaminants may produce poisonous metabolites.
- Bioremediation is a scientifically severe procedure that should be customized to site-specific situations.

REFERENCES

Alexander, M. 1999. *Biodegradation and Bioremediation*, 2nd Edition. Academic Press, San Diego.

Allard, A.S., and A.H. Neilson. 1997. Bioremediation of organic waste sites: A critical review of microbiological aspects. *Int Biodeter Biodegr* 39:253–285.

Norris, R.D. 1994. In-situ bioremediation of soils and groundwater contaminated with petroleum hydrocarbons. In *Handbook of Bioremediation* (J.E. Mathews, proj. officer). Lewis Publishers, Boca Raton, FL, pp. 17–37.

Verstraete, W. 2002. Environmental biotechnology for sustainability. *J Biotechnol* 94:93–100.

SUGGESTED READINGS

Adam, G., and H. Duncan. 2001. Development of a sensitive and rapid method for the measurement of total microbial activity using fluorescein diacetate (FDA) in a range of soils. *Soil Biol Biochem* 33:943–951.

Alexander, M. 1995. How toxic are toxic chemicals in soil? *Environ Sci Technol* 29:2713–2717.

Casida, L.E. 1977. Microbial metabolic activity in soil as measured by dehydrogenase determinations. *Appl Environ Microbiol* 34:630–636.

Claassens, S., Van Rensburg, L., Riedel, K.J., Bezuiddenhout, J.J., and P.J. Jansen Van. 2006. Evaluation of the efficiency of various commercial products for the bioremediation of hydrocarbon contaminated soil. *Environmentalist* 26:51–62.

Delille, D., Bassères, A., and A. Dessommess. 1998. Effectiveness of bioremediation for oil-polluted antarctic seawater. *Polar Biol* 19:237–241.

Gaspar, M.L., Cabello, M.N., Pollero, R., and M.A. Aon. 2001. Fluorescein diacetate hydrolysis as a measure of fungal biomass in soil. *Curr Microbiol* 42:339–344.

Gibson, D.T., and V. Subramanian. 1984. *Microbial Degradation of Organic Compounds*. Marcel Dekker Inc., New York, pp. 181–252.

Green, V.S., Stott, D.E., and M. Diack. 2006. Assay for fluorescein diacetate hydrolytic activity: Optimization for soil samples. *Soil Biol Biochem* 38:693–701.

Jackson, W.A., and J.H. Pardue. 1999. Potential for enhancement of biodegradation of crude oil in Louisiana salt marshes using nutrient amendments. *Water Air Soil Pollut* 109:343–355.

Kishino, T., and K. Kobayashi. 1995. Relation between toxicity and accumulation of chlorophenols at various pH, and their absorption mechanism in fish. *Water Res* 29:431–442.

Lee, S.H., Oh, B.I., and J. Kim. 2008. Effect of various amendments on heavy mineral oil bioremediation and soil microbial activity. *Bioresour Technol* 99:2578–2587.

Lundstedt, S., Haglund, P., and L. Öberg. 2003. Degradation and formation of polycyclic aromatic compounds during bio-slurry treatment of an aged gasworks soil. *Environ Toxicol Chem* 22:1413–1420.

Volkering, F., Quist, J.J., van Velsen, A.F.M., Thomassen, P.H.G., and M. Olijve. 1998. A rapid method for predicting the residual concentration after biological treatment of oil-polluted soil. In: Contaminated soil '98, Proceedings of the Sixth International FZK/TNO Conference on Contaminated Soil 17–21 May 1998, vol. 1. Edinburgh, UK, pp. 251–259.

Zilouei, H., Guieysse, B., and B. Mattiasson. 2008. Two-phase partitioning bioreactor for the biodegradation of high concentrations of pentachlorophenol using *Sphingobium chlorophenolicum* DSM 8671. *Chemosphere* 72:1788–1794.

12 Sustainable Manufacturing and Water Sustainability

Liwen Chen, Zexin Tian, and Helen H. Lou

CONTENTS

12.1 BACKGROUND

Manufacturing activities have been a crucial engine for US economy growth since the great recession in 2007–2009. With the steady increase of manufacturing productivity, it contributes to 26% of the total US economic growth and offers 11.8 million jobs, or 9% of total employment in 2011 [1]. Meanwhile, manufacturing accounts for a significant part of the world's consumption of resources and generation of waste. The energy consumption of manufacturing accounts for nearly a third of today's global usage, and it is responsible for 36% of global carbon dioxide emissions [2]. The scarcity of resources and environmental problems related to manufacturing triggers a significant paradigm shift from traditional manufacturing, which focused solely on the economic aspects, to a more integrated and holistic framework encompassing environmental and social aspects, such as green manufacturing and sustainable manufacturing.

12.2 HISTORY

There has been an increase in global attention toward environmental responsibility regarding manufacturing during the early 1990s, marked by the UN conference on environment and development, also known as "Earth Summit," held in 1992 in Rio de Janeiro, Brazil. This symbolic event underscored the role played by stakeholders with respect to environmental problems in the 300-page report known as Agenda 21 [3]. Another major achievement of this event is the birth of Kyoto Protocol, which sets binding obligations on industrialized countries to reduce the emissions of greenhouse gas. There are continuous developments from then on. In 1997, the Rio+5 held by the UN General Assembly appraised the status of Agenda 21 and identified the uneven progress. Rio+10, also known as the Johannesburg Declaration in 2002, affirmed full implementation of Agenda 21. In 2002, "Agenda 21 for culture" included culture as a new dimension in various subsections of Agenda 21 [4]. More recently, in 2012, Rio+20 reaffirmed their commitment to Agenda 21 in their outcome document called "The Future We Want" [5]. In this document, green economy in the context of sustainable development and poverty eradication was proposed as the theme of the conference.

On the other hand, US federal laws and regulations achieved three progressive stages regarding environmental protection since the early 1960s [6]:

Stage 1. Aims at end-of-pipe control of emissions and wastes (e.g., Clean Air Act [1963], Clean Water Act [1972], Resource Conservation and Recovery Act [1976])

Stage 2. Focuses on source reduction of pollution from industrial activities (Pollution Prevention Act [1990])

Stage 3. Encourages sustainable manufacturing by technological innovations to reduce the environmental and social impact of industrial production (National Technology Transfer and Advancement Act [1996])

Since the 1990s, new paradigms of manufacturing have also been raised. Lean manufacturing, green manufacturing, and sustainable manufacturing are the most frequently mentioned terms, which will be explained in the Section 12.3.

12.3 LEAN, GREEN, AND SUSTAINABLE MANUFACTURING

12.3.1 Lean Manufacturing

Lean manufacturing, which evolved from the Toyota Production System, was introduced to the United States in 1984. It was initially identified as a "Lean production" in the early 1990s, which aimed at creating more value for the customer with less generation of waste. Shah and Ward [7] developed a list of characteristics of Lean manufacturing:

1. *Supplier feedback*: provide regular feedback to suppliers about their performance
2. *Just-in-Time (JIT) delivery by suppliers*: ensure that suppliers deliver the right quantity at the right time in the right place
3. *Supplier development*: develop suppliers so that they can be more involved in the production process of the focal firm
4. *Customer involvement*: focus on a firm's customers and their needs
5. *Pull system*: facilitate JIT production including Kanban cards, which serve as a signal to start or stop production
6. *Continuous flow*: establish mechanisms that enable and ease the continuous flow of products
7. *Setup time reduction*: reduce process downtime between product changeovers
8. *Total productive/preventive maintenance*: address equipment downtime through total productive maintenance and thus achieve a high level of equipment availability
9. *Statistical process control*: ensure each process will supply defect-free units to subsequent process
10. *Employee involvement*: employees' role in problem solving, and their cross-functional character

These interrelated guidelines are designed for companies to achieve maximum profit, yet some of them may indirectly address the environment issue. For example, the pull system regulates the flow of resources in a manufacturing process by replacing only what has been consumed and only what is immediately deliverable. It was designed to reduce the inventory, work time, and human resources, which will also reduce waste released to the environment.

12.3.2 Green Manufacturing

Since 2000, the concept of green manufacturing, also called environmentally benign manufacturing or clean production, was introduced, explicitly incorporating environmental concern as part of the business model to enhance competitiveness, attributed to many external and internal reasons such as regulatory requirements, product stewardship, public image, and potential competitive advantages [8]. It requires continuous integration of environmental improvements of industrial processes or products to reduce or prevent the release of pollutants to the air, water, and land; to reduce and recycle waste; and to minimize health risks to human and other creatures [9]. Three major subsets of practices are pollution prevention [10], toxic use reduction [11], and design for environment [12].

1. *Pollution prevention*: to avoid or minimize waste and emissions through source reduction or on-site recycling
2. *Toxic use reduction*: to avoid or reduce the use of toxic substances during the process
3. *Design for environment*: to incorporate the environmental performance requirements in the product development, which involves analysis based on a life cycle perspective

12.3.3 Sustainable Manufacturing

Sustainable manufacturing, although often interchangeable with green manufacturing, incorporates economic efficiency, environmental sustainability, and social solidarity as the three dimensions commonly called the "triple bottom line" originated by John Elkington in 1994 [13]. Since then, this paradigm has been extensively enriched to incorporate economic, environmental, and societal evaluations into a holistic framework to analyze the sustainability at an enterprise level. According to Jawahir [14], manufacturing paradigms with more innovative elements would deliver more value to the stakeholder (Figure 12.1 [14]).

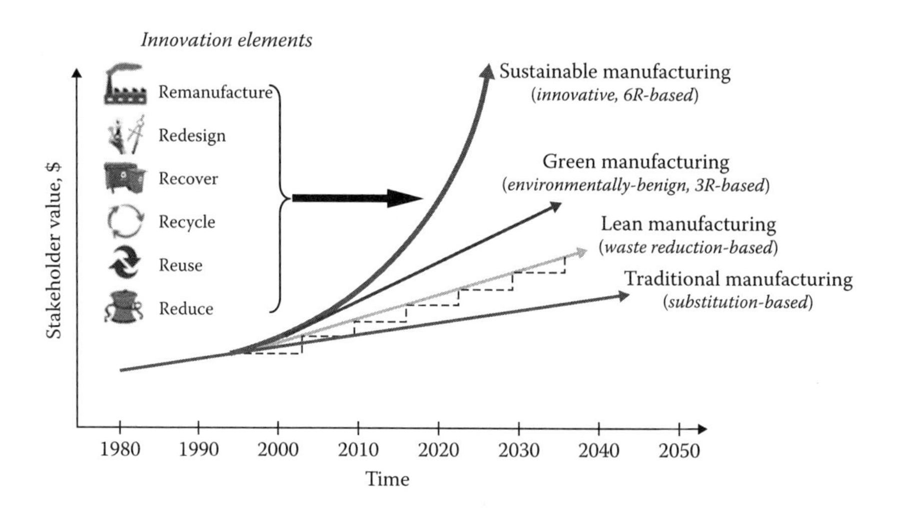

FIGURE 12.1 Comparisons of various manufacturing paradigms. (Reprinted with permission from Jawahir, I.S., *Sustainable manufacturing: The driving force for innovative products, processes and systems for next generation manufacturing*. 2011, College of Engineering, University of Kentucky: Lexington, KY.)

However, the implementation of sustainable manufacturing faced challenges from multiple perspectives such as economical, managerial, and technological. For instance, early sustainable practices that focused on the end-of-pipe pollution control were costly and thus lack economic incentives. The management's unwillingness to change together with the shortage of effective tools of measurement also hindered the progress toward sustainability. Fortunately, these challenges are gradually relieved by a shift in concept and business model, technological breakthroughs, and innovative tools to measure and facilitate sustainability.

Retrofit and renovation of existing manufacturing processes for sustainability purposes are normally considered to incur extra capital cost in the short term. However, research studies show that the extra cost can be well justified by the adoption of new sustainable practices. For example, Lou and Huang [15] proposed the "profitable pollution prevention (P3)" strategy focusing on source reduction of waste that can achieve pollution prevention and economic incentives as well. Also, many leading industrial companies have seen economic benefits by adopting sustainable manufacturing concepts. For example, the 3P Program (Pollution Prevention Pays) from 3M company, which aims for reduction of waste at the source, helped save more than $1 billion from 1975 to 2005 and prevented the release of more than 2.6 billion pounds of pollutants to the environment [16].

Managers from industrial manufacturers nowadays also raise increasing attention to sustainable practices. According to a global survey of 3000 companies in 2012 [17], 70% of respondents have placed sustainability permanently on their management agendas; two-thirds of them acknowledge that "sustainability is necessary for them to remain competitive on the marketplace." Also, in 2013, 72% of S&P 500 companies publish sustainability responsibility reports, while it was 52% and 20% in 2012 and 2011, respectively [18].

Process upgrade with less toxic raw material triggered by sustainable requirements may have huge potential for economic benefits. A striking example is DuPont's fluoro-product operation in the Netherlands. The novel thermal system could convert gaseous fluorocarbon waste to saleable aqueous hydrogen fluoride. The success helped DuPont avoid a $5 million end-of-pipe treatment fee, trimmed external disposal costs of $600,000, and decreased the equivalent of 12 billion pounds of carbon dioxide emissions per year [16].

The emergence of clear and consistent sustainability indicators and related software tools also strongly facilitate the evaluation and monitoring of sustainability for processes and products. Table 12.1 [19] lists the most common categories of sustainability indicators reported by the Organization for Economic Cooperation and Development [19]. Table 12.2 [20] lists the popular software tools on the market for sustainability measurements.

In the following sections, the water network (WN) synthesis technology will be introduced to demonstrate how it can be used to reduce water consumption. Then, several state-of-the-art water minimization strategies will be introduced for industrial cooling systems.

TABLE 12.1
A List of Categories of Sets of Indicators for Sustainable Manufacturing

Category	Description
Individual indicators	Measure single aspects individually
Key performance indicators (KPIs)	A limited number of indicators for measuring key aspects that are defined according to organizational goals
Composite indices	Synthesis of groups of individual indicators that is expressed by only a few indices
Material flow analysis (MFA)	A quantitative measure of the flows of materials and energy through a production process
Environmental accounting	Calculate environment-related costs and benefits in a similar way to financial accounting system
Eco-efficiency indicators	Ratio of environmental impacts to economic value created
Life cycle assessment (LCA) indicators	Measure environmental impacts from all stages of production and consumption of a product/service
Sustainability reporting indicators	A range of indicators for corporate nonfinancial performance to stakeholders
Socially responsible investment (SRI) indices	Indices set and used by the financial community to benchmark corporate sustainability performance

Source: *Sustainable manufacturing and eco-innovation synthesis report.* Organisation for Economic Co-Operation and Development (OECD) 2009 [cited February 24, 2013]; Available from: http://www.oecd.org/sti/inno/43423689.pdf.

12.4 WATER NETWORK

12.4.1 Introduction

In industrial processes, the quality and quantity of inlet and outlet water need to meet certain requirements in order to guarantee the quality of the product and meet the environmental requirement. The WN might become rather complicated when various units and interconnected streams are involved. Since the 1970s, research in the WN synthesis has been proven to be very useful to obtain the optimized WN that reduces freshwater intake and wastewater generation, especially in water-demanding industrial processes, such as electroplating, papermaking, refinery, and so on. The water intake ratio (i.e., theoretical demand of water needed without recycling divided by the actual consumption of water) has maintained a 10% growth each year because of the adoption of water-saving strategies among the US industries [21].

12.4.1.1 Reuse, Recycle, Regeneration

Before moving on, it is necessary to identify the difference between terminologies such as *saving*, *reuse*, *recycle*, and *regeneration* in the WN synthesis technology, although these terms more or less refer to the reduction of water use. *Saving* means the source reduction of water intake, while the rest means the repetitive utilization of water that already exists in the process. Although interchangeable in many

TABLE 12.2
A List of Software Tools for Sustainability Evaluation for Manufacturing

Software	Description	Source
ChemSTEER	Estimates occupational inhalation and dermal exposure to a chemical during industrial and commercial manufacturing, processing, and use operations involving the chemical	EPA
Eco-indicator	A damage-oriented method for Life Cycle Impact Assessment	PER
ECO-it	Uses Eco-indicator scores to express the environmental performance of a product's life cycle	PER
E-FAST	Provides screening-level estimates of the concentrations of chemicals released to air, surface water, landfills, and from consumer products	EPA
EPI SUITE	A Windows-based suite of physical/chemical property and environmental fate estimation models	EPA
Gabi4	Provides solutions for different problems regarding cost, environment, social and technical criteria, optimization of processes, and managing your external representation in these fields	PE Europe GMBH Life Cycle Engineering, IKP University of Stuttgart
IGEMS	Brings together in one system several EPA environmental fate and transport models and some of the environmental data needed to run them	EPA
LCA	Systematically describes and assesses all flows to and from nature, from a cradle-to-grave perspective	PER
MCCEM	Estimates average and peak indoor air concentrations of chemicals released from products or materials in houses, apartments, townhouses, or other residences	EPA and Versar Inc.
ReachScan	Estimates surface water chemical concentrations at drinking water utilities downstream from industrial facilities serving as a database for the identification of facilities and utilities	EPA and Versar Inc.
SDLC	Performs activities such as System/Information Engineering and Modeling, Software Requirements Analysis, Systems Analysis and Design, Code Generation, Testing and Maintenance	Stylus Systems Inc.
SimaPro	To collect, analyze, and monitor the environmental performance of products and services	PER
SRD	Performs a systematic screening-level review of more than 12,000 potential indoor pollution sources to identify high-priority product and material categories for further evaluation	EPA and Versar Inc.
UCSS	Identifies and screens clusters of chemicals ("use clusters") that are used to perform a particular task	EPA

(Continued)

TABLE 12.2 (CONTINUED)
A List of Software Tools for Sustainability Evaluation for Manufacturing

Software	Description	Source
Umberto	Visualizes material and energy flow systems for advanced process, flow, and cost modeling	German ifu Hamburg GmbH in cooperation with Ifeu. and PER
WPEM	Estimates the potential exposure of consumers and workers to the chemicals emitted from wall paint	EPA and Versar Inc.

Source: Reprinted from *Chem. Eng. J.*, **133**, J. García-Serna, L.P.-B., and M.J. Cocero, New trends for design towards sustainability in chemical engineering: Green engineering, 7–30, Copyright (2007), with permission from Elsevier Science Ltd.

occasions, *reuse* means that the water source that exits from a process can be reutilized in another process but does not allow reentry to the previous process. However, *recycle* allows the reentry of the water source to the process where it has previously been used. *Regeneration* means the use of treatment to make the water quality suitable for successive processes. Figure 12.2 [22] shows the differences of these terminologies in the WN synthesis.

The WN synthesis is generally categorized into two main classes: *insight-based* (graphical) method and *optimization-based* (mathematical) method [23]. The insight-based method is based on the water pinch analysis (WPA) from the seminal work by Wang and Smith in 1994 [24], which was further inspired by the similar pinch analysis for heat exchange [25] and mass exchange network synthesis [26].

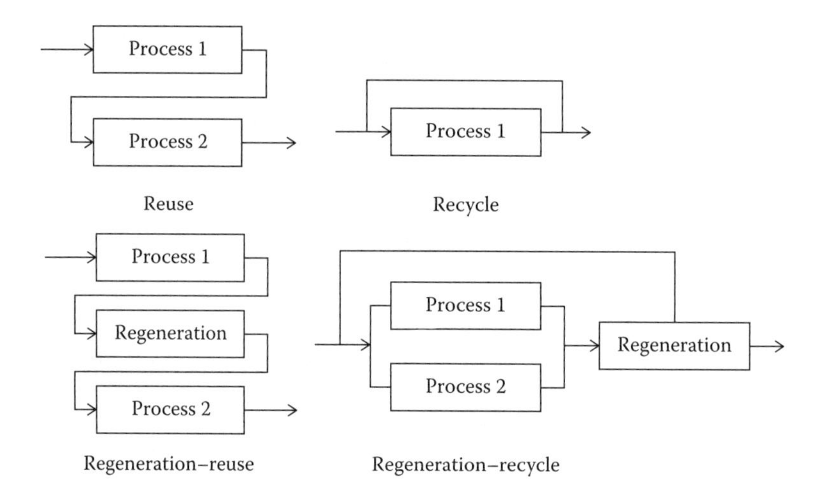

FIGURE 12.2 Reuse, recycle, regeneration–reuse, and regeneration–recycle. (Reprinted with permission from Foo, D.C.Y., State-of-the-art review of pinch analysis techniques for water network synthesis. *Ind. Eng. Chem. Res.*, 2009. 48: 5125–5159. Copyright [2009] American Chemical Society.)

A typical WPA consists of two sequential stages: the *flow rate targeting* stage where performance of the recovery system is predicted by first principle–based diagrams (e.g., limiting composite curve [24,27], surplus diagram [28], material recovery pinch diagram [29]) and the *network design* stage where resources are systematically allocated between process streams that contain them (sources) and process units that require these resources (sinks) [22]. The use of graphics such as the limiting composite diagram offers a visualization tool that is easy to master compared to the optimization-based method; however, many limitations of the former, such as the lack of computational effectiveness to guarantee global optimality and the system being confined with mostly one or two contaminants, prevent its application to more sophisticated real-world cases [23].

Optimization-based methods, on the other hand, are able to deal with both single and multiple contaminants, and targeting and design could be optimized simultaneously [30], taking advantage of the computational strength of modern computers. Among the early contributions, the seminal work by Takama et al. [31] in 1980 first demonstrated the usage of the nonlinear programming (NLP) model to solve the WN problem in the petroleum refinery, where both water-using processes and treatment units for a multicontaminant system were addressed. Since then, many authors presented linear or nonlinear optimization models with targeting objectives such as minimum freshwater usage for single [32] and multiple contaminants [33], simultaneous optimal freshwater consumption and wastewater treatment capacity [34], simultaneous heat and water integration for single [35] and multiple contaminants [36], simultaneous water targeting and flowsheet optimization [37], and so on. Although it was less popular among engineering practitioners because of the difficulty in mastering the technique, optimization-based methods overcome many shortcomings of insight-based methods and are becoming dominant in the WN synthesis field [38]. Furthermore, the aforementioned two approaches can be used together in a synergistic way to provide better engineering understanding through visualization and to handle complex real-world problems [39].

12.4.2 Insight-Based Method

12.4.2.1 Flow Rate Targeting

Many graphical and numerical targeting techniques have been developed since the origination of WPA. There are two main categories of WN synthesis cases: *fixed load* and *fixed flow rate* problems. In *fixed load* problems, water is regarded as a mass separating agent (i.e., the lean stream) to remove certain amount of impurity load from the rich steam in the mass transfer–based operations (Figure 12.3a [22]) such as surface cleaning in electroplating industries, pulp washing in paper mill, and desalter in refinery [40]. The major concern here is that the impurity load and the water flow rate are secondary. Hence, the inlet and outlet water flow rate in the process are assumed to be identical; in other words, no water loss or gain is considered in such processes.

On the other hand, in *fixed flow rate* problems, the inlet and outlet water flow rate may vary significantly in the non–mass transfer–based operations such as chemical reactors, boilers, or cooling towers, where water is being used for other functions besides as a mass separating agent, such as chemical reactions, evaporation, and condensation. In such cases, the flow rate becomes the major interest and it can be calculated by

$$F_p = \frac{\Delta m_p}{(C_{out} - C_{in})}. \tag{12.1}$$

In Equation 12.1, Δm_p is the impurity removal load and C_{in} and C_{out} are the maximum impurity concentrations of water inlet (sink) and outlet (source) as shown in Figure 12.3b [22]. This equation can also be used to convert the fixed load problem to a fixed flow rate problem from the limiting water data. It is worth mentioning that

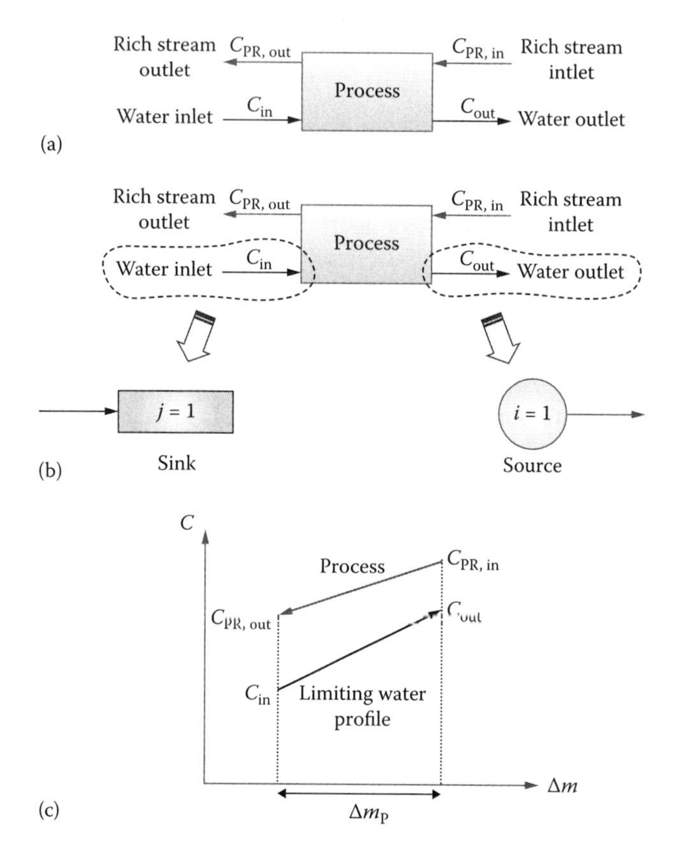

FIGURE 12.3 (a) The fixed load problem where water is the mass separating agent. (b) Conversion of a fixed load problem to a fixed flow rate problem. (c) Limiting water profile constructed by the maximum water inlet and outlet concentrations of the process. (Reprinted with permission from Foo, D.C.Y., State-of-the-art review of pinch analysis techniques for water network synthesis. *Ind. Eng. Chem. Res.*, 2009. 48: 5125–5159. Copyright [2009] American Chemical Society.)

the objective could be the same for both fixed load and fixed flow rate problems, that is, to minimize the flow rate of the freshwater source(s).

Then, we would show how to construct the limiting composite curve for a fixed load problem with a single freshwater source, which is the simplest of its kind. First, the limiting water data for all the processes are arranged in ascending order of quality level in a table as shown in Table 12.3 (data taken from Wang and Smith 1994 [24]). Second, the limiting water profile diagram for each individual process is plotted as contaminant concentration (C, ppm) versus. contaminant load diagram (Δm, kg/h) with ascending order of their concentration levels (Figure 12.4a). Third, connect the arrow of the previous process to the tail of the next process in each concentration intervals to form the limiting composite curve (Figure 12.4b [22]) that represents the overall WN system. Finally, draw the water supply line, which begins

TABLE 12.3
The Limiting Water Data

Process, Pp	Δm_p (kg/h)	C_{in} (ppm)	C_{out} (ppm)	F_p (ton/h)
1	2	0	100	20
2	5	50	100	100
3	30	50	800	40
4	4	400	800	10

Source: Reprinted with permission from Wang, Y.P. and R. Smith, Wastewater minimisation. *Chem. Eng. Sci.*, 1994. **49**: 981–1006. Copyright (1994) American Chemical Society.

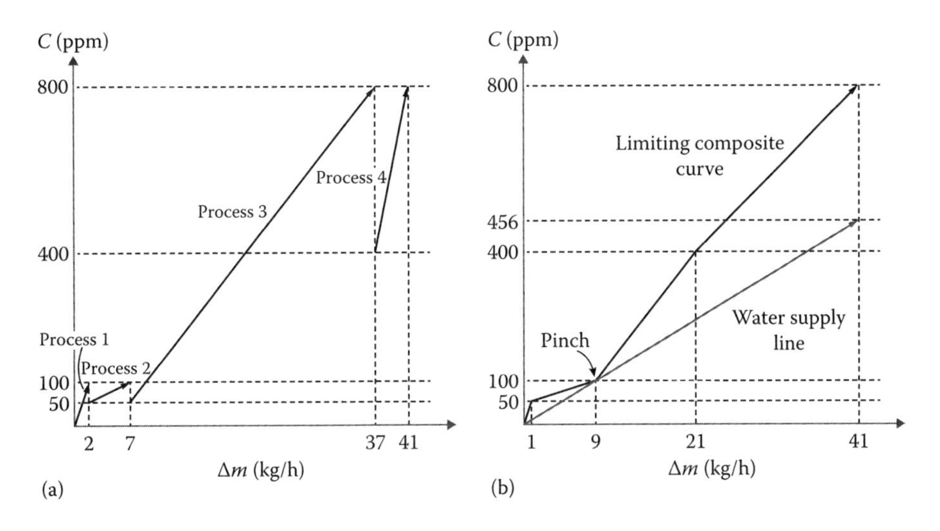

FIGURE 12.4 (a) Limiting water profile diagram. (b) Limiting composite curve. (Reprinted with permission from Foo, D.C.Y., State-of-the-art review of pinch analysis techniques for water network synthesis. *Ind. Eng. Chem. Res.*, 2009. 48: 5125–5159. Copyright [2009] American Chemical Society.)

from the y intercept and through the pinch point, that is, the lowest point on the limiting composite curve. The minimum flow rate of freshwater feed can be targeted by the inverse slope of the water supply line, that is, 90×10^3 kg/h. Also, the amount of water that can be directly reused or recycled is determined by the subtraction of the inverse slope value from the total cumulative water, that is, 80×10^3 kg/h. The corresponding composition at the pinch point, that is, 100 ppm, represents the location where freshwater is needed within the network to meet the target. For more complicated problems, such as targeting for multiple freshwater sources, impure freshwater sources, recycle with water regeneration, and so on, readers can refer to the comprehensive review by Foo [22].

12.4.2.2 Network Design

The second stage for the insight-based method is to design the WN once the minimum freshwater flow rate is determined by the pinch analysis. Among the various network design techniques, approximately half of them are based on the minimum flow rates established in the previous targeting stage, such as the water grid diagram [24], the water main method [41], the mass content diagram [42], the nearest neighbor algorithm (NAA) [43], and the network allocation diagram [44]. The other half are independent from the minimum flow rate targets and thus can be used without knowing the minimum flow rate target, such as the load table [45], the water source diagram [46], the source–sink mapping diagram [47,48], the source demand approach [49], and so on. Because of space constraints, only the NAA will be briefly introduced.

As we mentioned before, the *source* stream refers to any stream that exits any unit operation, and the *sink* stream refers to those that enter the operation. The basic principle of NAA is stated as, "To satisfy a sink, the source to be chosen are the nearest available neighbors to the sink in terms of contaminant concentration" [43]. In other words, two sources having the concentration level just higher and just lower than the sink (neighbors) are used to satisfy the flow rate and load requirement of the sink determined by the material balance equations. When the amount of a neighboring source is not sufficient for the sink, the next nearby source is used to satisfy the sink.

We now use the mathematical language to describe the NNA algorithm. Suppose a fixed load problem has n sources (S_1–S_n) and m sinks (D_1–D_m) numbered in the order of increasing contaminant concentration. Freshwater is identified as a source and is numbered S_0 accordingly. To fulfill the sink D_p with the principle of nearest neighbors, S_k and $S_{(k+1)}$ with contaminant concentration just below and above the concentration of D_p are selected. The flow rates of the sources are determined by solving the overall material balance and the contaminant material balance equations given below (Equations 12.2 and 12.3) simultaneously:

$$F_{Sk,Dp} + F_{S(k+1),Dp} = F_{Dp} \tag{12.2}$$

$$F_{Sk,Dp}C_{Sk} + F_{S(k+1),Dp}C_{S(k+1)} = F_{Dp}C_{Dp} \tag{12.3}$$

For situations where the nearest sources are not sufficient to fulfill the sink requirement, that is, if $F_{Sk,Dp} \geq F_{Sk}$, then the next nearest sources $S_{(k-1)}$ is used to satisfy the remaining requirement; similarly, if $F_{S(k+1),Dp} \geq F_{S(k+1)}$, then $S_{(k+2)}$ is used. In general, if S_s is the cleanest source to be used and S_t is the dirtiest source to be used, then the required flow rate of S_s and S_t for the sink D_p is given by

$$F_{Ss,Dp} + F_{St,Dp} = F_{Dp} - \sum_{i=s+1}^{i=t-1} F_{Si,Dp} \tag{12.4}$$

$$F_{Ss,Dp}C_{Ss} + F_{St,Dp}C_{St} = F_{Dp}C_{Dp} - \sum_{i=s+1}^{i=t-1} F_{Si,Dp}C_{Si} \tag{12.5}$$

Steps for synthesizing a maximum recovery network using the NNA algorithm are summarized in Figure 12.5 [43] and listed as follows:

1. Arrange the sources (S_1–S_m) and the sinks (D_1–D_n) (including the targeted minimum freshwater flow rate obtained from the previous water supply line) in an ascending order of contaminant level (lowest contaminant first), as shown in Table 12.4, which is based on the previous limiting water data (Table 12.3), and start the evaluation from D_1 ($p = 1$).
2. Find the source S_k with the same concentration as the sink D_p, that is, $C_{Sk} = C_{Dp}$, If yes, go to step 3; if no, go to Step 4.
3. Feed the source to the sink when $C_{Sk} = C_{Dp}$:
   a. If $F_{sk} \geq F_{Dp}$, the source is sufficient to satisfy the sink; update $F_{Sk} = F_{Sk} - F_{Dp}$, and go to Step 2 for the next sink ($p = p + 1$).
   b. If $F_{sk} < F_{Dp}$, feed the whole source to the sink (update $F_{Sk} = 0$), and replace s by $(k - 1)$ and t by $(k + 1)$ in Equations 12.4 and 12.5. Calculate $F_{Ss,Dp}$ and $F_{St,Dp}$. Go to Step 5.
4. Select S_k with contaminant concentration just below that of the D_p. Replace s by k and t by $(k + 1)$ in Equations 12.4 and 12.5. Calculate $F_{Ss,Dp}$ and $F_{St,Dp}$.
5. If both $F_{Ss,Dp}$ and $F_{St,Dp}$ are less than F_{Ss} and F_{St}, respectively, then the whole sink is satisfied. Update $F_{Ss} = F_{Ss} - F_{Ss,Dp}$, $F_{St} = F_{St} - F_{St,Dp}$, and $p = p + 1$. Go to Step 2.

   If $F_{Ss,Dp} > F_{Ss}$, then use the whole S_s ($F_{Ss} = 0$) and replace s by $(s - 1)$. If $F_{St,Dp} > F_{St}$, then use the whole S_t ($F_{St} = 0$) and replace t by $(t + 1)$. Solve Equations 12.4 and 12.5 again with the new s and t. Repeat this step until the whole sink is met. Update $p = p + 1$, and go to Step 2.

Stop when all the sinks are satisfied (i.e., $p = m$). The corresponding minimum WN is shown in Figure 12.6.

12.4.3 Optimization-Based Method

To use the optimization-based method, a superstructure encompassing all possible flow configurations within the process boundary needs to be set up as the first step.

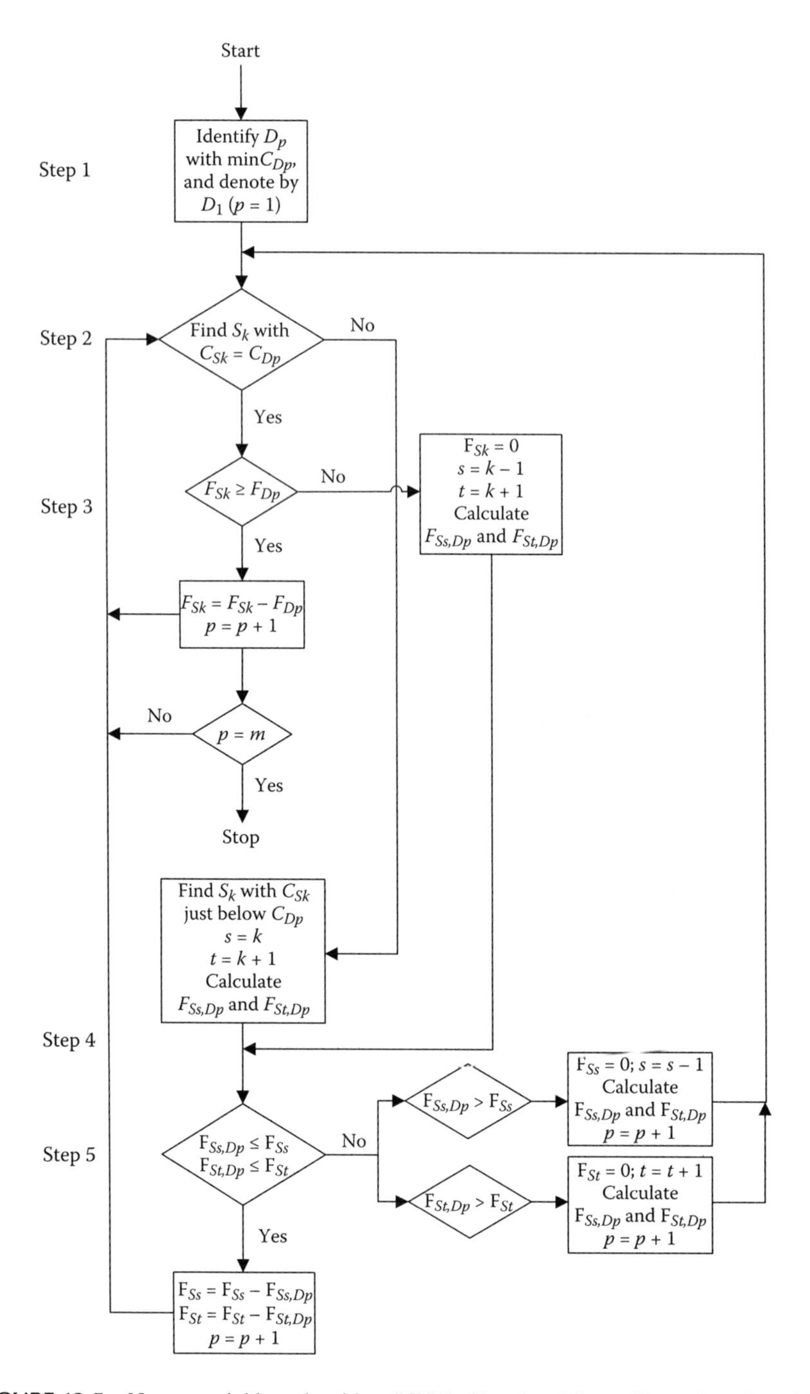

FIGURE 12.5 Nearest neighbor algorithm (NNA). (Reprinted from *Chem. Eng. Sci.*, 60(1), Prakash, R. and U.V. Shenoy, Targeting and design of water networks for fixed flowrate and fixed contaminant load operations, 255–268, Copyright [2005], with permission from Elsevier Science Ltd.)

TABLE 12.4
The Source and Sink Data Converted from the Limiting Water Data

	Contaminant Concentration (ppm)	Flow Rate (t/h)	Contaminant Load (kg/h)	Cumulative Flow Rate (t/h)	Cumulative Load (kg/h)
Sources					
Freshwater	0	90	0	90	0
S1	100	20	2	110	2
S2	100	100	10	210	12
S3	800	40	32	250	44
S4	800	10	8	260	52
Sinks					
D1	0	20	0	20	0
D2	50	100	5	120	5
D3	50	40	2	160	7
D4	400	10	4	170	11

Source: Reprinted from *Chem. Eng. Sci.*, **60**(1), Prakash, R. and U.V. Shenoy, Targeting and design of water networks for fixed flowrate and fixed contaminant load operations, 255–268, Copyright (2005), with permission from Elsevier Science Ltd.

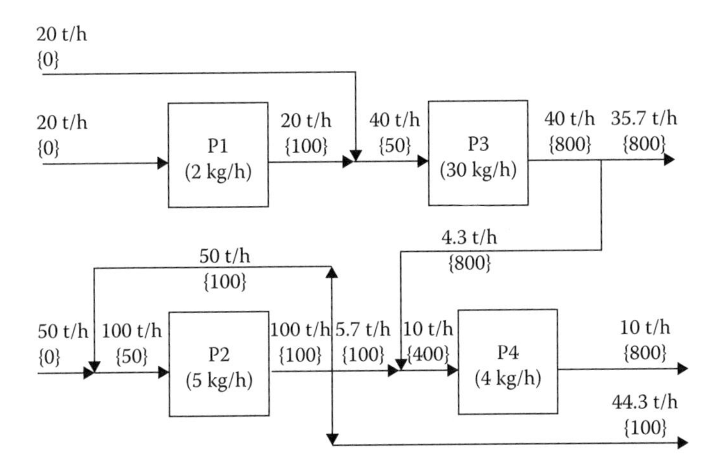

FIGURE 12.6 A minimum freshwater network by NNA (contaminant concentrations [ppm], contaminant loads [kg/h], and flow rates [t/h]). (Reprinted from *Chem. Eng. Sci.*, 60(1), Prakash, R. and U.V. Shenoy, Targeting and design of water networks for fixed flowrate and fixed contaminant load operations, 255–268, Copyright [2005], with permission from Elsevier Science Ltd.)

As an example, a general superstructure constructed by Huang et al. [34], which can deal with water losses as well as multiple sources and sinks commonly encountered in real-world cases, will be introduced. A general superstructure for water usage and treatment networks is shown in Figure 12.7 [34]. The related procedure for constructing such a superstructure is listed below:

1. Place a mixer node (M) at the inlet of every water-using unit (U) and water-treatment unit (T).
2. Place a mixer node before discharge to each of the type A sink (M^A, wastewater can be discharged to the environment).
3. Place a mixer node before discharge to each of the type B sink (M^B, wastewater require treatment).
4. Place two mixer nodes to collect loss streams. The operation losses from all water-using units are connected to one, and those from the water-treatment units are connected to the other.
5. Place a splitter node (S^P) after each primary source (most dependable, such as rivers, lakes). The split branches of every such node are connected to all mixer nodes established in Step 1.
6. Place a splitter node (S^S) after each secondary source (less dependable, such as generated by reaction). The split branches of every such node are connected to all mixer nodes established in Steps 1–3.
7. Place a splitter node (S) at the exit of every water-using and water-treatment unit. The split branches of every such node are connected to all of the mixer nodes installed in Steps 1–3 except the one before the same unit.

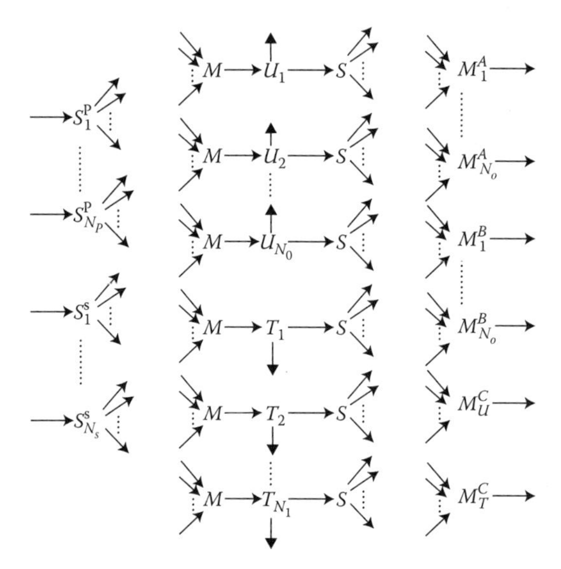

FIGURE 12.7 A general superstructure for water usage and treatment networks. (Reprinted with permission from Huang, C.H. et al., A mathematical programming model for water usage and treatment network design. *Ind. Eng. Chem. Res.*, 1999. 38: 2666–2679. Copyright [1999] American Chemical Society.)

The second step is to build an NLP model to find the optimal WN design. Equality constraints used in the model are simply water and solute mass balances, and inequality constraints are set for wastewater flow rate or pollutant concentrations after the mixers (M^As and M^Bs) in order to satisfy environmental requirements. It is assumed that the solute concentrations in each primary and secondary source are already given. The objective function for minimum freshwater use is represented as

$$F^W = \sum_{p \in I} W_p^P, \tag{12.6}$$

where W_p^P is the consumption rate of the pth primary water. The resulting NLP model can be solved with commercial software packages, such as GAMS [50]. For the aforementioned example from Wang and Smith (Table 12.3), the minimum water consumption rate is found to be 90 tons/h, which is the same as obtained through the insight-based method. The corresponding superstructure and alternative network structure are shown in Figures 12.8 and 12.9 [34]. It is worth mentioning that the equally acceptable alternative structures (Figure 12.9) can be easily generated by introducing perturbations to the initial guesses and solving the problem repeatedly.

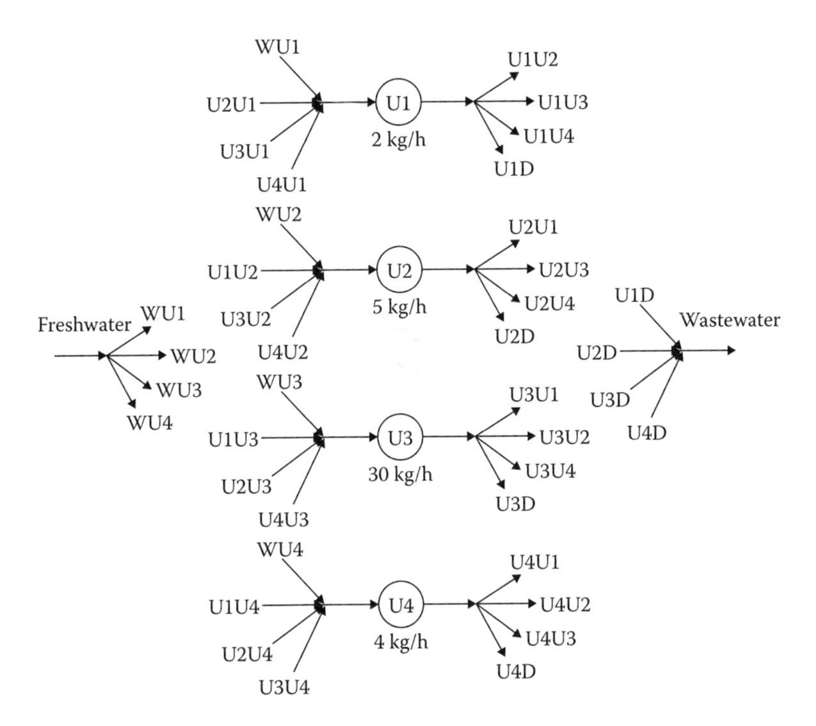

FIGURE 12.8 The superstructure for the example from Wang and Smith. (Reprinted with permission from Huang, C.H. et al., A mathematical programming model for water usage and treatment network design. *Ind. Eng. Chem. Res.*, 1999. 38: 2666–2679. Copyright [1999] American Chemical Society.)

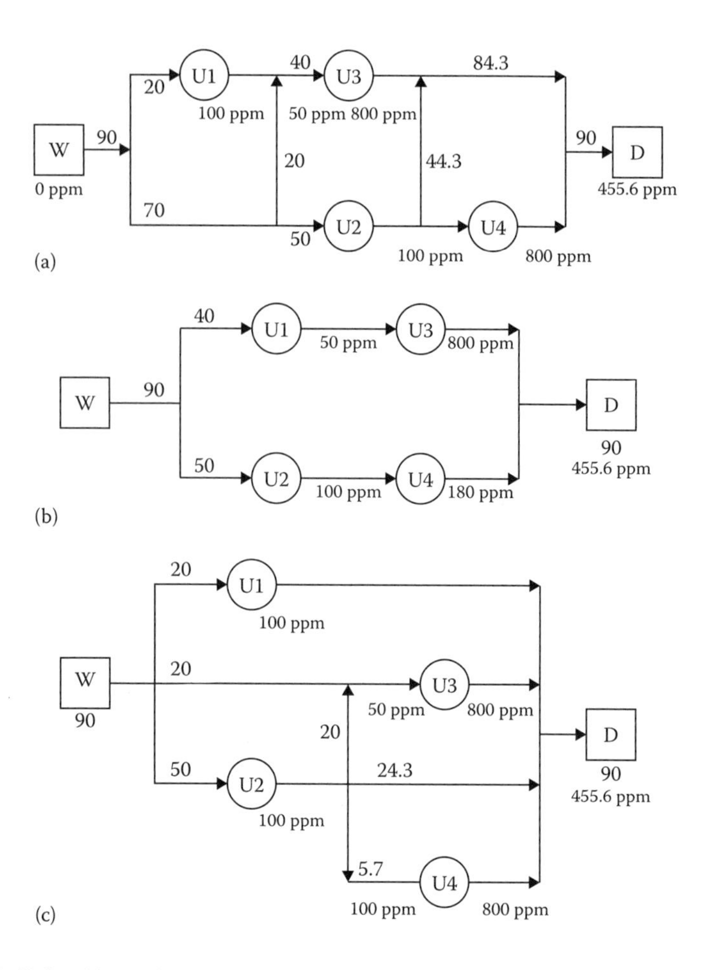

FIGURE 12.9 Alternative options for the optimized water usage and treatment network (WUTN) for the example from Wang and Smith: (a) option 1, (b) option 2, (c) option 3. (Reprinted with permission from Huang, C.H. et al., A mathematical programming model for water usage and treatment network design. *Ind. Eng. Chem. Res.*, 1999. 38: 2666–2679. Copyright [1999] American Chemical Society.)

12.4.4 Switchable Water Allocation Network

Some industrial processes involve characteristic plant dynamics that may be incorporated to the WN design as well. For example, in the water-demanding electroplating industry, the rinse tank runs in two operating modes and repeated in cycles: (i) a rinse mode in which parts are rinsed in tanks and (ii) an idle mode in which the rinse water in the tank is replenished to ensure that the rinse water quality meets the requirements for the next rinse job [51]. However, the duration of the idle mode for some rinse tanks may be too long and may cause excessive freshwater intake. The introduction of a dynamic model to control the cutoff and resume strategy may help save freshwater intake significantly.

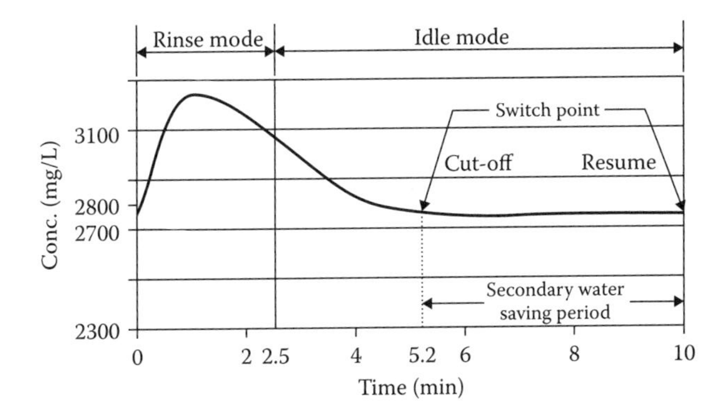

FIGURE 12.10 The switch of two water allocation networks (SWAN). (Reprinted with permission from Zhou, Q., H.H. Lou, and Y.L. Huang, Design of a switchable water allocation network based on process dynamics. *Ind. Eng. Chem. Res.*, 2001. 40: 4866–4873. Copyright [2001] American Chemical Society.)

On the basis of the optimization-based method, Zhou et al. [51] proposed a "switchable water allocation network (SWAN)" addressing the process dynamics by switching from a primary WAN design to a secondary WAN when operations for rinse tanks switch from the rinse mode to the idle mode as shown in Figure 12.10 [51]. The secondary WAN design is used because of the exclusion of those tanks whose inlet water could be safely cut off during the secondary water-saving period, while all rinse tanks are considered in the primary WAN. Both the primary and secondary WAN are modeled as mixed-integer nonlinear programming problems and solved by GAMS.

Even though the solution looks complicated to be implemented in the real world, as shown in the demonstrated electroplating line with six rinse tanks, the implementation of SWAN is surprisingly easy: just add four valves to control two rinse tanks [51]. Overall, the structural modification according to the primary WAN leads to a 33.3% of freshwater savings (from 960 to 640 gal/h). The inclusion of the full SWAN (primary and secondary WAN) further leads to a savings of 40.6%. The economic comparison also shows that the total annualized costs can be reduced by 39.3% through the implementation of SWAN.

12.5 WATER MINIMIZATION IN COOLING SYSTEMS

12.5.1 Background

The bulk of freshwater use in the United States is for irrigation (39%) and thermoelectric power generation (38%–39%), and 85%–90% of the power plant's freshwater is used for cooling [52]. Because of the steady increase in population and the corresponding energy demand, the freshwater demand is estimated to increase by 50% by 2030 [53]. The resulting fierce water competition among different users results in increasingly stringent restriction on cooling water use in thermoelectric power plants, which can be represented by the cooling constraint index as shown in

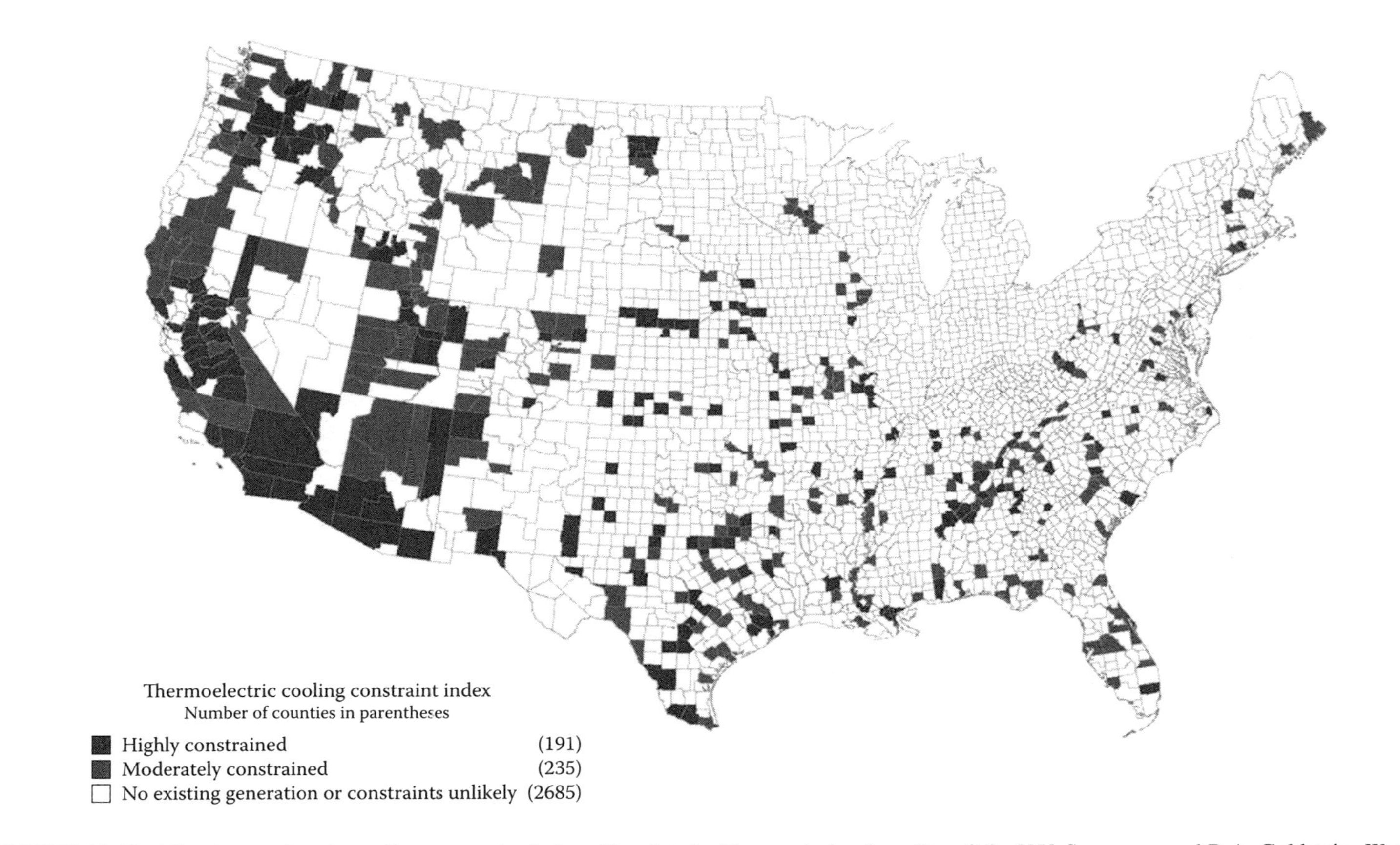

FIGURE 12.11 The thermoelectric cooling constraint Index. (Reprinted with permission from Roy, S.B., K.V. Summers, and R.A. Goldstein, *Water Resources Update*, 2003 (126): 94–99.)

Figure 12.11 [54]. Indeed, the lack of available water has already prevented the siting and permitting of new power plants in some regions [55]. In addition, Section 316(b) of the Clean Water Act [56] limits the amount of freshwater withdrawal by power plants, thereby putting challenges to explore new cooling water saving strategies.

As the working fluid in a typical steam power plant, water is converted to steam to drive the turbine and generate electricity, and is then condensed in the steam condenser by the cooling media, such as air, water, or other fluid [57]. The colder inlet temperature of the cooling water to the condenser results in lower steam condensation temperature, lower turbine back-pressure, and, consequently, higher power generation efficiency. The cooling water inlet temperature to the condenser is normally limited by the ambient wet or dry bulb temperature. The general types of cooling systems are listed below:

- Once-through cooling
- Wet recirculating cooling (cooling tower)
- Dry cooling
- Hybrid cooling

12.5.2 Once-Through Cooling

The once-through cooling systems draw surface water from lakes, rivers, or the ocean for one-time cooling and then discharge the heated water back to the water body. It is gradually phased out because of environmental problems, such as the increase in local water temperature. Furthermore, the construction of a once-through cooling system is highly restricted in many states in the United States according to the Clean Water Act 316(b) [58]. In addition, the intensity of water consumption for once-through cooling is pretty high (20–50 gal/kWh).

12.5.3 Wet Recirculating Cooling

Different from once-through cooling, warm water from the steam condenser is transferred to the wet recirculating cooling systems, often known as cooling towers, and exposed to ambient air for cooling. The classification of the cooling towers can be based on the direction of air flow (counterflow or cross-flow) or the type of draft (mechanical or natural). In a typical counterflow cooling tower, warm water is sprayed downward and evaporates at the ambient wet bulb temperature, while its heat is absorbed by the upward air flow through evaporation (Figure 12.12 [59]). Because of the water loss by evaporation or drift (mist or small droplets), the dissolved solids and suspended particles in the water may fall out of solution and cause scaling and fouling that reduce the thermal efficiency. Therefore, freshwater makeup and concentrate blowdown are needed to compensate for the loss and maintain the quality. Since water continues to recycle in the system, the intensity of water consumption in the cooling tower is only 0.3–0.6 gal/kWh [57]. When sufficient land is available, a cooling pond is constructed based on a similar mechanism to the cooling tower but relies on the natural heat transfer from the water to the atmosphere.

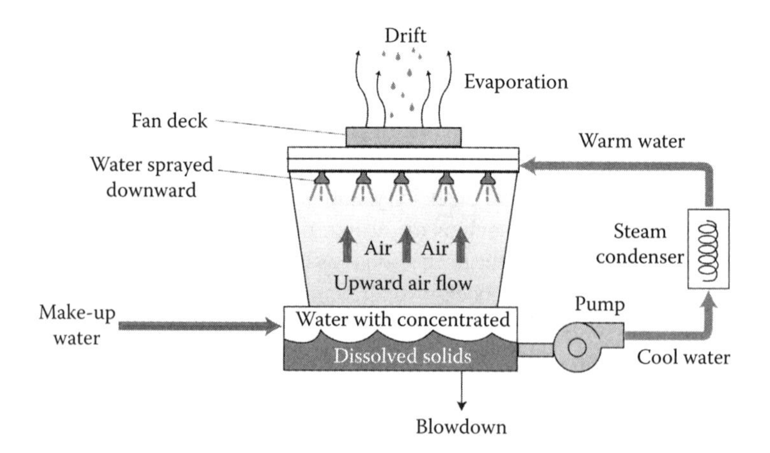

FIGURE 12.12 Illustration of the water flow in a typical cooling tower. (From Federal Energy Management Program, *Cooling towers: Understanding key components of cooling towers and how to improve water efficiency*. 2011, U.S. Department of Energy.)

12.5.3.1 Advanced Dew-Point Cooling Tower Fill

To improve the power efficiency and save cooling water, the fill is normally used below the spraying nozzles in the cooling tower to expand the water–air surface area, which can lower the inlet temperature of the steam condenser. An older type of fill is called splash bars, which serves to break the falling water into tiny droplets. In recent years, different forms of closely packed film fill were introduced to make water travel in thin streams, and it demonstrated superior thermal efficiency and evaporation rate compared to splash bars [59]. For example, the Gas Technology Institute (GTI) and its partners recently developed an advanced dew-point cooling tower fill that allows the cooling of water at lower than the current

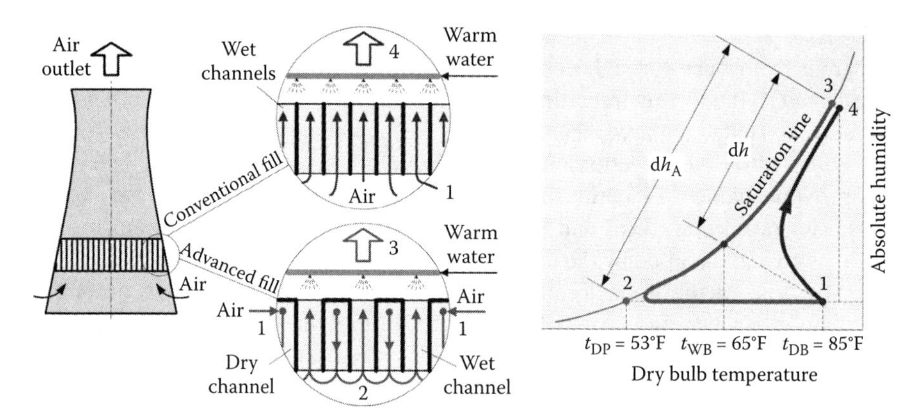

FIGURE 12.13 The advanced cooling tower fill to enable the dew point cooling. (Reprinted with permission from Electric Power Research Institute (EPRI), *Program on technology innovation: New concepts of water conservation cooling and water treatment technologies*. 2012.)

limit—the ambient wet bulb temperature—and even down to the dew point temperature using the patented M-Cycle approach [60]. To achieve this, the air flow in the fill is specially arranged in the dry channels adjacent to the wet channels in order to be indirectly precooled by evaporating water as shown in Figure 12.13 [61]. The conventional cooling process follows lines 1–4 on the psychrometric chart, which is from 100°F to 75°F when ambient dry bulb temperature (t_{DB}) is 85°F and web bulb temperature (t_{WB}) is 65°F. However, the advanced process (lines 1–2–3) is started at the same initial conditions (air inlet temperature and humidity, water inlet temperature) as conventional cooling but cooled to a lower temperature (55°F). Meanwhile, the cooling capacity (dh_A) of the advanced cooling tower fill is much higher than that (dh) of the conventional cooling tower. According to GTI, by using the advanced dew-point cooling tower fill, the plant power production can be increased by up to 4%, and the cooling water use could be decreased by 15%–20%, and the corresponding savings is more than $1,150,000 per year in a typical 500-MW power plant [62].

12.5.3.2 Multifunctional Nanofluid Development for Cooling Tower Evaporation Loss Reduction

Argonne National Laboratory proposes to develop a water-based nanofluid with multifunctional nanoparticles to be used in wet recirculating cooling systems that can reduce evaporation loss and improve thermal performance without requiring significant capital cost. As shown in Figure 12.14 [61], the nanoparticles are designed to have several thermophysical properties, such as higher heating capacity and thermal conductivity so that less amount of water is needed to achieve the given level of cooling and the amount of evaporation and drift loss can be reduced. It is estimated that the overall water consumption can be reduced by 20% because of the higher latent heat of nanofluid, and the coolant flow rate can be reduced by 15% because of the improved thermophysical properties.

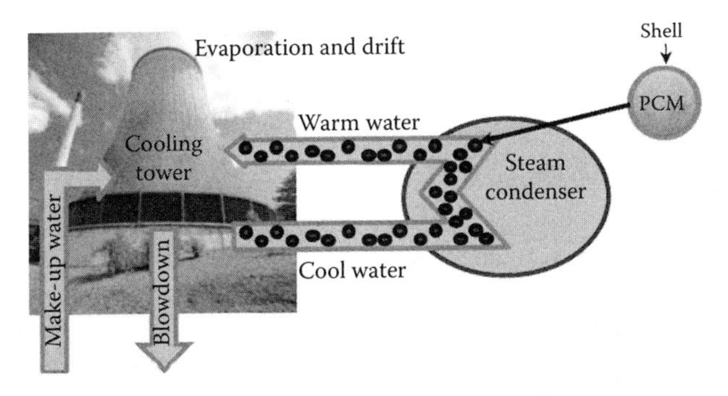

FIGURE 12.14 Schematic of circulating nanoparticles in a cooling loop. (Reprinted with permission from Electric Power Research Institute [EPRI], *Program on technology innovation: New concepts of water conservation cooling and water treatment technologies*. 2012.)

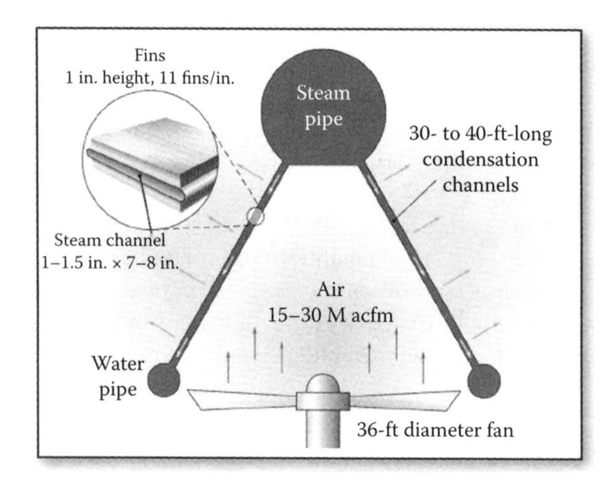

FIGURE 12.15 Illustration of a direct dry cooling system. (Reprinted with permission from Electric Power Research Institute (EPRI), *Program on technology innovation: New concepts of water conservation cooling and water treatment technologies*. 2012.)

12.5.4 Dry Cooling

There are basically two types of dry cooling: direct and indirect. In direct dry cooling, a high flow rate of air is blown on the surface of the condenser, which consists of thousands of banks of finned tubes to take the heat away from the turbine exhaust steam via convective heat transfer (Figure 12.15 [61]). Indirect dry cooling uses a water-cooled condenser for turbine exhaust steam, and the heated condenser cooling water is then recirculated to an air-cooled condenser before returning to the water-cooled condenser. Both direct and indirect dry cooling have no loss of cooling water, rendering almost zero withdrawal and consumption of freshwater. However, the performance of dry cooling is much poorer than that of wet cooling since it relies on local dry bulb temperature rather than on a much lower web bulb temperature [63], and since the former is more likely to fluctuate than the latter, cooling performance could also deteriorate during temperature peaks. In addition, the heat transferred in dry cooling is sensible instead of latent heat, rendering much larger size and higher (three to four times) capital cost [64].

12.5.4.1 Desiccant Dry Cooling

To overcome the shortcomings of traditional dry cooling technology, the Energy and Environmental Research Center proposed a novel indirect dry cooling technology, which is called "desiccant dry cooling (DDC)" [65]. The key feature of DDC is the use of a hygroscopic working fluid as the heat-transfer media between the steam condenser and the atmosphere. The hygroscopic working fluid contains a desiccant such as sodium chloride, calcium chloride, magnesium chloride, and lithium chloride that can retain moisture content until equilibrium with ambient moisture content. Therefore, unlike water that needs to be replenished as a result of evaporation

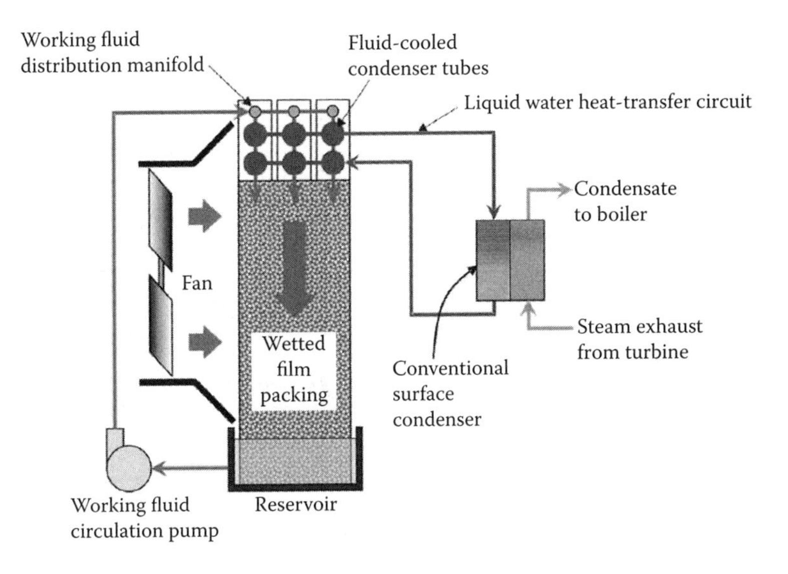

FIGURE 12.16 Schematic of a desiccant dry cooling (DDC) system. (Reprinted with permission from Martin, C.L., *Novel Dry Cooling Technology for Power Plants.* 2013, Energy and Environmental Research Center [EERC]: Grand Forks, North Dakota.)

or drift loss during cooling, the hygroscopic fluid is expected to last the life of the system after the initial charge. The process is similar to a conventional indirect dry cooling system, as shown in Figure 12.16 [66]. Case studies indicate that DDC maintains a much lower (40%) annual cost than the conventional dry cooling system and is comparable to the wet recirculating cooling system, but with much lower drift rate (<0.00006%).

12.5.5 Hybrid Cooling

The hybrid cooling system, which combines the wet and dry cooling systems, aims to achieve the best features of each: the wet cooling performance during hot days and the water conservation benefits of dry cooling during cold days [61]. It is also designed to reduce or eliminate plume formation, which may be regarded as environmentally or aesthetically objectionable. On the basis of the direction of air flow, there are two configurations for the arrangement of wet and dry cooling, in series or in parallel [63]. Figure 12.17 [61] shows the parallel arrangement of the wet and dry sections. During hot days, the dry section is isolated and the tower functions as a pure wet cooling tower with all air flow to the wet section, while during cold days, the dry cooler can transfer the majority of the heat duty, with the air flow mostly to the dry section; meanwhile, water flow rate to the wet section is diminished. However, hybrid cooling imposes challenges to control engineers owing to the wide range of conditions, and the corresponding control settings of valves and air control louvers have, in part, to be empirically determined [63].

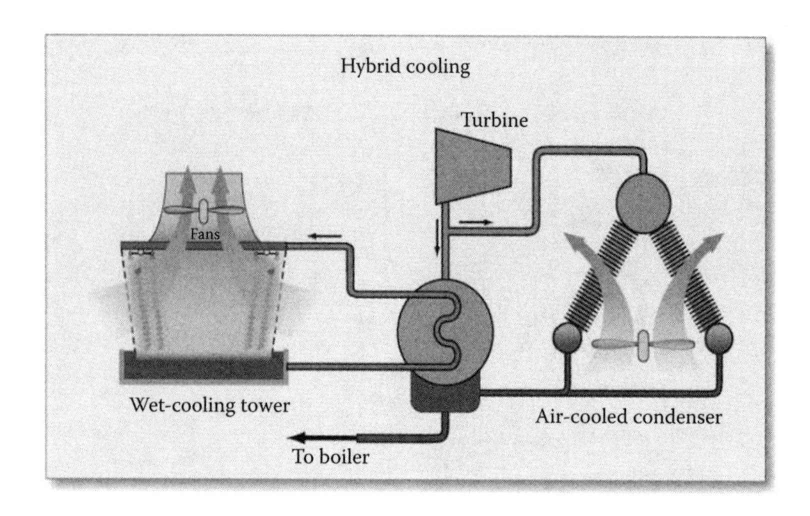

FIGURE 12.17 A hybrid cooling with parallel arrangement of wet and dry sections. (Reprinted with permission from Electric Power Research Institute [EPRI], *Program on technology innovation: New concepts of water conservation cooling and water treatment technologies*. 2012.)

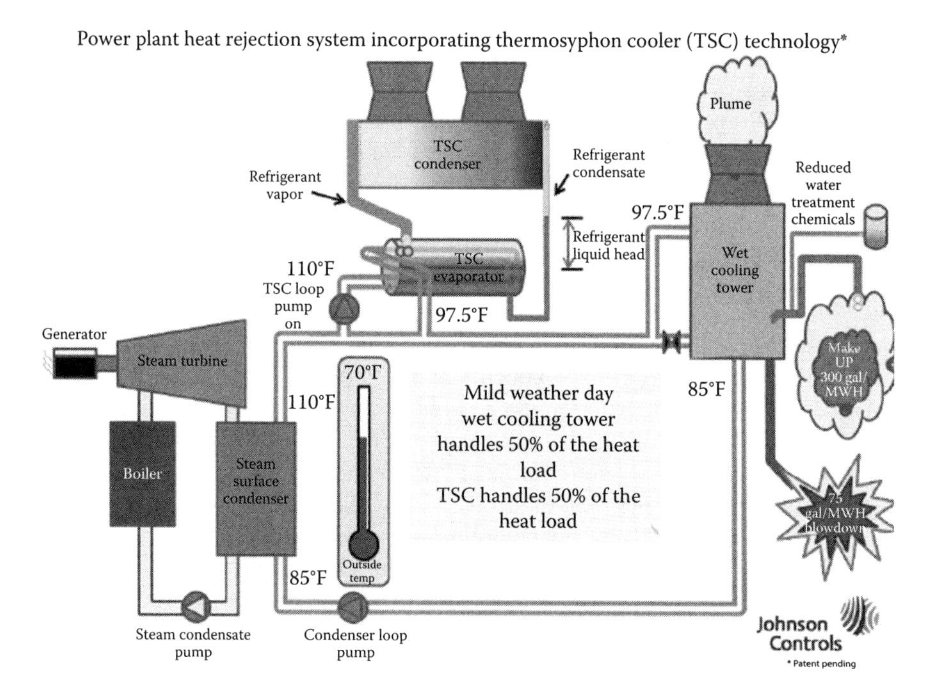

FIGURE 12.18 A cooling system with the thermosyphon cooler. (Reprinted with permission from Electric Power Research Institute [EPRI], *Program on technology innovation: New concepts of water conservation cooling and water treatment technologies*. 2012.)

12.5.5.1 Thermosyphon Hybrid Cooling System

Johnson Controls, Inc. recently developed a thermosyphon hybrid cooling system [61] that combines the thermosyphon dry heat rejection device (i.e., the thermosyphon cooler) with the wet recirculating cooling in series that would reduce the annual evaporative water loss by 30%–80% compared to the traditional wet recirculating cooling tower (Figure 12.18). The system proposes to use innovative control strategies for both the dry and wet sections to ensure the most economical balance between water savings and parasitic fan energy. It can also be applied to existing power plants with minimal piping modifications because the dry cooling component works with the traditional condenser water loop. A small-scale laboratory prototype of the thermosyphon system has been developed and tested by the manufacturer, Johnson Controls, Inc. A 1-MW pilot system has been installed and is currently being tested at the water research center located in Euharlee, Georgia.

12.5.6 Comparisons

Comparisons among the aforementioned four types of cooling towers (once-through, wet recirculating, dry direct, and hybrid) are listed in Table 12.5 [61] in terms of system cost, cost ratio relative to wet recirculating, evaporative loss, steam condensation temperature, and coolant flow rate. Dramatically higher (2.5–5 and 2–4) system costs for dry and hybrid equipment, respectively, is observed compared to the wet recirculating counterpart. The cost of water is assumed to be free, but a realistic assessment of this cost, especially because of the increasing water cost and more stringent water use regulations, may favor the dry and hybrid cooling in the future.

TABLE 12.5
The Cost and Operation Data Comparison among Various Cooling Systems for a 500-MW, Coal-Fired Steam Power Plant

Cooling System	System Cost (Million US$)	Cost Ratio Relative to Wet	Evaporative Loss (kgal/MWh)	Steam Condensation Temperature[a] (°F)	Coolant Flow Rate (kg/min)
Wet cooling tower and condenser	20–25	1	0.5–0.7	116	100–250
Dry direct	60–100	2.5–5	0	140–155	0
Once-through cooling	10–15	0.4–0.75	0.2–0.3	100	150–350
Hybrid	40–75	4-Feb	0.1–0.5	116	50–150

Source: Reprinted with permission from Electric Power Research Institute (EPRI), *Program on technology innovation: New concepts of water conservation cooling and water treatment technologies*. 2012.

[a] Steam condensation temperatures are based on ambient air dry-bulb temperature of 100°F and ambient air wet-bulb temperature of 78°F.

12.5.7 Other Recent Developments

The Electric Power Research Institute has conducted a worldwide solicitation between February 2011 and June 2012 and received 114 proposals of innovative power plant water-conserving technologies for cooling, waste heat utilization, and water treatment [61]. Some of the innovative concepts are briefly introduced in Sections 12.5.3 through 12.5.5. For further information, readers may check their website (http://www.epri.com).

REFERENCES

1. LeCompte, J. *Infographic: A few facts about manufacturing.* 2012 [cited January 25, 2013]; Available from: http://www.treasury.gov/connect/blog/Pages/Infographic-A-Few-Facts-about-Manufacturing.aspx.
2. *Tracking Industrial Energy Efficiency and CO_2 Emissions.* 2007, International Energy Agency (IEA).
3. *Agenda 21 United Nations Conference on Environment & Development.* 1992; Available from: https://sustainabledevelopment.un.org/content/documents/Agenda21.pdf.
4. *Culture: fourth pillar of sustainable development.* Available from: http://www.agenda21culture.net/index.php/docman/-1/393-zzculture4pillarsden/file.
5. *The future we want: Outcome document of the United Nations conference on sustainable development.* 2012; Available from: https://sustainabledevelopment.un.org/content/documents/733FutureWeWant.pdf.
6. Dornfeld, D.A., *Green Manufacturing: Fundamentals and Applications.* 2012, New York: Springer.
7. Shah, R. and P.T. Ward, Defining and developing measures of lean production. *Journal of Operations Management*, 2007. **25**: 785–805.
8. Sarkis, J., Manufacturing strategy and environmental consciousness. *Technovation*, 1995. **15**(2): 79–97.
9. Berkel, v.R., E. Willems, and M. Lafleur, The relationship between cleaner production and industrial ecology. *Journal of Industrial Ecology*, 1997. **1**(1): 51–66.
10. Hanna, M., W.R. Newman, and P. Johnson, Linking operational and environmental improvement through employee involvement. *International Journal of Operations & Production Management*, 2000. **20**(2): 148–165.
11. Bergendahl, C.-G. et al., Environmental and economic implications of a shift to halogen-free printed wiring boards. *Circuit World*, 2005. **31**(3): 26–31.
12. Johansson, G., A. Greif, and G. Fleischer, Managing the design/dnvironment interface: Studies of integration mechanisms. *International Journal of Production Research*, 2007. **45**(18–19): 4041–4055.
13. Triple bottom line. *The Economist* [cited November 17, 2009]; Available from: http://www.economist.com/node/14301663.
14. Jawahir, I.S., *Sustainable manufacturing: The driving force for innovative products, processes and systems for next generation manufacturing.* 2011, College of Engineering, University of Kentucky: Lexington, KY.
15. Lou, H.R. and Y.L. Huang, Profitable pollution prevention: Concept, fundamentals & development. *J. of Plating and Surface Finishing*, 2000. **87**(11): 59–66.
16. Association for Manufacturing Excellence, *Green Manufacturing: Case Studies in Leadership and Improvement.* Enterprise Excellence. 2007, Productivity Press: New York.
17. Kiron, D., Sustainability nears a tipping point. *MIT Sloan Management Review*, 2012. **53**(2): 69–74.

18. Governance & Accountability Institute. *Flash report.* 2014; Available from: http://www.ga-institute.com/nc/issue-master-system/news-details/article/seventy-two-percent-72-of-the-sp-index-published-corporate-sustainability-reports-in-2013-dram.html.
19. *Sustainable manufacturing and eco-innovation synthesis report.* Organisation for Economic Co-Operation and Development (OECD) 2009 [cited February 24, 2013]; Available from: http://www.oecd.org/sti/inno/43423689.pdf.
20. J. García-Serna, L.P.-B., and M.J. Cocero, New trends for design towards sustainability in chemical engineering: Green engineering. *Chem. Eng. J.*, 2007. **133**: 7–30.
21. Electric Power Research Institute (EPRI), *Maximizing wasterwater reduction for the process industries final report.* 1999: Palo Alto, CA.
22. Foo, D.C.Y., State-of-the-art review of pinch analysis techniques for water network synthesis. *Ind. Eng. Chem. Res.*, 2009. **48**: 5125–5159.
23. Jezowski, J., Review of water network design methods with literature annotations. *Ind. Eng. Chem. Res.*, 2010. **49**: 4475–4516.
24. Wang, Y.P. and R. Smith, Wastewater minimisation. *Chem. Eng. Sci.*, 1994. **49**: 981–1006.
25. Linnhoff, B. and E. Hindmarsh, The pinch design method of heat exchanger networks. *Chem. Eng. Sci.*, 1983. **38**: 745–763.
26. El-Halwagi, M.M. and V. Manousiouthakis, Synthesis of mass exchange networks. *AICHE J.*, 1989. **35**(8): 1233–1244.
27. Wang, Y.P. and R. Smith, Wastewater minimization with flowrate constraints. *Trans. Inst. Chem. Eng. A*, 1995. **73**: 889–904.
28. Hallale, N., A new graphical targeting method for water minimisation. *Adv. Environ. Res.*, 2002. **6**(3): 377–390.
29. El-Halwagi, M.M., F. Gabriel, and D. Harell, Rigorous graphical targeting for resource conservation via material recycle/reuse networks. *Ind. Eng. Chem. Res.*, 2003. **42**: 4319–4328.
30. Grossmann, I.E., M. Martin, and L. Yang, Review of optimization models for integrated process water networks and their application to biofuel processes. *Current Opinion in Chemical Engineering*, 2014. **5**: 101–109.
31. Takama, N. et al., Optimal water allocation in a petroleum refinery. *Comput. Chem. Eng.*, 1980. **4**: 251–258.
32. Savelski, M.J. and M.J. Bagajewicz, On the optimality conditions of water utilization systems in process plants with single contaminants. *Chem. Eng. Sci.*, 2000. **55**(21): 5035–5048.
33. Savelski, M.J. and M.J. Bagajewicz, On the necessary conditions of optimality of water utilization systems in process plants with multiple contaminants. *Chem. Eng. Sci.*, 2003. **58**: 5349–5362.
34. Huang, C.H. et al., A mathematical programming model for water usage and treatment network design. *Ind. Eng. Chem. Res.*, 1999. **38**(2666–2679).
35. Savulescu, L.E. and R. Smith, Simultaneous energy and water minimization, in *AICHE Annual Meeting.* 1998: Miami Beach, FL.
36. Bogataj, M. and M.J. Bagajewicz, Synthesis of non-isothermal heat integrated water networks in chemical processes. *Comput. Chem. Eng.*, 2008. **32**: 3130–3142.
37. Yang, L. and I.E. Grossmann, Water targeting models for simultaneous flowsheet optimization. *Ind. Eng. Chem. Res.*, 2013. **52**(9): 3209–3224.
38. Bagajewicz, M.J., A review of recent design procedures for water networks in refineries and process plants. *Comput. Chem. Eng.*, 2000. **24**: 2093–2113.
39. Manan, Z.A. and S.R.W. Alwi, Water pinch analysis evolution towards a holistic approach for water minimization. *Asia-Pac. J. Chem. Eng.*, 2007. **2**: 544–553.
40. Tan, R.R., D.C.Y. Foo, and Z.A. Manan, Assessing the sensitivity of water networks to noisy mass loads using Monte Carlo simulation. *Comput. Chem. Eng.*, 2007. **31**: 1355–1363.

41. Kuo, W.C.J. and R. Smith, Designing for the interactions between water-use and effluent treatment. *Trans. IChemE (Part A)*, 1998. **76**: 287–301.
42. Mann, J.G. and Y.A. Liu, *Industrial Water Reuse and Wastewater Minimization*. 1999, New York: McGraw Hill.
43. Prakash, R. and U.V. Shenoy, Targeting and design of water networks for fixed flowrate and fixed contaminant load operations. *Chem. Eng. Sci.*, 2005. **60**(1): 255–268.
44. Alwi, S.R.W. and Z.A. Manan, Generic graphical technique for simultaneous targeting and design of water networks. *Ind. Eng. Chem. Res.*, 2008. **47**: 2762–2777.
45. Aly, S., S. Abeer, and M. Awad, A new systematic approach for water network design. *Clean Technol. Environ. Pol.*, 2005. **7**(3): 154–161.
46. Gomes, J.F.S., E.M. Queiroz, and F.L.P. Pessoa, Design procedure for water/wastewater minimization: Single contaminant. *J. Clean. Prod.*, 2006. **15**: 474–485.
47. El-Halwagi, M.M., *Pollution Prevention through Process Integration: Systematic Design Tools*. 1997, San Diego, U.S.: Academic Press.
48. Dunn, R.F. and G. Bush, Process integration technology for cleaner production. *J. Clean. Prod.*, 2001. **9**: 1–23.
49. Polley, G.T. and H.L. Polley, Design better water networks. *Chem. Eng. Progress*, 2000. **96**(2): 47–52.
50. Rosenthal, R.E., *GAMS—A User's Guide*. 2013, GAMS Development Corporation: Washington, D.C.
51. Zhou, Q., H.H. Lou, and Y.L. Huang, Design of a switchable water allocation network based on process dynamics. *Ind. Eng. Chem. Res.*, 2001. **40**: 4866–4873.
52. *Water Requirements for Existing and Emerging Thermoelectric Plant Technologies*. 2009, U.S. Department of Energy National Energy Technology Laboratory.
53. U.S. Department of Energy, *Estimating Freshwater Needs to Meet Future Thermoelectric Generation Requirement*. 2008, National Energy Technology Laboratory: Pittsburgh, PA.
54. Roy, S.B., K.V. Summers, and R.A. Goldstein, Water sustainability in the united states and cooling water requirements for power generation. *Water Resources Update*, 2003(126): 94–99.
55. Feeley, T.J. and M. Ramazan, Electric utilities and water: Emerging issues and R&D needs. In *Proceedings of 9th Annual Industrial Wastes Technical and Regulatory Conference*. 2003. San Antonio, TX: Water Environment Federation.
56. Clean Water Act. *Section 316(b) Cooling Water Intakes*. Available from: http://water.epa.gov/lawsregs/lawsguidance/cwa/316b/.
57. R. D. Vidic, D.A.D., *Final technical report: Reuse of treated internal or external wastewaters in the cooling systems of coal-based thermoelectric power plants*. 2009, U.S. Department of Energy.
58. Federal Water Pollution Control Act (Clean Water Act). *Thermal discharge—Cooling tower intake structures*. 2002 [cited February 25, 2013]; Available from: http://epw.senate.gov/water.pdf.
59. Federal Energy Management Program, *Cooling towers: Understanding key components of cooling towers and how to improve water efficiency*. 2011, U.S. Department of Energy.
60. *Energy- and water-efficient dew point cooling tower fill*. 2012, Electric Power Research Institute.
61. Electric Power Research Institute (EPRI), *Program on technology innovation: New concepts of water conservation cooling and water treatment technologies*. 2012.
62. Libert, J.-P., Advanced dew point cooling tower concept for industrial and commercial applications, in *EPRI 2012 Cooling Tower Technology Conference*. 2012: Pensacola Beach, FL.

63. Electric Power Research Institute (EPRI). Proceedings: Cooling tower and advanced cooling systems conference, in *Cooling Tower and Advanced Cooling Systems Conference*. 1995. St. Petersburg, Florida.
64. Maulbetsch, J., *Comparison of Alternate Cooling Technologies for California Power Plants: Economic, Environmental and Other Tradeoffs*. 2004, Electric Power Research Institute: Palo Alto, CA.
65. Martin, C.L., *Final Report for Subtask 5.10 Testing of an Advanced Dry Cooling Technology for Power Plants*. 2013, Energy and Environmental Research Center.
66. Martin, C.L., *Novel Dry Cooling Technology for Power Plants*. 2013, Energy and Environmental Research Center (EERC): Grand Forks, North Dakota.

Index

Page numbers followed by f and t indicate figures and tables, respectively.

C

D

E

J

K

N

O

Q

R

S

U

V

W

X

Z